# New Results in High Energy Physics–1978

### (Vanderbilt Conference)

## AIP Conference Proceedings
Series Editor: Hugh C. Wolfe
### Number 45
Particles and Fields Subseries No. 14

# New Results in High Energy Physics–1978
### (Vanderbilt Conference)

Editors
## R.S. Panvini and S.E. Csorna
Vanderbilt University

# American Institute of Physics
New York                                    1978

L.C. Catalog Card No. 78-67196
ISBN 0–88318–144–4
DOE CONF- 780338

v

# FOREWORD

This was the 3$^{rd}$ International Conference at Vanderbilt University emphasizing the latest significant results in High Energy Physics. The conference was held during March 6-8, 1978; the first two took place March 26-28, 1973 and March 1-3, 1976, respectively. Each of these three conferences have been attended by about 100 participants with good representation from high energy groups throughout the United States and Europe. This conference was open to anyone who wanted to attend.

Comparing these proceedings with those of the first two conferences dramatically reveals the rapid pace of high energy physics in the last five years. In March 1973, the CERN Intersecting Storage Rings and the National Accelerator Laboratory in Batavia, Illinois, had been operating for only a short time. The most interesting new results at that time were the large cross sections for particle production at high transverse momentum and the rise in the pp total cross section. In March 1976, the J/$\psi$ particle had been discovered and spectroscopy was the rage again. The chief goal of experiments in 1976 was to find charmed particles, which were found at SPEAR a few months after our second conference. Now, in 1978, charmed particles are often considered a background; the possibilities for new heavy particles (or quarks) being discovered seems endless after the discovery of the Upsilon, with mass of 9.5 GeV, at Fermilab and DESY.

The 3$^{rd}$ Vanderbilt Conference, like the first two, was organized by consulting with acknowledged leaders in the areas of greatest current interest in high energy physics. Some of these people also served as session chairmen. C. Baltay and B. Barish were our Neutrino Session Advisors, E. L. Berger and K. Gottfried advised us on Theory, K. Berkelman and B. Richter suggested names for the e$^+$e$^-$ program, J. Cronin and A. J. S. Smith were the experts on lepton and lepton pair production experiments, and R. Diebold and T. Ferbel were the individuals who recommended contributors in the general area of hadron physics. Invitations were issued to spokesmen of various experimental groups who were asked to suggest a representative from their group or collaboration to present their results. This is in line with our policy that, with few exceptions, new experimental results are to be presented by representatives of the individual experiments rather than by rapporteurs. However, we invite a few theorists/phenomenologists who help put the material discussed in some perspective.

The local physicists who worked with me in planning and running the conference were S. Csorna and J. S. Poucher. However, I must add that I am deeply indebted to my wife Doria who tirelessly assisted me in a variety of clerical tasks, who took charge of the registration desk, and who organized the social program. She, Judy Csorna, and Lois Poucher were chiefly responsible for making the arrangements for our successful banquet.

Financial support for the 1978 Vanderbilt Conference was provided by the National Science Foundation, the Department of Energy, and by the Vanderbilt University Research Council.

R. S. Panvini
Conference Chairman

June, 1978

TABLE OF CONTENTS

III.  $e^+e^-$ Interactions

IV.  Neutrino Interactions

## SEARCH FOR A STABLE DIHYPERON

M. S. Witherell
Princeton University, Princeton, N. J. 08540

### ABSTRACT

A Brookhaven National Laboratory-Princeton collaboration[*] has searched for a stable six-quark state in the reaction $p + p \rightarrow K^+ + K^+ +$ Missing Mass. Upper limits are presented for production of a narrow resonance in the mass range 2.1 to 2.4 GeV.

### MOTIVATION

The most spectacular success of the quark model is the production of all known mesons and baryons using only the $Q\bar{Q}$ and $Q^3$ states of SU(3). The absence of free quarks and other combinations such as diquark states is explained by the requirement that physical hadrons are color singlets. Some other exotic combinations are allowed, however, such as $Q^6$ and $Q^2\bar{Q}^2$. The fact that we have not seen such states does not argue strongly against their existence, since many such states are likely to be unstable, and others have non-exotic quantum numbers.[1]

Recently it has been suggested by Jaffe[2] that there may be six-quark states which are bound to strong decay into two baryons. We may take some guidance as to which states should have lowest mass by studying the ordinary hadrons. In particular there are strong spin-dependent forces which are responsible for large mass splittings, such as the N-Δ mass difference of 300 MeV. This spin-dependent interaction together with the requirement of a color singlet leads to a flavor antisymmetry principle: the most strongly bound state has quarks in the most antisymmetric flavor state allowed.[3] This is true even though it is assumed that the strength of the interaction does not depend on flavor. According to this argument, the "most bound" six-quark state is that composed of (uuddss). In fact the lowest state is the SU(3) in which the six quarks are in a L=0, S=0 state, and has the quantum number of two lambdas in the s-wave.

These ideas give a qualitative description of the procedure used by Jaffe[2], who applied the MIT bag model to the six-quark system. He found two stable states in the (uuddss) system. Lowest in mass of these was the H, the SU(3) singlet state with $J^P = 0^+$ and M-2150 MeV. Since $2M_\Lambda=2230$ MeV, a state with the mass listed for the H would be bound by 80 MeV. Jaffe also predicted the H[*], the I=0 member of the SU(3) octet with $J^P=1^+$, to have a mass of 2335 MeV. The exact masses are of course suspect, since it is an extrapolation of the model well beyond the region in which it has been tested. Still it seems rea-

*The experimenters were A. S. Carroll, I. H. Chiang, R. A. Johnson, T. F. Kycia, K. K. Li, L. S. Littenberg, and M. D. Marx from BNL and R. Cester, R. C. Webb, and M. W. Witherell from Princeton.

ISSN: 0094-243X/78/001/$1.50 Copyright 1978 American Institute of Physics

sonable to assume that certain qualitative features, such as the ordering of states, might be more general results.

It should be noted that, as Jaffe emphasizes, these are single hadrons and not deuteron-like composite objects. The model deals not with the residual forces that bind neutron and proton with 2 MeV of binding energy, but rather with spin-dependent gluon exchange interactions responsible for mass differences on the order of 100 MeV. The only dibaryon state other than the deuteron which is now well-known is the enhancement in the $\Lambda p$ system at M=2129 MeV. This is probably a deuteron-like bound state of $\Sigma N$, since its mass is about 2 MeV below that threshold.

If a state such as the H exists, the decay properties depend on mass. If the mass is below 2055 MeV, only the double weak decay to two neutrons is allowed. For a mass in the range 2055-2230 MeV the decay goes weakly to $\Lambda N$ and $\Sigma N$ states. Finally, for mass greater than 2230 MeV strong decay into two hyperons is allowed. The middle range easily includes the masses which might result in the bag model. A dihyperon state would be narrow, if its mass were not much greater than twice the lambda mass. The $H^*$, a $J^P=1^+$ state, could not decay into two lambdas because of statistics, so it would be narrow if $M_{H^*} < M_\Lambda + M_\Sigma$.

There are a number of reactions in which the H might be produced, mostly those such as $K^-p \rightarrow \Lambda H$, in which there is one unit of strangeness in the initial state. Our experiment, however, looked for the reaction $pp \rightarrow K^+K^+H$ because of some experimental advantages which might offset the lower cross section expected. The $K^+K^+$ final state gives an extremely clean signature, and there are intense proton beams which are free of contamination by other particles. We chose to use the missing-mass technique because it doesn't depend on lifetime or decay mode. The feature of this experiment which enables a such small cross sections to be probed is the absence of physical background. If we properly identify $K^+K^+$ events there should be no background with missing mass below twice the lambda mass.

It is useful to try to calculate an expected cross section for the reaction. The best estimate was made by relating $\sigma(pp \rightarrow K^+K^+H)$ to $\sigma(K^-p \rightarrow K^+K^0\Omega^-)$, which also goes by $\Xi$ exchange. The $\Omega^-$ production reaction has a cross section of somewhat more than 1 $\mu$b. Putting in reasonable estimates of vertex couplings, etc., we arrived at an estimate of 50-100 nb.

## EXPERIMENT

There were two basic experimental problems to face in doing this experiment. The first was to obtain a beam with intensity sufficient to do an experiment of this sensitivity. The experimental apparatus was a modified version of the double-arm spectrometer described if reference 4, located in the High Energy Unseparated Beam at the Brookhaven AGS. Although this is the highest intensity secondary beam at Brookhaven, it had never been operated at an intensity of greater than $3 \times 10^7$ particles per pulse, which would give somewhat less than $10^{13}$ particles in a two week run. Also, with a primary proton beam of 30 GeV/c, the secondary beam is predominantly pions at the desired momentum of 5 GeV/c. For comparison, a total of approximately $10^{14}$ protons on the 9" liquid hydrogen target were required for this experiment. The only solution to the problem was to run the AGS as a 5 GeV/c accelerator. We received the full extracted beam of $2 \times 10^8$ protons per second for spill lengths as long as 6 seconds.

The $K^+$ selection was the most difficult experimental problem. It was necessary to reduce the double-arm trigger rate by a factor of $10^3$ to be able to run at these intensities. Further off-line rejection was required, for a total suppression factor of $10^6$ for events with no K in the final state and $10^3$ for events with one K. Four Cherenkov counters were used in the trigger and for off-line particle identification, and a time-of flight system with resolution of 0.32 nsec, was used for further suppression of non-K events. The suppression factors for pions and protons in the momentum range 0.5–1.6 GeV/c were $(3-10) \times 10^3$, for each arm.

RESULTS

In our early data we saw a number of events with apparent missing mass below the threshold at $2M_\Lambda$. There were two expected

sources of background, accidental coincidences and misidentified pions and protons. Both of these contributions could be estimated quite easily. The worst was from misidentified πK events, but together they accounted for less than ½ of all events below threshold. After some effort we concluded that these events were due to double-scattering sources. For example the reaction pp→$K^+\Lambda$p, with the outgoing proton interacting to give a second $K^+$, produces two $K^+$ with an apparent missing mass below threshold. Although we were only sensitive to such a second scatter within a few cm of hydrogen, such events still produce backgrounds that are nonnegligible at these cross sections. Of course there are many similar processes which contribute at the same level. One other type of event which formed an even more surprising background is represented by the reaction pp→$K^+\bar{K}^0$pn, with the $\bar{K}^0$ undergoing a transition to $K^0$ and charge exchange scattering, producing a $K^+$. We were able to suppress such backgrounds by cutting rather tightly on the vertex cuts, at some loss to real events.

The total experiment ran in less than three weeks of dedicated AGS time. The mass range accepted by the experiment depended upon the beam momentum. Approximately one-half of the beam was taken at a momentum of 5.33 GeV/c, where the acceptance spanned the range 2.1–2.4 GeV. Beam momenta of 5.06, 5.33, and 5.89 GeV/c were used, with a total of $3 \times 10^{13}$, $8 \times 10^{13}$, and $4 \times 10^{13}$ protons respectively. There were a total of $4 \times 10^9$ interactions with a particle in each arm, which after all cuts was reduced to 75 events. The 13 events with mass below $2M_\Lambda$ are consistent with the number expected from accidentals and misidentified events. Above threshold there are 62 $K^+K^+$ events, of which only about 20% are from these same background sources. Fig. 1 shows the mass plots at each beam momentum. The mass resolution is approximately $\pm 8$ MeV. No striking structure is seen in any of the plots. Although the sensitivity is lower in the 5.89 GeV/c sample, it has by far the largest number of events, because most of the acceptance lies above threshold.

The sensitivity of the experiment is shown in Table I. Listed are upper limits on the cross section for producing a narrow resonance in the reaction $pp \rightarrow K^+K^+X$. The important mass region is 2.1-2.35 GeV, where the limits are about 30 nb for the 5.33 GeV/c data.

Table I. Upper Limits on production cross section (in nanobarns)

| $P_{BEAM}$ (GeV/c) | 5.06 | 5.33 | 5.89 |
|---|---|---|---|
| Mass (GeV) | | | |
| 2.0 – 2.1 | 80 | | |
| 2.1 – 2.23 | 60 | 30 | |
| 2.23 – 2.35 | | 30 | 90 |
| 2.35 – 2.45 | | | 60 |

This experiment was performed in a minimum of time with an apparatus most of which already existed. The upper limit of 30 nb was below our best estimate for H production. It seems difficult to design a similar experiment which could improve the present limit by more than a factor of 10, partly because the effect of secondary interactions would be more difficult to suppress. There is an experiment scheduled for the CERN PS by a Madrid-Rome-Saclay-Vanderbilt collaboration, which plans to look at the reaction $K^-d \rightarrow K^+X$ at the level of 1-10 event/nb. The cross section is probably more favorable than the proton-induced reaction, but the background is worse. It only sees I=1 objects, and would concentrate on the region above $\Lambda\Lambda$ threshold.

## REFERENCES

1. R. L. Jaffe, Phys. Rev. D 15, 267 (1977).
2. R. L. Jaffe, Phys. Rev. Lett. 38, 195 (1977).
3. For a clear, qualitative discussion of these matters, see the contribution of Harry J. Lipkin in Prospects for Strong Interaction

Physics at ISABELLE, edited by D. P. Sidhu and T. L. Trueman
(Brookhaven National Laboratory, Upton, N. Y., 1977).
4.  R. Cester et al., Phys. Rev. Lett. 37, 1178 (1976)

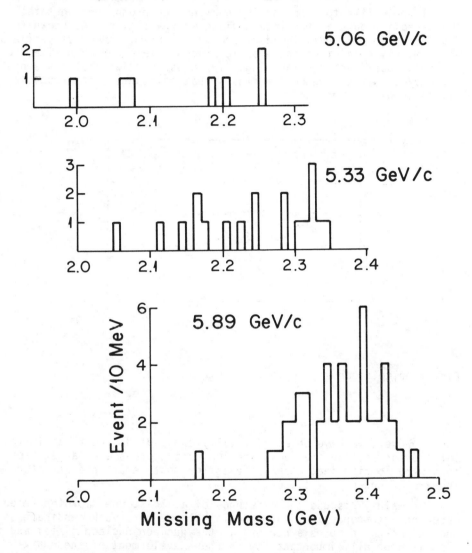

Figure 1.  Missing mass spectra for the three data samples.

# RESONANT-LIKE STRUCTURES IN PP SYSTEM
# IN THE MASS REGION 2100 to 2800 MeV*

Akihiko Yokosawa
Argonne National Laboratory, Argonne Il. 60439

I would like to start out by showing the proton-proton total cross-section data at the intermediate-energy region. As shown in Fig. 1, up to 1.2-GeV/c incident proton momentum, the cross sections, which mainly consist of the elastic process, fall and then rise due to the inelastic-channel opening. The cross section flattens above 1.5 GeV/c. We observe no structures that may suggest the possible existence of a resonance.

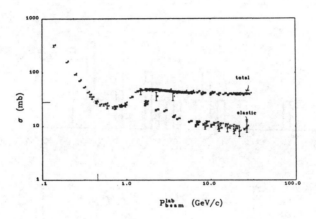

Fig. 1 pp total cross section.

However, we have observed totally unexpected structures in the total cross section when both the incident protons and target protons were longitudinally polarized. The most remarkable structure appears around $p_{lab}$ = 1.5 GeV/c.

We mainly discuss the existence of at least one diproton resonance and its properties, and speculate three more such candidates. Such a resonance opens a new era in the nucleon-nucleon system and also is crucially important for further development of the quark models that require six quarks in a bag.[1,2,3]

* Work supported by the U.S. Department of Energy.

First, we describe experimental observables in terms of the helicity amplitudes, and then in terms of singlet and triplet partial-wave amplitudes. There are three s-channel helicity amplitudes at $\theta_{c.m.} = 0$:

$$\phi_1 = \langle ++ | ++ \rangle,$$

$$\phi_2 = \langle -- | ++ \rangle, \text{ and}$$

$$\phi_3 = \langle +- | +- \rangle.$$

These amplitudes are related to total cross section as follows:

   i)  Total cross section

$$\sigma^{Tot} = (2\pi/k)\,\mathrm{Im}\left\{\phi_1(0) + \phi_3(0)\right\} = (\tfrac{1}{2})\left\{\sigma^{Tot}(\rightleftarrows) + \sigma^{Tot}(\rightrightarrows)\right\} \quad (1)$$

   ii)  Difference between total cross section for parallel and
        antiparallel spin states (longitudinal)

$$\Delta\sigma_L = (4\pi/k)\,\mathrm{Im}\left\{\phi_1(0) - \phi_3(0)\right\} = \left\{\sigma^{Tot}(\rightleftarrows) - \sigma^{Tot}(\rightrightarrows)\right\} \quad (2)$$

   iii)  Difference between total cross section for transverse and
         antitransverse spin states

$$\Delta\sigma_T = -(4\pi/k)\,\mathrm{Im}\ \phi_2(0) = \sigma^{Tot}(\uparrow\downarrow) - \sigma^{Tot}(\uparrow\uparrow). \quad (3)$$

Argonne ZGS facilities provide various spin directions of incident beam and target. Spin directions are illustrated in Fig. 2. To express observables in elastic scattering, we adopt the notation (Beam, Target; Scattered, Recoil); (0, N; 0, 0) for polarization, (N, N; 0, 0) and (L, L; 0, 0) for spin correlation parameters, etc. A typical experimental setup for $\Delta\sigma_L$ is shown in Fig. 3. The measurements were done in standard transmission experiment.

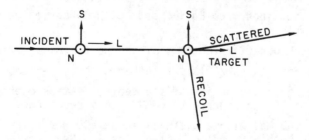

N: NORMAL TO THE SCATTERING PLANE
L: LONGITUDINAL DIRECTION
S = N x L IN THE SCATTERING PLANE

Fig. 2  Unit vectors N, L, and S.

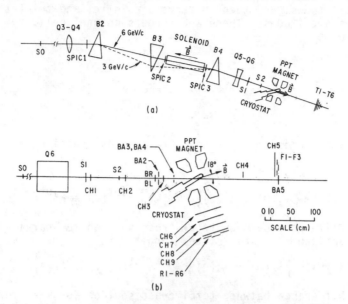

Fig. 3  (a) Beam line and experimental setup for the $\Delta\sigma_L$ measurement.
(b) Experimental setup for the $C_{LL} = (L,L;0,0)$ measurement.

We show the results of $\Delta\sigma_L$ measurements from 1.0 to 6.0 GeV/c (Fig. 4a)[4-6] There is a sharp peak near 1.2 GeV/c and a dip near 1.5 GeV/c.  From Eq. 2, a structure in $\phi_1(0)$ and $\phi_3(0)$ should appear as a peak and dip, respectively, in $\Delta\sigma_L$.  Figure 4b shows $\sigma^{Tot}(\rightleftarrows)$ and $\sigma^{Tot}(\rightrightarrows)$ as obtained from Eqs. 1 and 2.  We observe the third structure in $\sigma^{Tot}(\rightleftarrows)$ near 2.0 GeV/c.

To study the behavior in terms of the partial scattering amplitude, the data on $(k^2/4\pi)\Delta\sigma_L$ together with $(k^2/4\pi)\Delta\sigma_T$ are plotted in Fig. 5 as a function of the center-of-mass energy.[7] If the dip in $\Delta\sigma_L$ is considered to be due to a resonance, the mass is about 2260 MeV with a width of about 200 MeV.  The 1.2 GeV/c peak is seen in both $\Delta\sigma_L$ and $\Delta\sigma_T$ data.  The 2.0 GeV/c peak is clearly visible in $\Delta\sigma_T$.

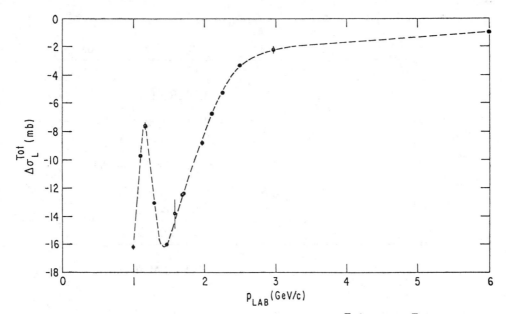

Fig. 4a  Total cross-section difference $\Delta\sigma_L = \sigma^{Tot}(\overrightarrow{\leftarrow}) - \sigma^{Tot}(\overrightarrow{\rightarrow})$.

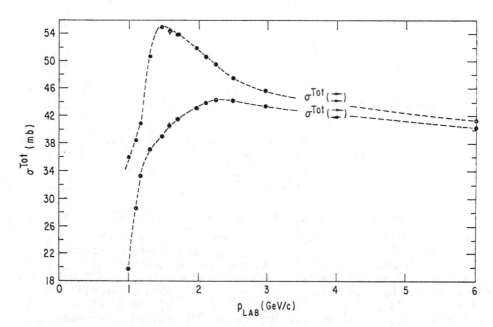

Fig. 4b  Total cross sections for pure initial spin states.  The
         dotted curves are only to guide the eye.

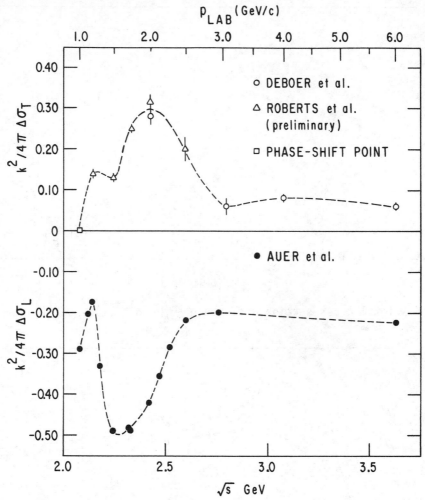

Fig. 5 $(k^2/4\pi)\ \Delta\sigma_L$ together with $(k^2/4\pi)\ \Delta\sigma_T$.

Using the data on $\Delta\sigma_L$, Grein and Kroll have calculated the real part of $\left[\phi_1(0) - \phi_3(0)\right]$ by applying dispersion relations.[8] In the Argand plot of the $\left[\phi_1(0) - \phi_3(0)\right]$ amplitude, we observe a clear resonance-like behavior around the incident proton momentum of 1.5 GeV/c (mass $\sim$ 2260 MeV) and possibly at 1.2 GeV/c (mass $\sim$ 2100 MeV) as shown in Fig. 6. So far, such a behavior is discussed only at $|t| = 0$, and the properties of the possible resonance (e.g., spin and parity) are not determined. One needs to study the angular distributions of the observables in pp scattering.

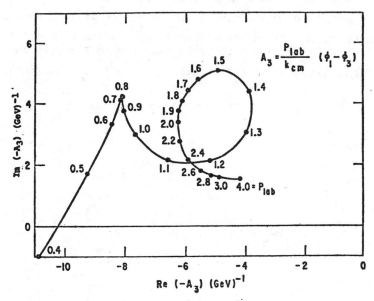

Fig. 6  Argand plot of the $\left[\phi_1(0) - \phi_3(0)\right]$ .

When the helicity amplitudes are decomposed into partial waves,

$$\text{Im } \phi_1(0) = \frac{1}{k} \sum_J \text{Im}\left\{(2J + 1)R_J + (J + 1)R_{J+1,J} + JR_{J-1,J} + 2\left[J(J+1)\right]^{\frac{1}{2}}R^J\right\}, \quad (4)$$

$$\text{Im } \phi_2(0) = \frac{1}{k} \sum_J \text{Im}\left\{-(2J + 1)R_J + (J + 1)R_{J+1,J} + JR_{J-1,J} + 2\left[J(J+1)\right]^{\frac{1}{2}}R^J\right\}, \quad (5)$$

and

$$\text{Im } \phi_3(0) = \frac{1}{k} \sum_J \text{Im}\left\{(2J + 1)R_{JJ} + JR_{J+1,J} + (J + 1)R_{J-1,J} - 2\left[J(J + 1)\right]^{\frac{1}{2}}R^J\right\}, (6)$$

where k is the center-of-mass momentum, $R_J$ is the spin-singlet
partial-wave amplitude with J = L = even, $R_{JJ}$ and $R_{J\pm1,J}$ are spin-
triplet waves with J = L = odd and J = L $\mp$ 1 = even, respectively,
and $R^J$ is the mixing term.  The partial wave characteristics to
$\phi_1(0)$ is $R_J$ and to $\phi_3(0)$ is $R_{JJ}$; the peak in Fig 4a is due to
one of the singlets $^1S_0$, $^1D_2$,..., and the dip is due to one of the
triplets $^3P_1$, $^3F_3$,... .
    Let us see if we observe a similar structure in other channels.

Figure 7 shows the elastic total cross section, $\sigma_{el}^{Tot}$.[9,10] There is also a structure in the plot of polarization against incident momentum at fixed $|t|$, as shown in Fig. 8.[11] We note that the structure in polarization has nothing to do with the peak in $\Delta\sigma_L$ at 1.2 GeV/c, because the polarization does not include singlet terms.

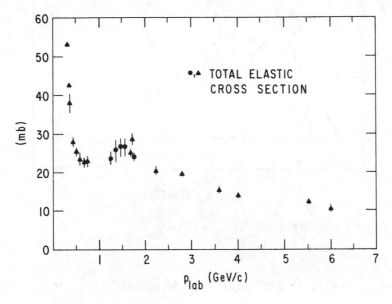

Fig. 7 Elastic total cross section.

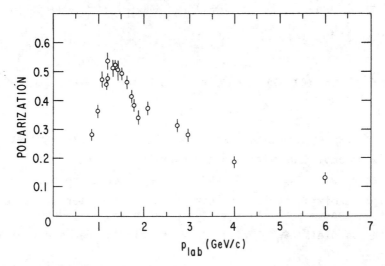

Fig. 8 Polarization at $0.1 < |t| < 0.2$.

We pursue the possibility of the existence of such a resonance and investigate the nature of the possible resonance by
studying differential cross-section and polarization data in pp
elastic scattering around 1.5 GeV/c. Our interest then is to investigate if an $R_{JJ}$ partial wave has the behavior of a Breit-Wigner

formula. The effect of resonances can be studied through the
energy dependence of the Legendre expansion coefficients obtained
from the differential cross-section and polarization data.[10,11] We
report the results of such an analysis using data at 1.0 - 2.0
GeV/c.[12] The analysis was carried out by looking at the energy
dependence of the coefficients $a_n$ and $b_n$ in the expansions:

$$d\sigma/d\Omega = \lambda^2 \sum_{n=0}^{N} a_n P_n (\cos\theta) \tag{7}$$

$$P d\sigma/d\Omega = \lambda^2 \sum_{n=2}^{N} b_n P_n^{(1)} (\cos\theta). \tag{8}$$

These coefficients are related to various partial waves, and we
show only relevant relations here.

The coefficients $a_n$, obtained by fitting differential cross-
section data to Eq. 7, mainly tell us that the highest significant
value of J is four; $a_8$ and higher coefficients are nearly zero, and
we ignore those terms with $J > 4$ and $L > 4$. Figure 9 shows the
coefficients $b_n$ obtained by fitting the product of differential
cross-section and polarization data, plotted against incident
momentum and energy. All coefficients up to and including $b_6$ have
a remarkable energy dependence around $p_{lab} \sim 1.5$ GeV/c. We need
to know if such a rapid change is due to one or two resonant partial
waves while the other amplitudes vary slowly with energy; our
particular attention is on the $R_{JJ}$ partial waves, $^3P_1$ and $^3F_3$. We
determine how the behavior compares with the Breit-Wigner formula,

$$A_{res} = (\varepsilon + i) (\Gamma_{el}/\Gamma)/(\varepsilon^2+1), \text{ where } \varepsilon = 2(E_0 - E)/\Gamma. \tag{9}$$

The energy dependence of this formula is illustrated in Fig. 9.

In general, the coefficient with higher order is easier to
interpret because fewer terms are involved. The coefficient $b_6$
is related to the partial-wave amplitude by

$$b_6 = 1.8 (Im^3F_3 Re^3F_4 - Re^3F_3 Im^3F_4) + ..., \tag{10}$$

where residual terms (...) include neither $^3F_3$ nor $^3P_1$. A rise in
$b_6$ with respect to energy is consistent with $^3F_3$ following the

Breit-Wigner formula while other amplitudes vary slowly with energy; the value of the first term in Eq. 10 is the same both before and after resonance, say at 2110 and 2410 MeV, respectively, and the difference in $b_6$ at these energies is due to the second term, which has $-\text{ReA}_{res}$ changing from minus to plus ($\text{Im}^3F_4 > 0$ by unitarity). We note that $^3P_1$, another possible resonance candidate in $R_{JJ}$, is absent in $b_6$.

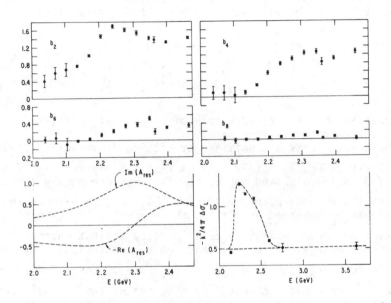

Fig. 9    Results of Legendre coefficient analysis.

To look for other possible resonances and also to accumulate more data around the resonance region, we have measured the spin-spin correlation parameter $C_{LL}(\theta_{c.m.}) = (L,L:0,0)$, which is expressed in terms of s-channel helicity amplitudes.

$$C_{LL}(d\sigma/d\Omega) = -\tfrac{1}{2}\left(|\phi_1|^2 + |\phi_2|^2 - |\phi_3|^2 - |\phi_4|^2\right), \qquad (11)$$

$$(1 + C_{LL})(d\sigma/d\Omega) = |\phi_3|^2 + |\phi_4|^2 + 2|\phi_5|^2, \qquad (12)$$

and

$$(1 - C_{LL}) \, (d\sigma/d\Omega) = |\phi_1|^2 + |\phi_2|^2 + 2|\phi_5|^2, \tag{13}$$

where $d\sigma/d\Omega = \frac{1}{2} (|\phi_1|^2 + |\phi_2|^2 + |\phi_3|^2 + |\phi_4|^2 + 4|\phi_5|^2)$ is a differential cross section. We note here that Eqs. 11 and 13 contain singlet partial waves with different sign. No $R_{JJ}$ waves are in Eq. 13 and only triplet waves are in Eq. 12.

The parameter $C_{LL}$ is determined by measuring $I^{++}$, $I^{+-}$, $I^{-+}$, and $I^{--}$.

$$C_{LL} = \frac{1}{P_B P_T} \frac{(I^{++} + I^{--}) - (I^{+-} + {}^{-+})}{(I^{++} + I^{--}) + (I^{+-} + {}^{-+})} \, .$$

The measurements covered the angular region of $\theta_{c.m.}$ = 60 to $90^\circ$.[13] The results at $p_{lab}$ = 1.0 to 2.5 GeV/c are shown in Fig. 10. We observe dip structures at 1.2 and 2.0 GeV/c in Fig. 11a. These structures already appeared in Figs. 4a and 4b. In Fig 11a we examine the behavior of the ${}^3F_3$ (2260) diproton resonance around 1.5 GeV/c. The predicted curve, obtained by using the results of partial-wave analyses for nonresonant waves and the resonance is consistent with the data. The dip at 1.2 GeV/c is due to a singlet term because we observe a peak in Fig. 11c at 1.2 GeV/c. This is the first evidence of spin singlet structure that appeared in the elastic-scattering channel. Because of a structure in pp→πd at 1.2 GeV/c, there was speculation that resonance existed (${}^1D_2$ according to the phase-shift analyses). Also, the structure was considered to be an s-wave NΔ production, which is fed from ${}^1D_2$ state in the pp system. A new phase-shift analysis using this result should help to clarify the nature of the structure.

We would like to point out other independent work on the diproton resonances. Possible candidates in the ${}^1D_2$ (2100 MeV) and ${}^3F_3$ (2260 MeV) are shown in Hoshizaki's phase-shift analysis.[14] Earlier the possible existence of ${}^1D_2$ resonance has been discussed by several authors.[15]

Do we observe any structure above a mass of 2500 MeV? Measurements of $\Delta\sigma_L$ and $\Delta\sigma_T$ are yet to be made at small momentum interval. But we observe a remarkable energy dependence in $C_{NN}$ = (N,N;0,0) data at all angles[16] and also at $\theta_{c.m.}$ = $90^\circ$.[17] As shown in Fig. 10 $C_{LL}$ = (L,L;0,0) data are all positive at large angle up to 2.5 GeV/c, but we observe a value as large as -35% at 6 GeV/c.[13]

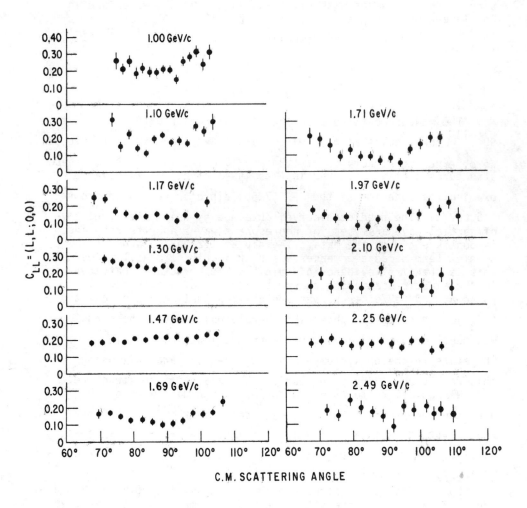

Fig. 10  $C_{LL}$ = (L,L;0,0) data at 1.0 to 2.5 GeV/c.

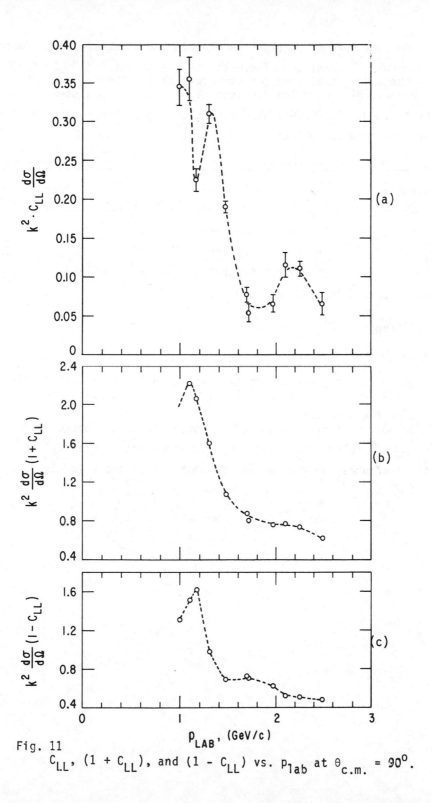

Fig. 11
$C_{LL}$, $(1 + C_{LL})$, and $(1 - C_{LL})$ vs. $p_{lab}$ at $\theta_{c.m.} = 90^o$.

There are several attempts to explain the structure in $\Delta\sigma_L$ data by Deck models,[18] opening of inelastic channels, etc.[19,20] So far, none of the models explained the structure at 1.5 GeV/c.

We have discussed the four structures in the pp system including one of them, $^3F_3$ (2260), as a strong candidate for resonance. We summarize them as follows:

| Lab Momentum GeV/c | Mass, Mev | Spin State | Spin Parity | Measured Observables* |
|---|---|---|---|---|
| 1.2 | 2100 | Singlet | $2^+$ | $\Delta\sigma_L$, $\Delta\sigma_T$, $C_{LL}$ |
| 1.5 | 2260 | Triplet | $3^-$ | $\Delta\sigma_L$, $\Delta\sigma_T$, P, $C_{LL}$ |
| 2.0 | 2500 | Singlet | ? | $\Delta\sigma_L$, $\Delta\sigma_T$, $C_{LL}$ |
| >3.0 | 2800-3000 | Singlet | ? | $C_{NN}$, $C_{LL}$ |

* In addition to $\sigma^{Tot}$ and $d\sigma/d\Omega$.

We tentatively assign the singlet state to the fourth candidate (Mass >2800), because there is no unusual energy dependence observed in the polarization data.[16]

We note that predictions of diproton resonances based on the MIT bag model include $^3F_3$ (2260). It is important to clarify the rest of structures for the further development of quark models.

REFERENCES

1. R. L. Jaffee, Phys. Rev. Lett. <u>38</u>, 195 (1977) and Errata <u>38</u>, 617 (1977).
2. P. J. G. Mulders, A. Th. M. Aerts, and J. J. DeSwart, THEF-NYM-78.1.
3. H. Lipkin, private communication.
4. I. P. Auer et al., Phys. Lett. <u>67B</u>, 113 (1977).
5. I. P. Auer et al., Phys. Lett. <u>70B</u>, 475 (1977).
6. I. P. Auer et al., to be published.
7. For $\Delta\sigma_T$ data see W. deBoer et al., Phys. Rev. Lett. <u>34</u>, 558 (1975); E. K. Biegert et al., to be published in Phys. Rev. Lett.
8. P. Kroll, private communication; for earlier work see N. Grein and P. Kroll, Wuppertal report, WU B77-6.
9. A compilation of NN and ND interactions, UCRL-20000 NN (1970).
10. To obtain the total elastic cross section from 1.2 to 1.7 GeV/c, we have integrated the differential cross-section data by B. A. Ryan et al., Phys. Rev. <u>3</u>, 1 (1971). These data were used because of internal consistency.
11. M. G. Albrow et al., Nucl. Phys. <u>B23</u>, 445 (1970).
12. K. Hidaka et al., Phys. Lett. <u>70B</u>, 479 (1977).
13. I. P. Auer et al., to be published.
14. N. Hoshizaki, private communication; N. Hoshizaki, Prog. Theor. Phys. <u>57</u>, 1099 (1977).
15. R. A. Arndt, Phys. Rev. <u>165</u>, 1834 (1968); G. L. Kane and G. H. Thomas, Phys. Rev. <u>D13</u>, 2944 (1976); L. M. Libby and E. Predazz, Lettere Al Nuovo Cimento, Vol. II, N. 18, 881 (1969); J. H. Hall et al., Nucl. Phys. <u>B12</u>, 573 (1969); H. Suzuki, Prog. Theor. Phys. <u>54</u>, 143 (1975) and earlier references therein.
16. D. Miller et al., Phys. Rev. Lett. <u>36</u>, 763 (1976); Phys. Rev. <u>16D</u>, 2016 (1977).
17. A. Lin et al., UM HE 78-3 (1978).
18. E. L. Berger, P. Pirila and G. H. Thomas, ANL-HEP-75-72.
19. W. M. Koet et al., LA-UR-77-2321.
20. M. Arik and P. G. Williams, preprint, Westfield College (1977).

# STATUS OF BARYONIUM EXPERIMENTS*

DAVID H. MILLER

PURDUE UNIVERSITY

WEST LAFAYETTE, INDIANA  47907

## ABSTRACT

The experimental evidence for resonant states coupled to the baryon - antibaryon system is reviewed.  The evidence for broad states known as the S(1963), T(2190) and U(2350) is compelling, however, there is no evidence for I = 2 states and the observation of narrow states needs confirmation.

## INTRODUCTION

It has long been conjectured that meson resonances should exist with strong couplings to the baryon - antibaryon system. The original proposals were made in connection with duality[1,2]. Over the last year or so there has been renewed interest in such states and numerous theoretical papers have resulted.  The sub structure of such states is normally that of a diquark and an anti diquark which naturally leads to preferred decays to $B\bar{B}$.  There is considerable disagreement on the predicted widths and spins of such objects although there is general agreement that I = 0, 1, 2 states should exist, closely degenerate in mass.  It remains, however, an experimental question as to wether such states exist, and it is to this that we will address ourselves.

* Work supported in part by the U.S. Department of Energy.

## STATUS OF BARYONIUM EXPERIMENTS

### TYPES OF EXPERIMENTS

Since it is clear that conventional mesons lying on leading and daughter Regge Trajectories exist in nature, experiments to detect baryonium states look at reactions which are favourable to $B\bar{B}$ coupling. The main classes of such experiments are $\bar{p}N$ formation, u channel production via $N$ or $\Delta$ exchange and t channel production with decays involving $B\bar{B}$. This review will divide this data into exotic searches, production of narrow states, production of broad states and the S meson followed by a summary and conclusions. Of necessity this review will concentrate on recent results so the reader is recommended to also read the excellent review by L. Montanet.[3]

### EXOTICS

Much data has been published over the last decade in the search for charge 2 meson resonances. To date there is no firm evidence that such states exist. Two recent papers have presented results of a search for charge 2 mesons produced by baryon exchange.

The first[4] has studied the reactions:

$$\pi^-_- p \rightarrow p_f X^-_{--}$$
$$\pi^- p \rightarrow p_f X^-$$

at 12 GeV/c

with a mass resolution for X between 12 and 18 MeV. They find no evidence for resonant structures in X between 2 Mn and 3 GeV, with upper limits in the 10-50 nano barn range. The final states of the X system were $p\bar{p}\pi^-$, $p\bar{p}\pi^-\pi^-$, $p\bar{p}\pi^-\pi^o$, $\pi^+\pi^-\pi^-\pi^-$, $\pi^+\pi^-\pi^-\pi^o$. A Typical histogram is shown in Fig 1.

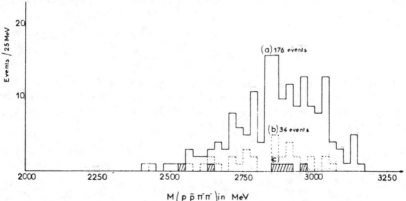

Fig. 6. Invariant mass distributions of the $(p\bar{p}\pi^-\pi^-)$ system in the reaction $\pi^-d \rightarrow$ Pspect Pforward $pp\pi^-\pi^-$. (a) 2C fits with Pforward between 7 and 10 GeV/c. (b) Events of (a) with a $p_{slow}$ forward in the $(pp\pi^-\pi^-)$ rest system. (c) Events of (b) with both $\Delta^0$ in $(p\pi^-)$ and $(\Delta)^{--}$ in $(\bar{p}\pi^-)$ mass. One event in histogram (a) represents a cross section of 1.7 nb.

### Fig 1. The data of Ref 4.

STATUS OF BARYONIUM ECPERIMENTS

A similar experiment has produced results from:

$$\pi^- d \to N_s P_f X^-$$
$$\pi^- d \to N_s P_f X^{--}$$

at 13.2 GeV/c

in the SLAC streamer chamber.[5]  No structure is observed with limits $\sim$ 250 nano barns.  The mass resolution is 25 $\to$ 50 MeV and a typical histogram is shown in Fig 2.

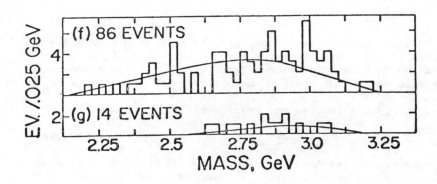

Fig 2.  The data of reference 5.
Shown is $X^- \to p\bar{p}\pi^-$ and $X^{--} \to p\bar{p}\pi^-\pi^-$ in (f) and (g) respectively.

NARROW ($\Gamma$ < 30 MeV) STATES

Over the years there have been many claims for narrow states in the 2 $\to$ 3 GeV mass region.[6]  Recently there have been two experiments which have presented strong evidence for such states. The first [7] has studied the reaction:

$$\pi^- p \to (p\pi^-)_f \; p\bar{p}$$

at 9 and 12 GeV/c

and find evidence for two states at:

$$M = 2020 \pm 3 \qquad \Gamma = 24 \pm 12$$
$$2204 \pm 5 \qquad \Gamma = 16 \pm^{20}_{16}$$

in the $p\bar{p}$ system produced opposite $\Delta$ (1238) and N (1520).  The observed cross section is $\sim$ 15 $\to$ 30 nano barns.  These results are shown in Fig 3.

STATUS OF BARYONIUM EXPERIMENTS

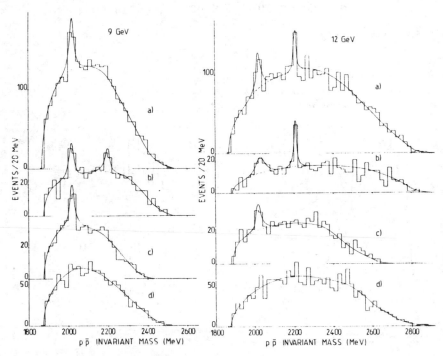

Fig. 1. The distribution of the p̄p invariant mass
(a) All the events. (b) Events with invariant mass p̄pπ⁻ in
Δ°(1232) region 1175 < $M(p_F\pi^-)$ < 1300 MeV. (c) Events
with invariant mass p̄pπ⁻ in the N°(1520) region 1450 <
$M(p_F\pi^-)$ < 1600 MeV. (d) Events with invariant mass p̄pπ⁻
outside the regions of b and c. The full curves represent the
fit of the data with a smooth background and one or two
Breit Wigner resonances. The dotted curve under the peak represent the contribution of the background.

## Fig 3.   The data of Ref 7.

One problem not fully resolved is that these states have not
been observed in p̄p formation although the data does not rule them
out either. The second experiment[8] published recently observes a
peak in the p̄pπ⁻ system produced in the reactions:

$$\pi^- p \to (p\bar{p}\pi^-)_f \; P$$
$$p\pi^o$$
$$p\pi^+$$
$$p\pi^+\pi^-$$

at 16 GeV/c

with properties:

$$M = 2950 \pm 10 \qquad \Gamma \leq 32 \qquad \sigma \sim 1\mu b$$

This is shown in Fig 4.

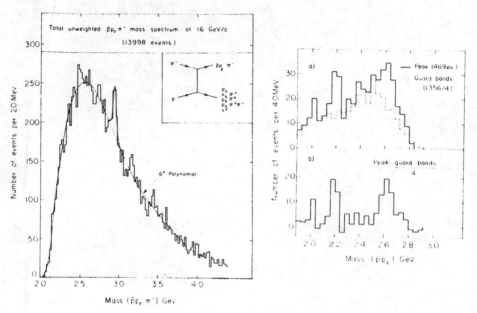

Fig. 3. Total $\bar{p}p_f\pi^-$ effective mass distribution in 20 MeV bins (13 998 events).

Fig 4.   The data of Reference 8.

Of interest is that the $\bar{p}p$ mass spectrum associated with the 2.95 Peak itself has peaks at 2.02 and 2.20, this is also shown in Fig 4. The data is clearly consistent with intermediate decays involving the states discussed previously. There is no evidence in the data for an intermediate decay involving $\Delta$, $\bar{\Delta}$, N* or $\bar{N}$*. Since the width is small it suggests that decays to other channels are inhibited which means that the production mechanism is probably not conventional meson exchange. In fact the data could be interpreted as production via baryonium exchange.

In summary, although the peaks discussed have good statistical significance, confirmation with more data is clearly desirable.

STATUS OF BARYONIUM EXPERIMENTS

## BROAD STATES $\Gamma > 30$

Structures in total and partial cross sections were first observed several years ago.[9] They consist of broad bumps with masses $\sim$ 2190 and 2350 with cross sections of several milli-barns. The 2190 has isospin 1 but the 2350 is a mixture of isospin 0 and 1. A recent paper [10] contains results from a $\bar{p}p$ elastic scattering experiment and the structures can clearly be seen (Fig 5).

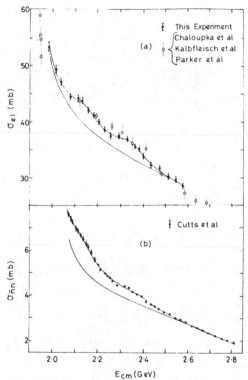

Fig. 1. (a) $\bar{p}p$ total elastic cross sections from our experiment and from a sample of bubble chamber experiments. (b) Cross sections for $\bar{p}p \rightarrow \bar{n}n$ from [12]. In both parts the upper curve is the two-resonance fit described in the text. The lower curve is the background contribution to it.

Fig 5.   The data of Ref 10.

The same institutions have also published results[11] from the reaction $\bar{p}p \rightarrow \pi^+\pi^-$ using a polarised target. These results give strong evidence for spin 3, 4, and 5 mesons in this overall mass region. A summary of these results taken from Ref 10 is shown in Table 1 where in addition the old results on $\sigma$TOT have been fitted by the

# STATUS OF BARYONIUM EXPERIMENTS

same procedure. The determination of masses and widths show good agreement within the errors. One problem is wether the p̄p → π π resonances at 2480 and 2310 are in fact part of the U bump at 2350 and 2385. Another is the low cross section for p̄p → n̄n relative to p̄p → p̄p. In order to explain this one can invoke strong interferences with states or background of the same $J^P$, however, one does need to do this for all the observed structures. The T and U do have the required property for baryonium in that their compling to BB̄ is large, the coupling to π π is ~ 10%.

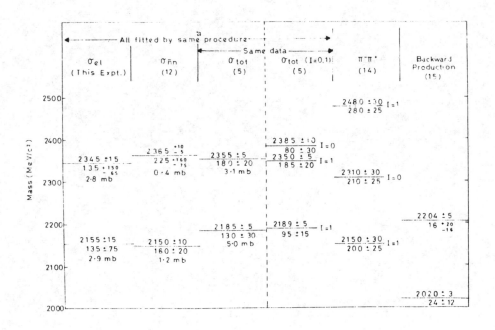

Table 1. Taken from Ref 10.

## S MESON

This state is also seen in the total cross section of p̄p[12] with a magnitude ~ 10 mb and in the elastic channel with σ ~ 7 mb.[13] It appears to be dominantly I = 1 but an I = 0 component is also possible. It differs from the T and U in that its width is small. Its properties averaged over several experiments are :

$$M = 1963 \pm 1 \qquad \Gamma \ 4 \rightarrow 8$$

STATUS OF BARYONIUM EXPERIMENTS

Once again the n̄n decay is very small[14] meaning that strong interferences have to be involved to explain the data.  Some typical
data[15] is shown in Fig 6.  The peak is seen clearly in σT and σel
but is not obvious in σo which is dominantly n̄n.

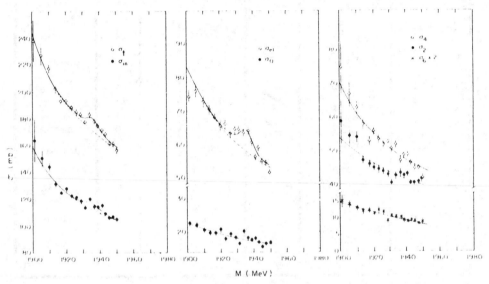

Fig 6.  The data of Ref 15.

## BOUND STATES

A recent experiment on p̄p at rest[16] has observed discrete rays
which they interpret as evidence for bound states of the p̄p system.
An older experiment[17] has also seen evidence for a bound state
just below p̄p threshold by looking at p̄n interactions with the
neutron off mass shell.  The masses of these states do not coincide with states previously observed in other decay modes in conventional spectroscopy.  It is possible that they belong to a class
analogous to the deuteron rather than having constituent diquarks.
More experimental data is clearly needed before we reach an understanding of these states.

## SUMMARY

Very few states exist which couple dominantly to BB̄ and have
been confirmed experimentally.  Table 2 shows a list of the states
we have reviewed with an indication of their status.  Only the S,
T, U are on a firm footing and even then the small n̄n decays need
to be understood in more detail.

## STATUS OF BARYONIUM EXPERIMENTS

| MASS | MODE | STATUS | $\Gamma$ | J | I | COMMENTS |
|---|---|---|---|---|---|---|
| 1395<br>1646<br>1684 | $\bar{p}p \to \gamma$ | | small | | | |
| 1897 ± 1 | $\bar{p}n$ | | 25 ± 6 | > 0 | 1 | |
| 1936 | $\bar{p}p$, $\bar{p}n$ | ✓ | 4 → 8 | | 1,0 | $n\bar{n}$ small |
| 2190 ± 10<br>2350 ± 15 | $\bar{p}p$<br>$\sigma_T$, $\sigma_{el}$, $\pi\pi$ | ✓<br>✓ | 150<br>180 | 3 | 1 | $n\bar{n}$ small |
| 2385 ± 10 | $\sigma_T$ | ✓ | 100 | 4 | 0 | $n\bar{n}$ small |
| 2480 | $\bar{p}p \to \pi\pi$ | | 280 | 5 | 1 | part of U? |
| 2020 ± 3<br>2204 ± 5 | U channel<br>with $\bar{p}p$<br>decay | | 24 ± 12<br>16 ± $^{20}_{16}$ | | | $\sigma \sim$ 15 nb<br>$\sigma \sim$ 19 nb |
| 2950 ± 10 | t channel<br>$p\bar{p}\pi$ decay | | ≤ 20 | | ≥ 1 | ≤ 1μb |
| 2850<br>3050 | $\bar{p}p$<br>↓<br>$\pi X$ | | ≤ 39<br>≤ 15 | | | $\sigma \sim$ 83 μb<br>$\sigma \sim$ 22 μb |

### FUTURE

Experiments looking for baryonium states are long and therefore results are difficult to confirm.  There are, however, several
experiments now under way which will help to clarify the situation.
First the groups[8] which observed the 2.95 have now completed a new
experiment at the same energy with a slightly different trigger
which will yield a factor of ten increase in some of the channels
in which the enhancement was observed.  This data is being analyzed
and should be available within six months.  Two experiments are
utilizing the line reversed reaction to the u channel processes
discussed, that is, using incident antiprotons with a fast forward
meson ($\pi^+$ for example)[18].  Preliminary results from one of these
were presented at this conference.

The crucial information needed apart from confirmation of the
states listed is the spin and parity and decay modes of such states.
In particular the spin helps to differentiate between a number of
models[19]. In addition the observation of a manifestly exotic state
is vital to the  theory of baryonium.  All these questions can be
answered by good experimental data, some of which should be available within a year.

# STATUS OF BARYONIUM EXPERIMENTS

## REFERENCES

1.  J. Rosner, Phys. Rev. Letters 21, 950, (1968).
    J. Rosner, Physics Reports 11C, 189, (1974),

2.  P. G. O. Freund et al., Nucl. Phys. B13, 237, (1969).

3.  L. Montanet, Experimental Meson Spectroscopy, 260, (1977).

4.  J. Boucrot et al., Nucl. Phys. B121, 251, (1977).

5.  M. S. Alam et al., Indiana, Purdue, SLAC, Vanderbilt preprint.

6.  For a complete survey see the latest Particle Data Group
    compilations.

7.  P. Benkheiri et al., Phys. Letters 68B, 483, (1977).

8.  C. Evangelista et al., Phys. Letters 72B, 139, (1977).

9.  R. J. Abrams et al., Phys. Rev. D1, 1917, (1970).

10. M. Coupland et al., Phys. Letters 71B, 460, (1977).

11. A. A. Carter et al., Phys. Letters 67B, 117, (1977)

12. A. S. Carroll et al., Phys. Rev. Letters 32, 247, (1974).

13. W. Bruckner et al., Phys. Letters 67B, 222, (1977)

14. M. Alston - Garnjost et al., Phys. Rev. Letters 35,1685(1975).
    D. Cutts et al., Sub. to Phys. Rev.

15. V. Chaloupka et al., Phys. Letters 61B, 487, (1976).

16. P. Pavlopoulos et al., Phys. Letters 72B, 415, (1978).

17. L. Gray et al., Phys. Rev. Letters 26, 1491, (1971).
    T. E. Kalogeropoulos et al., Phys. Rev. Letters 34, 1047,
    (1975).

18. One experiment will utilize the SLAC Hybrid Faculity, the
    other the Brookhaven MPS.

19. A recent theoretical review was given by C. Rosenzweig at the
    1977 D.P.F. meeting.

# SEARCH FOR NARROW STATES IN THE
## REACTION $\bar{p}p \rightarrow \pi^+_f X^-$ AT 6 GeV/c[†]

D. R. Green

Carnegie-Mellon University, Pittsburgh, Pennsylvania 15213

## ABSTRACT

The reaction $\bar{p}p \rightarrow \pi^+_f X^-$ at 6 GeV/c has been studied in a search for narrow states in the $X^-$ decays into $\bar{p}n$, $\pi^-\pi^+\pi^-$, $\pi^-K^+K^-$, and $\pi^-\bar{p}p$. Narrow enhancements are seen at masses of 1940, 2040 and 2175 MeV. The cross sections for these enhancements are typically $\sigma B \sim 50$ nb and the widths are $\Gamma \lesssim 25$ MeV.

## INTRODUCTION

We describe here a subset of the data from an experiment performed at the Brookhaven National Laboratory (B.N.L.) Multiparticle Spectrometer (MPS). The experiment was designed to study a wide variety of low mass two-body and quasi-two-body baryon exchange reactions[1,2,3] initiated by an incident $\bar{p}$ beam. The sensitivity of the experiment was typically 1 event/nb. The missing mass resolution was determined from elastic scattering data to be ±15 MeV (σ) at 2 GeV missing mass. The missing mass range was MM ≤ 2.5 GeV while the momentum transfer range was -t' ≤ 1.0(GeV)². The solid angle coverage was essentially 4π. Data were taken at 4, 6, and 8 GeV/c. Only the 6 GeV/c data have been fully analyzed. A comparable data set at 8 GeV/c will be completed momentarily.

This data is complementary to that taken in an experiment at the C.E.R.N. Omega (Ω) spectrometer,[4] which reported narrow states in $\pi^-p \rightarrow p_f(\bar{p}p)$ at masses of 2020 and 2204 MeV. Note that this reaction is mediated by N and Δ exchange while $\bar{p}p \rightarrow \pi^+_f X^-$ is mediated by Δ exchange alone. If the same enhancements are seen in our data, then one can conclude that the isotopic spin is one, I=1. In addition, our data covers the mass of 1940 MeV where a narrow state is observed in $\bar{p}p$ formation.[5]

One can consider the implications of our data in the context of a four quark, diquark-anti-diquark model.[6] The SU(3) content of a 4-quark state is:

$$q^2 \bar{q}^2 \quad 3 \times 3 \times \bar{3} \times \bar{3} = (1+8) + (8+10) + (\bar{8}+\overline{10}) + (1+8+27)$$

Note that:

$$\bar{N} p \quad \bar{8} \times 8 \quad = (1+8) + (8+10) + (\quad \overline{10}) + (\quad 27)$$

While:

$$\bar{\Delta} p \quad \overline{10} \times 8 \quad = \quad\quad\quad (8+10) + (\quad 27) + 35$$

†Work performed under the auspices of the U.S. Department of Energy.

Thus, in a diquark model explicitly exotic states (contained in the 10 representation) will be fairly near in mass to the observed enhancements.

## APPARATUS

A schematic layout of the apparatus is shown in Fig. 1. The incident $\bar{p}$ beam from the Medium Energy Separated Beam (MESB) had a typical $\bar{p}$ flux of 40K/burst and a $\bar{p}/\pi^-$ ratio of 2. The incident beam was momentum analyzed to ±0.25% and tagged by 2 beam Cerenkov counters. The MPS magnet contains 10 kG over a volume 15' × 8' × 4'. The hydrogen target was surrounded by 7 cylindrical wire chambers viewing angles $\theta > 25^o$ and by 25 planar spark chamber gaps viewing $\theta < 25^o$.

The trigger for this data was $\bar{p} \cdot [\sum\limits_{i}^{45} \sum\limits_{j}^{112} (H_4)_i \cdot (H_5)_j] \cdot C_6$. $C_6$ was a one atmosphere $F^{114}$ Cerenkov counter with efficiency >96% over the active area. The main feature of the trigger was that the two scintillation counter hodoscopes $H_4$ (45 elements) and $H_5$ (112 elements) were used in a matrix coincidence. The target point, $H_4$, and $H_5$ then comprise a crude 3-point trajectory yielding charge momentum and angle selection corresponding to $0 \le M_{X^-} \le 2.5$ GeV and $-t'| \le 1.0 (GeV)^2$.

The spark chamber module labeled as E provided improved momentum resolution for the $\pi_f^+$ track. Our momentum resolution was monitored by taking subsidiary forward elastic data. The momentum resolution for the elastic data was ±0.85% at 6 GeV/c corresponding to ±15 MeV missing mass resolution at $M_{X^-} = 2.0$ GeV. This improvement is crucial in a search for narrow states since the scale of $\Gamma \sim 15$ MeV is set by the widths reported by the $\Omega$ experiment.[4]

## EVENT RECONSTRUCTION

The data was first analyzed through pattern recognition programs to convert sparks to tracks. Then extra beam tracks were deleted and the $\pi_f^+$ track was found by requiring that it satisfy all the trigger track requirements. All other tracks were defined to be recoil tracks.

A vertex was found which consisted of the intersection of the beam track $(\bar{p})$, the trigger track $(\pi_f^+)$, and all other recoil tracks within some limits. The non-vertex associated tracks were then deleted. The $\bar{p}$ and $\pi_f^+$ were propagated to the vertex and missing mass $M_{X^-}$ and momentum transfer $(t' \equiv t - t_{min})$ were formed. The resulting missing mass spectra for 1 and 3 prong $X^-$ decays are shown in Figs. 2a) and 2b) respectively. The arrows are the locations of the previously mentioned 1940, 2020, and 2204 MeV states. Even at this level of analysis, an enhancement is seen at a mass of ~2175 MeV of 125 events above a background of 1,000 events.

The $n_{X^-} = 1,3$ events were then further kinematically processed. For $n_{X^-} = 1$ events the missing neutral momentum was solved for and the assignment of decay masses was made on the basis of the best energy balance. The resulting energy imbalance is shown in Fig. 3a).

A cut of ±25 MeV was imposed which yields a signal/background ratio
of ~1.

For typical $n_{\chi-} = 3$ decays the angles were well measured but the
lever arm in the field was insufficient to measure the momentum of
the recoils accurately. For this reason, the momenta were solved for
using momentum conservation and masses were assigned on the basis of
the best energy balance. The energy imbalance is shown in Fig. 3b).
A cut was imposed at ±50 MeV resulting in a signal/background ratio of
~3. It should be noted that essentially no misidentification of a
$\pi^-\pi^+\pi^-$ decay as a $\pi^-K^+K^-$ or $\pi^-p\bar{p}$ decay occurs since the recoil tracks
are slow. Therefore, the energy imbalance is ~0, 1 GeV, and 2 GeV
respectively.

## DATA ANALYSIS AND CUTS

For all of the subsequent data a cut of $\left|t'_{\bar{p}\to\pi_f^+}\right| < 0.5$ $(\text{GeV})^2$
was imposed. This cut enhances the signal due to peripheral baryon
exchange reactions. Since the logarithmic slopes we observed were
$b \sim 3$ to $5$ $(\text{GeV})^{-2}$, this cut passes $\geq 80\%$ of the baryon exchange
peak.

The missing mass spectrum for events identified as $\bar{p}p \to \pi_f^+ (\bar{p}n)$
is shown in Fig. 4a). An enhancement is observed at a mass of
2040 MeV consisting of 70 events above a background of 190 events.
No clear enhancements are seen at masses of 1940 or 2200 MeV.

A source of high mass background in this reaction arises from
events of the type $\bar{p}p \to (\pi_f^+ \bar{p})n$ where low mass $(\pi_f^+ \bar{p})$ causes a kinematic reflection into high mass $(\bar{p}n)$. This background was studied
by displaying $t'_{\bar{p}\to n}$ vs. $M_{\pi_f^+ \bar{p}}$. Indeed, low mass peripheral enhancements are observed which were then partially removed. Events were
rejected if $M_{\pi_f^+ \bar{p}} < 2.5$ GeV and $\left|t'_{\bar{p}\to n}\right| < 0.5$ $(\text{GeV})^2$. The resulting
mass spectrum is shown in Fig. 4b). The enhancement at 2040 MeV
persists and its signal/background ratio is improved, indicating
that kinematic reflection is not the source of this enhancement.
(60 events above a background of 80 events). A hint of an enhancement at 2175 MeV exists in Fig. 4b) but no firm conclusion can be
drawn.

The missing mass spectrum of events identified as $\bar{p}p \to \pi_f^+(q^-q^+q^-)$
with no missing neutral $(\left|\delta E_3\right| < 50$ MeV) is shown in Fig. 5a). Nar-
row enhancements are seen at masses of 1940 MeV (40 events above a
background of 125) and at 2175 MeV (70 events above a background of
180). The missing mass spectrum for the particular reaction
$\bar{p}p \to \pi_f^+(\pi^-K^+K^-)$ is shown in Fig. 5b). Again one sees enhancements at
1940 MeV (30 events above a background of 90) and at 1975 MeV (55
events above a background of 120). No convincing evidence exists
for the 2040 MeV enhancement seen in $\bar{p}n$ decays.

The shaded regions in Fig. 5 consist of events which pass additional cuts designed to remove kinematic reflections. The important
point is that these enhancements persist after the cuts are imposed.

Consider only the reaction $\bar{p}p \to \pi_f^+(\pi^-K^+K^-)$. A kinematic reflection occurs for two possible allowed baryon exchange quasi two-
body processes; $\bar{p}p \to (\pi_f^+ K^-)(K^+\pi^-)$ and $\bar{p}p \to (\pi_f^+\pi^-)(K^+K^-)$. The two-

body effective mass spectra for these two pairings are shown in
Fig. 6. In Fig. 6a) one sees a low mass ($\pi_f^+ K^-$) clustering while in
Fig. 6c) one sees a low mass ($\pi_f^+\pi^-$) clustering effect indicating that
events from both of the processes $\bar{p}p \to K^{*O}(K^+\pi^-)$ and $\bar{p}p \to \rho^O(K^+K^-)$
exist in the data sample. We reduce the backgrounds due to these
sources by removing events from the sample if $M_{\pi_f^+ K^-} < 1.7$ GeV and

$$|t'_{\bar{p}\to\pi_f^+ K^-}| < 0.25 \text{ (GeV)}^2 \text{ or if } M_{\pi_f^+ \pi^-} < 1.5 \text{ GeV and}$$

$$|t'_{\bar{p}\to\pi_f^+ \pi^-}| < 0.25 \text{ (GeV)}^2.$$

The recoil mass distributions $M_{K^+\pi^-}$ and $M_{K^+K^-}$ are shown in Figs.
6b) and 6d) respectively, where the shaded regions correspond to
events which pass the cuts on the forward two-particle cluster.
Clearly, events of the type $\bar{p}p \to K^{*O}K^{*O}$ and $\bar{p}p \to \rho^O\phi^O$ have been removed by these cuts. The remaining events show no evidence of two-
body recoil mass resonant substructure.

In Fig. 7a) is shown the distribution of events of the type
$\bar{p}p \to \pi_f^+ (\pi^-K^+K^-)$ which pass the cuts described above. The enhancement at 1940 MeV persists as 12 events on a background of 30. The
2175 MeV enhancement is 21 events on a background of 40. As mentioned previously no clear evidence of resonant two-body substructure
in the ($\pi^-K^+K^-$) system remains after the cuts. However, one can
attempt to see whether the decays lead preferentially to low or high
mass two-body subsystems. In Fig. 7b) is shown the missing mass
distribution for the events of Fig. 7a) which have $M_{K^+K^-} < 1.5$ GeV.
In this sample the 1940 MeV enhancement is roughly 10 events on a
background of 23 events while the 2175 MeV enhancement is roughly 16
events on a background of 24 events. Clearly the signal/background
ratio is improved.

## CONCLUSIONS

We have studied the reaction $\bar{p}p \to \pi_f^+ X^-$ at 6 GeV/c with $X^-$ decaying into $\bar{p}n$, $\pi^-\pi^+\pi^-$, $\pi^-K^+K^-$, and $\pi^-\bar{p}p$ with a sensitivity of
roughly 1 event / nb over a mass range $0 < M_{X^-} < 2.5$ GeV and t' range
of $|t'| < 1.0$ (GeV)$^2$. The signal/background ratio for $\bar{p}n$ decays is
about 1 while the same ratio for 3-body decays is about 3.

A peripherally produced enhancement is seen in the $\bar{p}n$ channel
at a mass of 2040 MeV as an excess of roughly 70 events above a
background of 190. In the 3-body decays enhancements exist at
masses of 1940 and 2175 MeV at the 40 events above 125 events background level and at the 70 events above 180 events background level
respectively.

These enhancements persist when cuts are made to remove known
sources of high mass kinematic reflections; in addition, the signal/
background level improves from ~1/3 to ~1/2. In the $\pi^-K^+K^-$ decay
mode there is weak evidence that low $K^+K^-$ masses are favored in the
decay.

The observation of these enhancements in the charge one final
state, if they are identified with the $\Omega$ results,[4] allow one to
conclude that I=1. Since $\bar{p}p \to \pi_f^+ X^-$ is mediated by pure $\Delta$ exchange,

in a diquark-antidiquark model,[6] one expects that these states are probably in the (8+10) representations of SU(3).

Since the branching ratios for $\pi^-\pi^+\pi^-$ and $\pi^-K^+K^-$ are not enormously different from that for $\bar{p}n$ seen in this experiment, and those reported for $\bar{p}p$,[4] one expects that the diquark is in a color sextet. These two facts then imply (in the context of this particular model) that explicitly exotic states in the 10 representation of SU(3) will be fairly near in mass, and that they will be reasonably narrow in width.

## REFERENCES

1. N. A. Stein et al., Phys. Rev. Letters 39, 378 (1977).
2. D. R. Green et al., Phys. Rev. Letters 39, 1243 (1977).
3. N. Sharfman et al., Phys. Rev. Letters 40, 681 (1978).
4. P. Benkheirt et al., Phys. Lett. 68B, 483 (1977).
5. A. S. Caroll et al., Phys. Rev. Letters 32, 247 (1974);
   V. Chaloupka et al., Phys. Lett. 61B, 487 (1976).
6. C. H. Mo et al., Phys. Lett. 72B, 121 (1977).

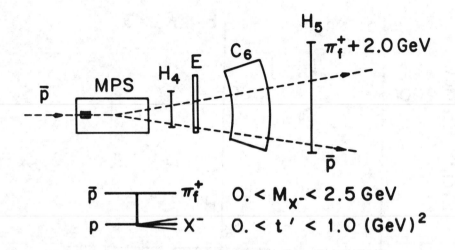

Fig. 1   Schematic layout of the apparatus.

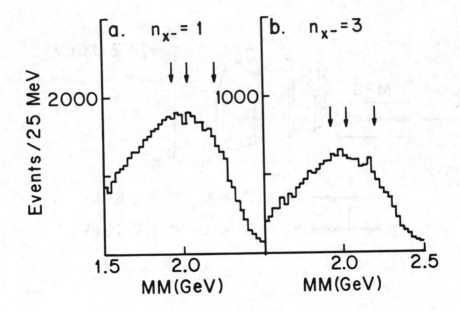

Fig. 2　Missing mass spectra for $\bar{p}p \rightarrow \pi_f^+ X^-$
a) $n_{X^-} = 1$　　b) $n_{X^-} = 3$
The arrows are at 1940, 2020, and 2200 MeV.

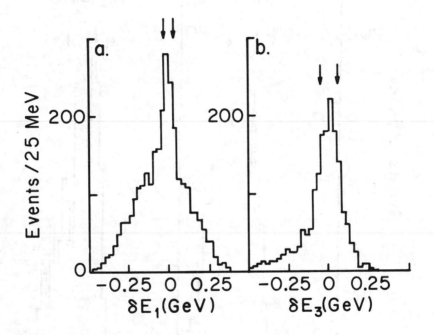

Fig. 3    Energy imbalance for a) $\bar{p}p \rightarrow \pi_f^+ (q^-q^0)$ and b) $\bar{p}p \rightarrow \pi_f^+(q^-q^+q^-)$. The arrows indicate the cut limits.

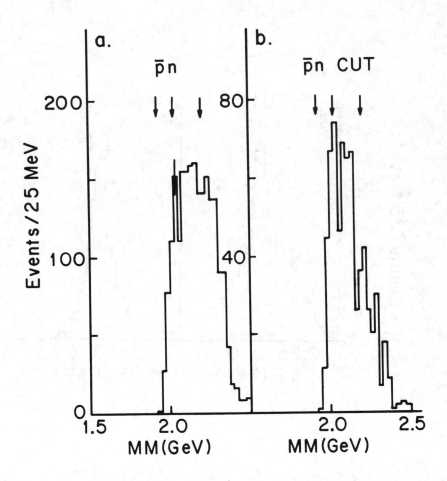

Fig. 4   Missing mass spectra for $\bar{p}p \rightarrow \pi_f^+ (\bar{p}n)$.  The arrows are at 1940, 2020, and 2200 MeV.  a) All events.  b) events passing additional kinematic reflection cuts.

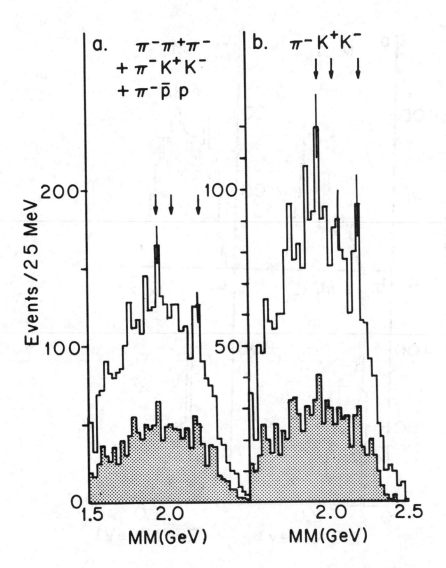

Fig. 5  Missing mass spectra for a) $\bar{p}p \rightarrow \pi_f^+(q^-q^+q^-)$ and $\bar{p}p \rightarrow \pi_f^+(\pi^-K^+K^-)$. Events in the shaded region pass additional kinematic reflection cuts.

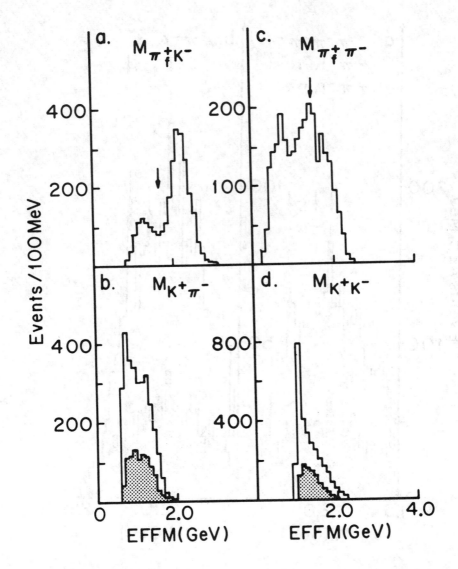

Fig. 6    Two-body submass spectra for events of the type $\bar{p}p \rightarrow \pi_f^+(\pi^- K^+ K^-)$, a) $\pi_f^+ K^-$ mass  b) $K^+\pi^-$ mass  c) $\pi_f^+ \pi^-$ mass  d) $K^+K^-$ mass.  The arrows in a) and c) indicate the cuts.  In b) and d) the shaded events pass the cuts.

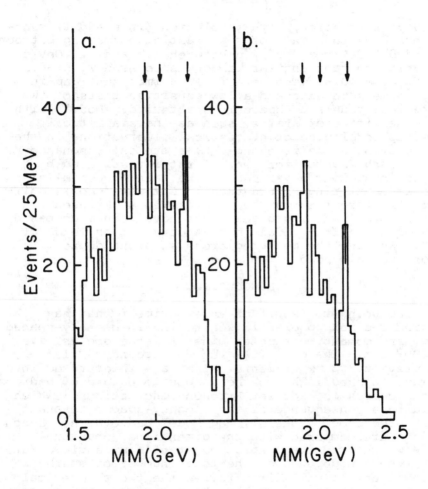

Fig. 7    Missing mass spectra for events of the type $\bar{p}p \rightarrow \pi_f^{\pm}(\pi^- K^+ K^-)$.
a) All events passing the kinematic reflection cuts.  b) Events
in a) with $K^+ K^-$ mass < 1.5 GeV.

# ELASTIC SCATTERING NEAR 90° c.m.:
## TESTS OF PARTON AND RESONANCE MODELS *

L. E. Price
Columbia University, New York, N.Y.   10027

## ABSTRACT

High statistics, high resolution (in s and t) measurements have been made of $\pi^-p$ elastic scattering between 1.9 and 9.5 GeV/c and of $\pi^+p$ between 1.9 and 6.3 GeV/c. A roughly constant angular range was covered with $-0.3 < \cos\theta_{c.m.} < +0.4$. The data fill an important gap in existing measurements and allow sensitive tests of two theoretical models of hadronic scattering. Models with rising densities of states, such as the statistical bootstrap model predict statistical fluctuations in the cross section as $\sqrt{s}$ is changed by a typical resonance width. Such fluctuations were first investigated by Ericson in nuclear physics and have been proposed for hadronic scattering by Frautschi. Our data show significant structure that can be interpreted as Ericson fluctuations. Scaling predictions have been made based on the quark parton model that at constant angle $d\sigma/dt \propto s^{-8}$. We find the measured cross sections do not conform to this prediction in detail.

## INTRODUCTION

I am going to report on an experiment that has measured $\pi^{\pm}p$ and pp elastic scattering in closely-spaced energy and momentum-transfer bins at large angles, i.e. near 90° c.m. The data fill the gap around 90° left by most previous $\pi$p experiments. They are interpreted in two ways to shed light on the composite nature of hadrons. First a search is made for Ericson Fluctuations,[1] which should be a consequence of the closely-spaced resonant states predicted by the MIT Bag Model and others. Second the data are compared with the dimensional counting rule,[2] which predicts that at constant c.m. angle $d\sigma/dt \propto s^{-n}$, where n depends on the total number of quarks in the reaction. I will first review the two theoretical predictions and then describe the experiment and present the data.

## FLUCTUATIONS

The analysis of fluctuations has been made by Ericson[1] in Nuclear Physics, where it has become a standard method of analysis, and extended to hadrons by

ISSN:  0094-243X/78/042/$1.50  Copyright 1978 American Institute of Physics

Frautschi.[3]  They show that when resonances overlap so
that they cannot be individually resolved, there will
still be significant changes in cross sections for c.m.
energy changes on the order of a resonance width, but
the changes will come from variations of the number and
properties of contributing resonances, rather than the
effects of single resonances.  Frautschi[3] and Carlson[4]
have shown that the amount of structure in forward and
backward $\pi^{\pm}p$ elastic scattering cross sections is consistent with this picture.  Such fluctuations should be
easiest to observe in the differential cross section
in regions where exchange contributions are small
relative to s-channel resonances, i.e. at large angles,
away from the forward and backward peaks.

Frautschi[3] has made specific predictions for $\pi^{\pm}p$
elastic scattering based on a statistical bootstrap
model.  He predicts that typical resonance widths, and
therefore the scale for significant cross section fluctuations, will be about equal to the pion mass.[5]  In
fact, Schmidt et al,[6] analyzing data from a CERN experiment[7] at 5 GeV/c, have reported changes in the differential cross section of up to a factor of 3 in $\pi^{+}p$ (but
not $\pi^{-}p$) elastic scattering at large angles, for a change
of only 30 MeV in $\sqrt{s}$.  In the model of multiple overlapping resonances, this result suggested that large
angle $\pi^{+}p$ scattering is dominated by small numbers of
resonances whose width might be quite small.  Confirmation of this conclusion, however, requires closely
spaced measurements at other energies.

## DIMENSIONAL COUNTING RULE

One attractive model for large angle hadron-hadron
scattering involves scattering of the constituents (e.g.
quarks) of the projectile and target particles.  An
early encouragement for this approach was the apparent
success of the dimensional counting rule, which predicts
the asymptotic energy dependence of differential cross
sections at large fixed angles.  The rule predicts that
at constant c.m. angle, $d\sigma/dt \propto s^{-n}$, where n = m-2 and
m is the sum of the number of quarks in the initial and
final states of the interaction, s is the square of the
center-of-mass energy, and t is the square of the four-
momentum transfer.  It has been claimed that previous
data agree with the predicted exponents (7 for photo-
production of pions, 8 for $\pi p$ and Kp elastic scattering,
and 10 for pp elastic scattering) for laboratory momenta
above 5 GeV/c.  However, in the $\pi p$ case the comparison
is based on very few data points with generally large
statistical uncertainties.  While the pp data have much

better statistics, the case for $s^{-10}$ is debated in
the literature.[8]

We are reporting here some results from a new
experiment which supplement and check the existing pp
data and provide enough additional $\pi^-p$ data for a
stringent test of the dimensional counting rule.

## DESCRIPTION OF EXPERIMENT

This high statistics experiment measured the differential cross section $d\sigma/dt$, for $-0.35 < \cos\theta_{c.m.} <$
$0.35$ from 1.9 to 9.7 GeV/c for $\pi^-p$, from 1.9 to 6.3 GeV/c
for $\pi^+p$, and from 1.9 to 9.0 GeV/c for pp interactions.
The momentum region was covered in 2% steps. The experiment was performed at the Argonne Zero Gradient Synchro-
tron (ZGS). The experimental apparatus, shown in Fig. 1,
was designed to allow large data taking rates, to have
a simple and smooth geometric acceptance, and to give
high angular and momentum resolution.

The high intensity, two stage, secondary beam had
a momentum spread of $\pm$ 5%. A 19 bin hodoscope at the
first focus permitted tagging of the incident particle
momenta with a resolution of $\pm$ 0.25%. The absolute momentum of the beam was determined to $\pm$ 0.5%. Two ethy-
lene filled threshold Cherenkov counters were used to
identify incident $\pi$'s, K's, and p's. The direction and
position of each beam particle were measured by two x-y
hodoscopes with resolutions of $\pm$ 3 mr and $\pm$ 3 mm.

The beam of a few million particles per 0.6 sec
spill was focused on a 30 cm long liquid hydrogen target.
The scattered projectile particle and the recoiling proton were detected by the double arm multiwire proportional chamber spectrometer. Each arm consisted of
three pairs of vertical and horizontal chambers, which
could be rotated to concentrate on c.m. angles of 90°.
The resolution of each arm was $\pm$ 3 mr in the laboratory
scattering angle. At the end of each arm was a threshold Cherenkov counter used to identify the scattered
particles when kinematics could not distinguish between
the pion and proton. There was no magnet in the detector.

The experiment trigger consisted of a coincidence
of the beam defining counters ($B_1$, $B_2$, $B_3$ and $\overline{BH}$) and
the four counters of the MWPC arms ($L_1$, $L_2$, $R_1$ and $R_2$).
In addition, to suppress background due to inelastic
scatters, the trigger required anticoincidence with
eight veto counters surrounding other areas. These
were multilayer lead-scintillator sandwiches, which
detected both charged particles and photons from $\pi^0$ decay.
An additional single layer anticoincidence counter rejected non-interacting beam particles. The trigger and
veto counters covered approximately 3.5 $\pi$ sr in the c.m.

solid angle.

The acceptance for elastic scattering was calculated by a Monte Carlo simulation, and is a smooth function of $\cos\theta_{c.m.}$ with a peak value typically 22%. Points with acceptance less than 25% of the peak value at a given momentum have been excluded from the data.

The data were recorded on magnetic tape and partially analyzed on line. The total pion and proton flux was $1.4 \times 10^{12}$ particles, resulting in $2.8 \times 10^7$ triggers, from which $1.3 \times 10^6$ elastic scatters were recognized. Elastic events were selected by using the coplanarity of the incoming, scattering, and recoiling tracks, and the angles of the scatter and recoil. The remaining inelastic background was subtracted by fitting the coplanarity distribution for each momentum and scattering angle bin to the sum of a Gaussian and a background polynomial. Backgrounds ranged from 1% at the lowest momentum to 25% at the highest. The data were also corrected for random accidental vetoing, nuclear absorption after scattering, ambiguous momentum determination, and chamber-hodoscope inefficiencies. A typical total correction factor was 1.5. We estimate that the normalization is uncertain to $\pm$ 10%. This error is not included in the data plotted below.

Data were taken in such a way that the momentum ranges overlapped for adjacent settings of the beam magnets. These overlapping bins provide an important test of the reproducibility of the data, since often the data at adjacent momentum settings were taken with the spectrometer arms at different angles and/or with weeks or months intervening. Point-by-point comparisons of these overlapping differential cross sections give a satisfactory distribution of $\chi^2$. In making these comparisons, we have allowed the relative normalizations to change within the estimated run-to-run systematic error of $\pm$ 3%. In the cross sections reported here, the overlap bins have been averaged together.

The absolute momentum calibration of the incident beam was determined by Monte Carlo simulation of the magnet system, wire orbit studies, and a separate experiment which detected elastic pp scattering.[9] A further check is provided by the kinematics of the elastically scattered particles. The overall uncertainty of the momentum calibration is estimated to be $\pm$ 0.5%.

## EVIDENCE FOR FLUCTUATIONS

The search for fluctuations has been conducted by plotting $d\sigma/dt$ at fixed t vs. s. Some of these graphs are shown in Fig. 2. Fixed t cross sections have been chosen because of the known presence of structures in the cross section at fixed t, in particular a sharp dip at t = -2.8 GeV/c$^2$. If the energy dependence of fixed angle cross sections were investigated, as might seem more appropriate in a search for resonance effects, extraneous structure would be produced as a fixed t feature moved past the particular scattering angle.

Lines have been drawn through the data points in Fig. 2 to guide the eye and to indicate the major structures that appear in the data. It is clear that the data cannot in general be represented by smooth curves and that significant, previously unknown structures with widths in s of 1-2 GeV$^2$/c$^2$ are revealed. Plots at intermediate values of t generally show a smooth transition between the graphs of Fig. 2.

Previous data have been omitted from Fig. 2 for lack of space. They are sparse above $p_{inc}$ = 3 GeV/c, but are generally in agreement with the present data within quoted errors.

We summarize here the characteristics of the observed structures:

a) Excursions about a smooth curve are as much as a factor of 2 in either direction.

b) Full widths are between 100 and 200 MeV in c.m. energy. These widths are significantly narrower than the observed widths of established nucleon resonances, which for s > 6 GeV$^2$/c$^2$ are at least 350 MeV.[10]

c) None of the structures is present in all of the constant-t graphs. About half of the structures are centered at a constant value of u, but are present only for a limited range of t, even though there are data for the particular u value in a wider range of t.

d) The number and relative amplitude of the structures are qualitatively constant as a function of s or t, and as a function of pion charge in the region where we have data for both $\pi^+p$ and $\pi^-p$. This behavior is in sharp contrast to similar plots of our pp elastic scattering data, where no significant narrow structure is observed, confirming the negative results of previous searches.[11]

e) Structures that appear in the same kinematic region for $\pi^+p$ and $\pi^-p$ are not more prominent in $\pi^+p$ by a large factor (the scattering amplitude should be larger by a factor of 3) as they would be if due to pure I = 3/2 states.

We have paid particular attention to $\pi^+p$ elastic scattering near 5.0 GeV/c incident momentum because of the sharp structure previously reported.[6] We have followed the analysis of Ref. 6 in computing an asymmetry parameter for adjacent momentum bins. If the two adjacent momenta are $p_1 < p_2$ ,

$$A(t) = [\frac{d\sigma}{dt} (p_2, t) - \frac{d\sigma}{dt} (p_1, t)]/[\frac{d\sigma}{dt} (p_2, t) + \frac{d\sigma}{dt} (p_1, t)] .$$

This asymmetry parameter is plotted in Fig. 3 for the two bins at 5.10 and 5.20 GeV/c ($\Delta\sqrt{s}$ = 31 MeV) along with the data from Ref. 6.

The two bins used have momenta about 3% higher than those of Ref. 6, but have been chosen because they show a structure similar to that of the previous experiment. No other pair of bins between 4.5 and 5.5 GeV/c shows as much structure. A comparison of the two sets of asymmetries shows that the present data substantially confirm the sharp drop to negative values for -t > 5 GeV$^2$/c$^2$. However, it does not show the large positive values of asymmetry reported near -t = 4 GeV$^2$/c$^2$. The structure revealed in Fig. 3 is evident in Fig. 2 at -t = 5.9 GeV$^2$/c$^2$, s $\approx$ 10 GeV$^2$/c$^2$.

We note that there is narrow structure in the non-exotic $\pi^{\pm}$p channels but not in the exotic pp channel. This suggests an origin in s-channel resonances. The structures are not well explained as constant u pheno-mena. (See point c) above.) However the narrow widths and the relationship between $\pi^+$p and $\pi^-$p structures cannot be explained by known s-channel resonances. Thus it is probable that the structure is due to either new individual resonances or to multiple-resonance fluctuations. In the latter case, the density of states must be different from the exponentially rising mass spectrum of the statistical bootstrap model, since in that model Frautschi[3] finds that the relative size of the structures must fall by about an order of magnitude across the region of s measured by this experiment.

## TEST OF DIMENSIONAL COUNTING RULE

To compare to the dimensional counting rule, the data are binned in intervals of 2% in the laboratory momentum and 0.1 in $\cos\theta_{c.m.}$. The $\pi^{\pm}$p and pp elastic differential cross sections at several angles are shown in Fig. 4, along with power law predictions of the dimensional counting rule, and all previous $\pi$p data[3-8] above 3 GeV/c at these angles.

The pp data are inconsistent with one value of n over the entire energy range. Several previous experiments have reported breaks in the slope of the 90$^\circ$ cross sections plotted against a variety of parameters.[19] The disagreement with s$^{-10}$ in our data is a reflection of these breaks. Plotted against -t, the logarithm of the 90$^\circ$ cross section shows changes of slope at t $\simeq$ -3 GeV$^2$/c$^2$ and t $\simeq$ -6.5 GeV$^2$/c$^2$, as reported earlier. Looking at other angles, we find that the breaks occur at approximately constant t. However, the energies at which these breaks occur may not be high enough for the dimensional counting rule to apply. We have therefore fitted the 90° data to $d\sigma/dt = a \, s^{-n}$ for $s > 12$ GeV². The result is $n = 10.07 \pm 0.11$ with a $\chi^2$ of 30 for 24 degrees of freedom. We note, however, that several other forms fit the data equally well at these energies.[20]

The $\pi^{\pm}$ data are more complicated. Both $\pi^{-}p$ and $\pi^{+}p$ have a dip (at $s \simeq 7$, 8 and 9 GeV² for 100°, 90°, and 80°, respectively) which is due to a constant dip at $t = -2.8$ GeV²/c². It is clear from the figure that there are significant departures from any power law $s^{-n}$, even beyond this constant dip. Some fits to the $\pi^{-}p$ data are shown in Table 1. Our data agree well with the data of Owen et al,[13] on which the claim of $s^{-8}$ is based, yet in the same energy range ($12 \leq s \leq 19$ GeV²), our more detailed data yield lower exponents. If the lower limit of the fit is made sufficiently high ($s > 16$ GeV²), the data are consistent with $s^{-8}$ at all angles, suggesting an "asymptotic" agreement with the theory. However, the errors on the exponents are then rather large, and the agreement may not be significant.

This experiment was done by a collaboration of groups from Argonne, Columbia University and the University of Minnesota. The people involved are shown in Fig. 5.

## REFERENCES

1. T.E.O. Ericson, T. Mayer-Kuckuck, Ann. Rev. Nucl. Sci. 16, 183 (1963).
2. S.J. Brodsky, G.R. Farrar, Phys. Rev. Lett. 31, 1153 (1973).
3. S. Frautschi, Nuovo Cimento 12A, 133 (1972).
4. P.J. Carlson, Phys. Lett. 45B, 161 (1973).
5. S. Frautschi, Nucl. Phys. B91, 125 (1975).
6. F.H. Schmidt et al, Phys. Lett. 45B, 157 (1973).
7. A. Eide et al, Nucl. Phys. B60, 173 (1973).
8. V. Barger, F. Halzen, J. Luthe, Phys. Lett. 42B, 428 (1972). B. Schrempp, F. Schrempp, Phys. Lett. 55B, 303 (1975). A.W. Hendry, Phys. Rev. D10, 2300 (1974).
9. R.D. Klem et al, Phys. Rev. D15, 602 (1977).
10. N. Barash-Schmidt et al, Rev. Mod. Phys. 48, 2, pt. 2 (1976).
11. J.V. Allaby et al, Phys. Lett. 23B, 389 (1966); 25B, 156 (1967).
12. C.T. Coffin et al, Phys. Rev. 159, 1169 (1967).

13. D.P. Owen et al, Phys. Rev. 181, 1794 (1969).
14. A. Eide et al, Nucl. Phys. B60, 173 (1973).
15. M. Fellinger et al, Phys. Rev. Lett. 23, 600 (1969).
16. P.L. Bastien et al, Phys. Rev. D3, 2047 (1971).
17. D.R. Rust et al, Phys. Rev. Lett. 24, 1361 (1970).
18. C.W. Akerlof et al, Phys. Rev. 159, 1138 (1967).
19. R.C. Kammerud et al, Phys. Rev. D4, 1309 (1971).
20. We get good fits to $d\sigma/dt = a \exp(-bp_{c.m.})$; $d\sigma/dt = a \exp(-b\sqrt{s})$; and $d\sigma/d\Omega = a/s \exp(-bp_t)$. These forms were motivated by the discussion in Ref. 9. The forms $\exp(-b\sqrt{s})$ and $\exp(-bp_{c.m.})$ appear to extrapolate well to lower energies.
21. J.F. Gunion, S.J. Brodsky, R. Blankenbecler, Phys. Rev. D8, 287 (1973).

Table I: Values of the Parameter n from Fits to the $\pi^- p$ Data Using the Form $d\sigma/dt = A\, s^{-n}$

| Energy Range | $\theta_{c.m.} = 80°$ | | $\theta_{c.m.} = 90°$ | | $\theta_{c.m.} = 100°$ | |
|---|---|---|---|---|---|---|
| | n | $\chi^2$/DF | n | $\chi^2$/DF | n | $\chi^2$/DF |
| $12 \leq s \leq 19$ GeV$^2$ | $6.1 \pm 0.2$ | 49/23 | $6.7 \pm 0.2$ | 45/23 | $7.6 \pm 0.3$ | 34/22 |
| $14 \leq s \leq 19$ GeV$^2$ | $7.3 \pm 0.4$ | 25/15 | $8.2 \pm 0.4$ | 19/15 | $6.2 \pm 0.6$ | 16/14 |

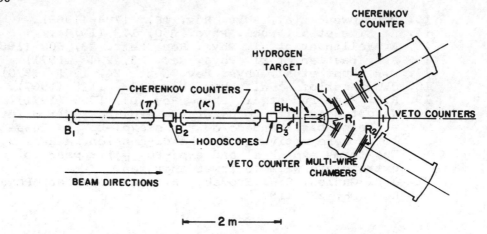

Fig. 1    Experimental apparatus.

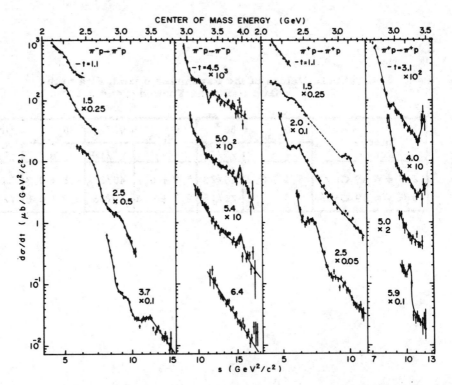

Fig. 2    Constant-t cross-sections to search for
fluctuations.  Lines are to guide the eye.

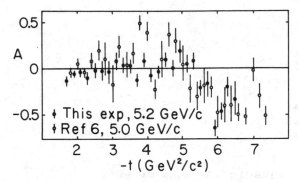

Fig. 3    Asymmetry for adjacent momentum bins near
          5 GeV/c.

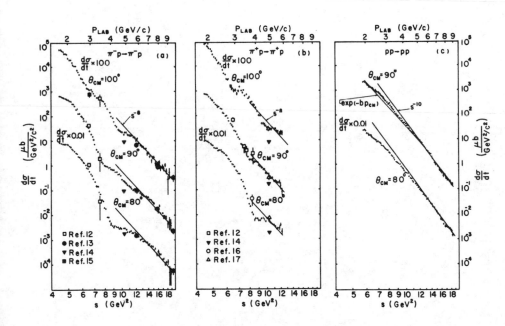

Fig. 4    Constant angle cross sections to compare with
          dimensional counting rule.

Columbia University
    K. A. Jenkins
    L. E. Price
Argonne National Laboratory
    R, Klem
    R. J. Miller
    P. A. Schreiner
University of Minnesota
    H. Courant
    Y. I. Makdisi
    M. L. Marshak
    E. A. Peterson
    K. Ruddick

Fig. 5    Participants in the experiment.

# Inclusive Scattering Results from the
# Fermilab Single Arm Spectrometer

D. Cutts, R. Dulude, R E. Lanou, and J.T. Massimo
Brown University, Providence, Rhode Island 02912

R. Meunier
CERN, Geneva, Switzerland

A.E. Brenner, D.C. Carey, J.E. Elias, P.H. Garbincius,
G. Mikenberg, V.A. Polychronakos, and G.A. Weitsch
Fermi National Accelerator Laboratory, Batavia, Illinois

M.D. Chiaradia, C. DeMarzo, C. Fazuzzi, G. Germinario,
L. Guerriero, P. Lavopa, G. Maggi, F. Posa, G. Selvaggi,
P. Spinelli and F. Waldner
Istituto di Fisica and INFN, Bari, Italy

W. Aitkenhead, D.S. Barton, G. Brandenburg, W. Busza,
T. Dobrowolski, J.I. Friedman, H.W. Kendall, T. Lyons,
B. Nelson, L. Rosenson, W. Toy, R. Verdier, and L. Votta
Massachusetts Institute of Technology
Cambridge, Massachusetts  02139

## ABSTRACT

Results are presented on the inclusive scattering
process $a + p \rightarrow c + X$ where a and c are any of $\pi^{\pm}$, $K^{\pm}$,
p or $\bar{p}$.  The data were taken at 100 and 175 GeV/c incident
momentum using the Fermilab Single Arm Spectrometer.  A
vertex detector was also used to measure the associated
charged multiplicity.  The x dependences of the inclusive
cross section exhibit the power law behavior expected from
various fragmentation models.

ISSN:  0094-243X/78/053/$1.50  Copyright 1978 American Institute of Physics

## I. INTRODUCTION

I'd like to present some preliminary results of Fermilab Experiment E118, a high precision measurement of inclusive and semi-inclusive hadron scattering.[1] The experiment has been carried out using the M6 beam line and the Single Arm Spectrometer (SAS) in FNAL's Meson Lab.[2] The purpose of our experimental program is to provide detailed information on the x and $p_\perp$ dependence of the inclusive cross section for positive values of x over a modest $p_\perp$ range ($p_\perp \lesssim 1$ GeV/c). We have taken data at both 100 and 175 GeV/c incident momenta.

Utilizing the Cerenkov power of the M6 beam and of the SAS we are able to separate $\pi$'s, K's and p's over the entire kinematic range of the experiment; by taking data with all beam-spectrometer charge permutations we potentially measure 6 x 6 = 36 separate reactions. These data provide severe tests for models which attempt to describe the fragmentation and diffractive regions of phase space; they are also an excellent place to look for violations of the factorization and scaling hypotheses.

In addition to the single arm inclusive measurements, we also count the number of associated charged secondary particles and measure their angular distribution with a non-magnetic vertex detector. Therefore we can study both the dependence of the associated multiplicity on x and $p_\perp$ and the inclusive cross section for various classes of event topology.

## II. THE EXPERIMENT

The high resolution M6 beam at Fermilab has three optical stages and is instrumented with four Cerenkov counters.[2] It provides excellent momentum and angular resolution as well as complete particle identification for fluxes as high as 5 Mhz. Its momentum range is from 50 to 200 GeV. Just before the final focus the beam passes through three bending magnets which bend the beam vertically. These magnets are used to pitch the beam through the target and thus vary the scattering angle relative to our fixed spectrometer.

The Single Arm Spectrometer (SAS) is equivalent to a fourth optical stage of the M6 line.[2] A schematic of the SAS is shown in Fig. 1. The momentum bend is provided by 5 main ring bending magnets. Since rates are generally lower in the spectrometer than in the beam, it is instrumented with MWPC's instead of hodoscopes. It also has four large Cerenkov counters allowing complete particle identification. At the end of the SAS are a pair of shower counters and a muon calorimeter which are used to measure electromagnetic backgrounds and particle decay rates. The entire length of the SAS is approximately 150 meters.

The vertex detector is a smaller scale ($\sim$2m. long), multiparticle spectrometer which is located in the area

just downstream of the target (see Fig. 2).  This device
is non-magnetic and consists of three sections, each designed to measure the charged particle multiplicity in a
different angular region.  The three sections together
cover the entire forward hemisphere.  The first section
is an eighteen element barrel hodoscope which surrounds
the target and counts the large angle hadrons.  The scintillation counter elements are double-layered to provide
δ-ray rejection and identification of slow recoil protons.
The second section consists of 9 MWPC planes, each with
288 wires, spaced by 2 mm.  The centers of these planes
through which the beam passes have been deadened to avoid
background tracks and chamber deterioration.  Finally, the
deadened region of the MWPC's is covered by a small, six
element lucite hodoscope.  The amount of Cerenkov light
collected from each element is a direct measure of the
number of relativistic particles passing through it.[3]
This hodoscope covers the angular region $\theta \gtrsim 30$ mrad.
    The data from both the SAS and the vertex detector
are logged onto magnetic tape by a PDP-11.  This computer
is also used as an on-line monitor of the performance of
both systems.

### III. DATA
    The analysis of the single arm inclusive cross sec-
tions is essentially complete.[4]  For the data shown here
each spectrometer setting is treated as a single kinematic
point, although in the high-rate diffractive region we
plan to subdivide the data further.  The momentum bite
of the SAS is $\sim\pm5\%$ and its angular acceptance is $\sim6\mu$ster.
Empty target rates are typically 25% of the full target
rates for small angle running dropping to 5% at the largest angles.
    Absorption and decay corrections have been determined
directly from the data, including runs where the beam is
run through the spectrometer.  As a check these corrections
have also been calculated from the known properties of the
system.  These corrections vary from 36% at x = .3 to 15%
at x = .9, and are known to $\pm2\%$.  The relative systematic
error between reaction types is estimated to be less than
5% while the overall normalization is known to better than
15%.  Effects due to sprectometer asymmetries and angle
offsets are minimized by always taking data at symmetrical
positive and negative scattering angle settings.  The errors in the data plots are statistical only.
    The   dependences of the cross sections for the re-
actions pp → a + x are shown in Fig. 3 for both 100 and
175 GeV/c incident momenta.  The data shown are for
$p_\perp$ = .3 GeV/c; we also have x -sweeps at $p_\perp$ = .5 and .75
GeV/c.  It is clear from this figure that the data scale
for these two energies.  With the exception of the dif-

fraction dominated reaction pp → pX, the data exhibit a power law behavior in 1- x. This is shown by the solid curves which are power law fits to each channel. This behavior is characteristic of fragmentation models as discussed in the next section.

Similar data for the $\pi^+$-induced reactions are displayed in Fig. 4, here for 100 GeV/c only. Again the power law behavior is evident, with the exception of the channels with an outgoing $\pi^+$ or $\pi^-$. The latter case is not diffractive, however the observed high x shoulder can be explained in terms of peripherally produced resonances (see section IV). It should be noted that the data in Figs. 3 and 4 represent only a fraction of the reaction channels and kinematic points which we have measured.

The analysis of the vertex detector data is currently in a more preliminary stage. The track decoding for the MWPC's and the pulse height analysis of the lucite radiator hodoscope are both complete. By taking data with two different target lengths we are able to correct for multiple interactions in the target, $\pi^0$ conversions, and δ-rays. We are currently in the process of Monte-Carloing this system to get a more accurate estimate of these effects. Until this computation is completed we believe our systematic error in the average charged multiplicity is not more than 15%.

Some preliminary data from the vertex detector are shown in Fig. 5. Here we display the average charged multiplicity of the system X in the reaction $\pi^+p \to \pi^+X$. The data are plotted versus $p_\perp$ for three different x settings. The error bars are statistical. For the high x data the multiplicity appears to be independent of $p_\perp$, whereas at lower x the multiplicity is larger and appears to rise with increasing $p_\perp$. Other experiments have also seen the multiplicity increase with $p_\perp$, but have observed this at both low and high x.[5] However, the constancy at high x of $\bar{n}_x$ as a function of $p_\perp$ or t at high x is consistent with an exchange picture where the multiplicity of system X is independent of the mass of the (virtual) projectile.

## IV. FRAGMENTATION MODEL COMPARISON

Numerous authors have in recent years applied the quark-parton model, which was developed to explain deep inelastic lepton scattering results, to hard scattering in hadronic processes, i.e., to scattering at large $p_\perp$.[6] More recently it has also been suggested that these ideas may also be applicable to fragmentation in low $p_\perp$ collisions.[7,8] Basically it is predicted that the quark (or anti-quark) content of the projectile determines the probability that it will fragment into a given hadron.

In particular Brodsky and Gunion predict that the invariant cross section for the production of such fragments will have the power law form $(1-x)^n$.[7] The exponent n is given by $2n_s-1$ where $n_s$ is the number of spectator quarks in the projectile-to-fragment reaction, assuming a quark exchange mediates the reaction. For example, if a proton fragments to a $\pi^+$, we have $n_s = 2$, so an exponent n = 3 is expected.

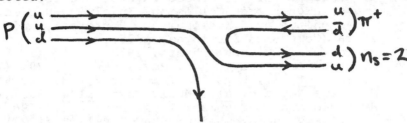

In Fig. 6 the invariant cross sections for four different reactions are shown as a function of x. In each case the data are well fitted by a power law (solid curve). For both $p \to \pi^+$ and $\pi^- \to p$ fragmentation an exponent x = 3 is expected. The data are in good agreement with this prediction in both cases; for $p \to \pi^+$ two different $p_\perp$ values are displayed both of which yield exponents consistent with n = 3. The power law is expected to be independent of $p_\perp$. For the other two cases shown, $\pi^- \to K^-$ and $p \to \bar{p}$, the predicted exponents are 1 and 9 respectively. Even with these extreme predictions, the data are in excellent agreement.

We find that this simple prescription is a good description of the majority of our data, as have other experiments with similar data.[9] As noted earlier, channels where the projectile and the outgoing particle are equivalent are dominated by a diffractive peak at x ≃1 and are clearly not relevant to this model. However, the reaction $\pi^+ p \to \pi^- X$ would seem to be a likely candidate with a predicted exponent n = 5. As shown in Fig. 3 the data instead have a shoulder at high x. We have shown that this shoulder is the result of peripherally produced $\rho^0$ and $f^0$ mesons which can then decay into a forward $\pi^-$ (and $\pi^+$).[1] Thus if one is careful to exclude diffractive channels and to account for possible resonance contributions, the quark-parton fragmentation picture is in good agreement with our data.

We plan to summarize the results on all the channels which we have measured soon. We will also correlate the results from the vertex detector with the SAS results to attempt to learn more about the fragmentation process. For example it may be possible to eliminate the diffraction and resonance production contributions by selecting events with large associated multiplicities.

REFERENCES

1. D. Cutts et al., Phys. Rev. Lett. 40, 141 (1978).
2. Fermilab Single Arm Spectrometer Group, Phys. Rev. D15, 3105 (1977).
3. W. Busza et al., in Proceedings of the Topical Meeting on High Energy Collisions Involving Nuclei, Trieste 1974, edited by G. Bellini, L. Bertocchi, and T. Rancoita (Compositori Bologna, 1975).
4. W. Toy, MIT Doctoral Thesis, 1978.
5. E. Anderson et al., Phys. Rev. Lett. 34, 294 (1975).
   D. Fong et al., Phys. Rev. Lett. 37, 736 (1976).
   M. Della Negra et al., Nucl. Phys. B104, 365 (1976).
6. R. Field and R. Feynman, Phys. Rev. D15, 2590 (1977).
7. S. Brodsky and J. Gunion, SLAC-PUB-1939 (1977).
8. W. Ochs, Nucl. Phys. B118, 397 (1977);
   K. Das and R. Hwa, Phys. Lett. 68B, 459 (1977);
   B. Andersson, G. Gustafson, and C. Peterson, Phys. Lett. 69B, 221 (1977).
9. J. Johnson et al., Phys. Rev. D17, 1292 (1978);
   R.T. Edwards, et al., University of Colorado Preprint (1978).

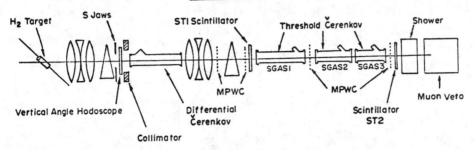

SPECTROMETER

Fig. 1.  Schematic of Single Arm Spectrometer.  Total length ∿ 150 m.

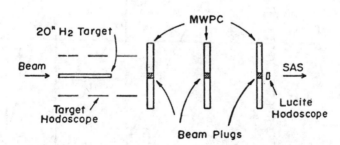

VERTEX DETECTOR

Fig. 2.  Schematic of Vertex Detector.  Length ∿ 2 m.

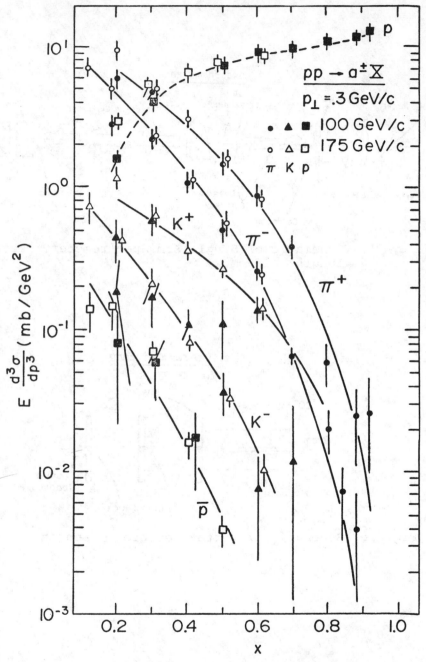

Fig. 3.  Invariant cross sections for
pp → a$^{\pm}$X at 100 and 175 GeV/c.
Solid curves are fits to $(1-x)^n$.

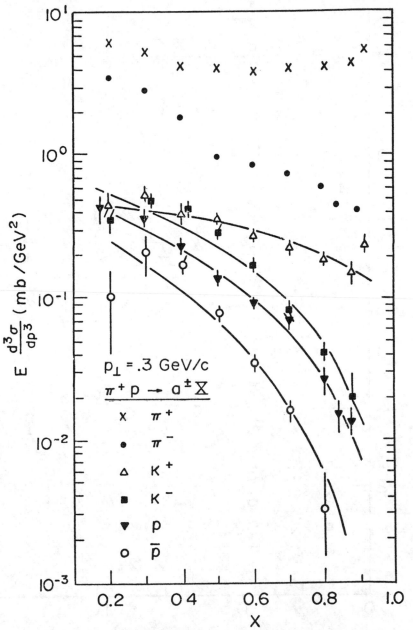

Fig. 4. Invariant cross sections for
$\pi^+p \rightarrow a^\pm X$ at 100 GeV/c with p
= .3 GeV/c. Solid curves are
fits to $(1-x)^n$.

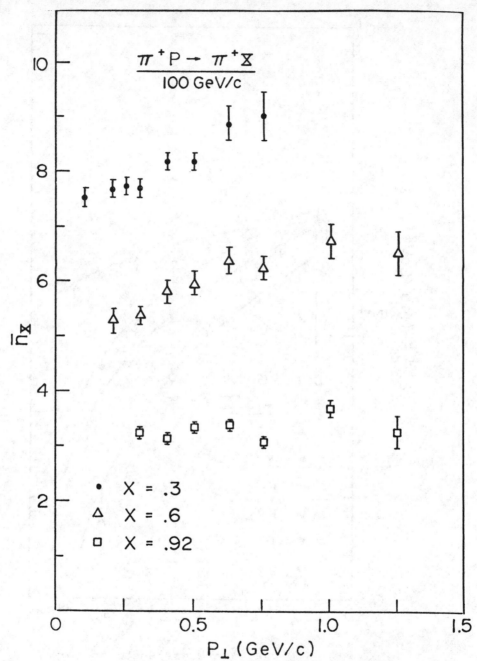

Fig. 5. Associated charged multiplicity in the reaction π+p → π+X plotted versus p⊥ at
100 GeV/c.

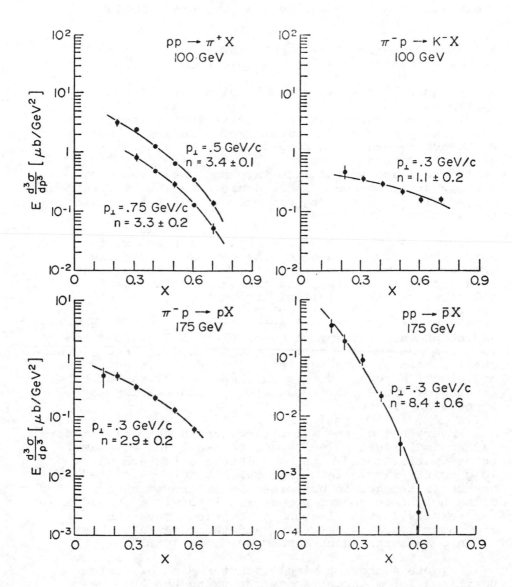

Fig. 6. Invariant cross sections and power law fits
for four different reaction channels.

# A Preliminary Look at Hadron Structure
## With a Double Arm Calorimeter

Fermilab, Lehigh, Pennsylvania, Wisconsin Collaboration

This experiment is a collaborative effort by four institutions involving people shown in figure 1.

Figure 2 is an approximate plan view of the apparatus. The steel was part of a 24 absorption length hadron shield which effectively prevented small angle scatters upstream from simulating large $P_T$ target events. A hydrogen target consisted of separate 12" and 6" segments for studying large $P_T$ accidentals due to multiple interactions. Two of the six drift chamber planes achieved unique x-y readout on individual hits using cathode delay lines in addition to electron drift time.

Both calorimeter arms rested on a steel plate which moved along the beam line on rails. The right arm could be moved independently by jacking it along on a teflon pad. The right arm was a delux model calorimeter with an initial layer of Pb (5 radiation lengths) for $\pi°$ identification followed by several layers of steel (each $\sim$ 2 absorption lengths) to observe hadron shower development. For financial reasons the left arm has smaller solid angle for hadrons and coarser spatial resolution. The left $\pi°$ detector is quite efficient, however, with a thickness of 21 radiation lengths of Pb. A beam's-eye view of the calorimeters appears in figure 3.

Every one of the $\sim$ 150 elements in the calorimeters had an individual pulse height read-out and an unambiguous spatial coordinate. Unlike calorimeters with long scintillator slats, our calorimeters permitted a true $P_T$ trigger to be formed by attenuating each pulse height an amount proportional to $\sin \theta$ and adding the results. Typical running positions gave the left arm about 1 sr in the c.m.s. and the right arm about 1.5 sr. Azimuthal acceptances were $\sim \pm 40°$.

Figure 4 shows a simple scatter plot of vertex positions for a target empty run using the 2 largest angle tracks in the drift chambers. One sees these events originate mostly in the superinsulation and the beam counters. Events upstream of the target area would have presented serious difficulty.

ISSN: 0094-243X/78/064/$1.50 Copyright 1978 American Institute of Physics

Figure 5 shows the target empty data to be of the same character in $P_T$ as the target full data. Up to 5 or 6 GeV/c in $P_T$ it would not even appear important to make an empty subtraction. Target empty subtractions were nevertheless made.

There were three basic types of triggers used for physics analysis which we will now discuss. (L) or (R) triggers recorded data if the total $P_T$ of one arm exceeded a preset threshold. Figure 6 shows a preliminary invariant cross section for jets at 400 GeV using an (R) trigger at two different thresholds. There is a rapid rise in cross section at threshold followed by the typical exponential fall off found in high $P_T$ experiments using single particle triggers.

Figure 7 shows an effect that has been discussed[1] for single arm triggers but never so clearly seen in data as here. In the top histogram (right-side-up) we see a selected bin of $P_T$ for events which produced a left arm (L) trigger. These events produce a spectrum of $P_T$ in the right arm (inverted histogram) which peaks somewhat lower in $P_T$ than the trigger arm. When the trigger $P_T$ is increased in the left arm, we see in the second half of figure 7 that the peak in the right arm also increases but is still smaller than the trigger $P_T$.

This near-balance of $P_T$ is summarized for several trigger values in figure 8. The fact that the away side has a systematically lower $P_T$ might be explained in the context of a parton model by noting that transverse Fermi momentum of the partons would naturally conspire to deposit more momentum on the trigger side because of the steepness of the $P_T$ spectrum.

In contrast to real events figure 8 also contains[2] 3 points generated by a peripheral phase space program for a 15 body final state. Lower multiplicities than 10 tend to give no high $P_T$ single arm triggers. We see that even though an occasional phase space event gives a right arm trigger, the amount of $P_T$ captured by the other arm is independent of the trigger $P_T$. Figure 9 shows peripheral phase space spectra for 20 body final states when the right side threshold is 1.5 GeV/c. Only 6.5% of these events make a right arm trigger, and only 25% of those give any $P_T$ into the left arm.

(SPR) and (SPL) triggers were attempts to observe a single particle which carried unusually large $p_T$ into the right trigger. A group of several contiguous segments was used for this trigger. Figure 10 shows the angular distribution of $P_T$ about the centroid for such triggers. The $P_T$ is indeed tightly collimated as one would expect for a single particle and is dramatically different from jet distributions.

An invariant cross section using single $\pi°$ triggers (SPL) in the left arm is shown in figure 11. It agrees quite well with more precise $\pi°$ data of others and thus gives us confidence in the energy calibration and interpretations of the single particle trigger.

The third trigger type (L + R) required the sum of the $P_T$ magnitudes in the left and right calorimeter to exceed a preset value. Such a trigger would have no bias toward one arm or the other insofar as Fermi momentum of partons is concerned.

Figure 12 is a scatter plot of raw data taken with such a trigger at 400 GeV using a logarithmic intensity scale. Note the sharp threshold and the tendency for transverse momentum to balance in the two arms.

If we take a slice of data perpendicular to the 45° diagonal, we see in figure 13 the real extent of the $P_T$ balance. The trigger itself does not a priori favor such a balance. To see this we can make a Monte Carlo model using what we do know from single particle high $P_T$ data[3].

i.e. $\sigma_I (P_T) \sim P_T^{-8} (1-x_T)^9$ where $x_T = 2 P_T/\sqrt{s}$

Our model will assume each arm obeys this distribution independently.

$$\sigma \sim \sigma_I (P_{TR}) \cdot \sigma_I (P_{TL}) \text{ where } P_{TR} (P_{TL}) \text{ is}$$
$P_T$ in the right (left) arm.

The result is shown as a dashed histogram in figure 13. Computational reasons prevent us from extending the Monte Carlo as far from the diagonal as the data, but the model clearly favors unbalanced $P_T$. Thus there appears to be a physics related correlation between the $P_T$ collected in the two arms.

The peripheral phase space events mentioned earlier are highly discriminated against by the (L + R) trigger. These events drop off as $e^{-8.6 P_T}$ and $e^{-7.6 P_T}$ respectively for 15 and 20 body final states compared to real data which drop off like $e^{-3 P_T}$.

Feynman and Field have suggested that the $P_T$ bias of a single arm trigger for high $P_T$ events might take a spectrum which would otherwise fall as $P_T^{-6}$ and change it to a spectrum which falls as $P_T^{-8}$, a change of approximately 2 in the power of $P_T$. Since our (L + R) trigger does not exhibit this bias we have made a preliminary effort to look into this difficult and somewhat controversial matter.

We make the assumption that our invariant, double arm cross section obeys the form suggested by dimensional counting.

$$\sigma_I = \frac{d\sigma}{\dfrac{d^3 p_1}{E_1}\dfrac{d^3 p_2}{E_2}} = \frac{1}{P_T^n} f(x_T, \theta_{cms}) \quad \text{where } x_T = 2 P_T/\sqrt{s}.$$

If scaling holds, one expects the power of n to be 6 for a double arm cross section instead of the usual n = 4 expected for a single arm cross section.[4] By comparing this cross section at two energies (200 GeV and 400 GeV) for the same $x_T$ and $\theta_{cms}$ we can extract the power $\underline{n}$ without having to know the form of $f(x_T, \theta_{cms})$.

We make an effort to use the same detector solid angles $d\Omega_{Left}$ and $d\Omega_{Right}$ at 200 and 400 GeV/c so that we can use $\sigma_I \sim \dfrac{1}{P_T^2} \dfrac{d\sigma}{dP_T}$ for the cross section ratio of our double arm triggers at the two energies. The "$P_T$" assigned to an invididual event is taken to be the same for all events in a slice such as the one shown in figure 12. The $P_T$ value is taken to be that of one arm when the $P_T$ balance is perfect. Figure 14 is a plot proportional to $d\sigma/dP_T$ for data at 200 and 400 GeV in a limited range of $P_T$. Since $\sqrt{s}_{400}/\sqrt{s}_{200} = 1.41$, the two curves have approximately the same slope when plotted versus $x_T$.

The power of n that fits the data of figure 14 best at the highest $P_T$ is n = 7.8. The power is two units higher than we expect from scaling but not the usual four units higher that one finds from single arm data. Using jets to obtain this power has some special difficulties, however, that need better understanding. The worst problem is that one does not know how to relate the amount of energy measured for the jet to the energy that it actually had except through the use of jet models.

Some of the problems one has with jets can be appreciated it one looks at a particular model for a jet. We will show one here which agrees with our data remarkably well so far. It is a suggestion by Selove based on an observation by Feynman that since dN/dy $\approx$ 2 for ordinary backward/forward jets, it might also be true for high $P_T$ jets. This leads to the conclusion (see figure 15) that the longitudinal momenta in an average jet drop off by factors of $\sim$ 2.

In figure 16 is a plot of the distribution of momenta observed in whatever segment of the right arm calorimeter which happens to have the largest $P_T$. It is supposed that this momentum is directly related to that of the particle in the jet with the highest longitudinal momentum. Also shown is the distribution of momenta for whatever segment has the second highest $P_T$. We see that on the average these two momenta are in the ratio 2:1 as the model suggested.

Converted to a suitable Monte Carlo program, this model can predict the particle multiplicity that should be captured by one of our calorimeter arms. This prediction is shown in figure 17 as a function of jet $P_T$. An algorithm based on the number of calorimeter segments with energy above some minimum value can also give a measure of observed particle multiplicity. Because of finite shower size and sharing of some showers by adjacent segments, this algorithm probably over estimates the particle count by $\sim$ 2X. The result seems to agree rather well with the jet model.

If what we see is not a jet but is instead a soft spray of average $P_T$ particles, Sivers, Brodsky, and Blankenbecler[5] suggest that multiplicity should depend on $P_T$ as $<n> \approx P_T/<P_T>$. The multiplicity algorithm does not agree with this at all as can be seen in figure 18.

By treating the individual particles in a jet as massless, one can obtain a mass spectrum for that part of a jet observed in a particular arm. Figure 19 is the mass spectrum for jets with $3.5 < P_T < 4.0$ GeV/c in the right arm which had a jet axis directed near the center of the arm. The average jet mass is about $2 \text{GeV/c}^2$. A width of $1 \text{ GeV/c}^2$ indicates the calorimeter mass resolution is at least that good. Once again the Monte Carlo jet model agrees well with this spectrum.

Because the Monte Carlo jet model agrees so well in cases where it can be checked, we are inclined to believe its predictions for matters which are less directly observed. For instance, the calorimeter on the average misses a fixed amount of $P_T$ ($\sim 1$ GeV/c) regardless of the size of the $P_T$. Among other things this results in better angular definition of jet direction as $P_T$ increases. Figure 20 shows the average difference expected between a measured jet direction and the true direction that would have been measured if all particles could have been collected by the calorimeter. Clearly, one cannot expect to obtain a sharp angular distribution in the c.m.s. for jets with $P_T = 2 \text{GeV/c}$, however jets with $P_T = 6$ GeV/c can be defined to 5 degrees.

One of the original objectives of this experiment was to measure the difference between the pion and proton structure functions by studying the angular distribtuions of jets in the overall center of mass for the colliding hadrons. Figure 21 shows a naive picture for what one expects on the average. Partons in pions are thought to carry a larger fraction x of the hadron momentum than partons in a proton. Thus one should find both jets from the fragmenting partons in the forward hemisphere more frequently in πp collisions than in pp collisions.

In order to achieve independence of possible experimental difficulties what we plot in figure 22 is the ratio $R_{JET} = \sigma(\text{pp} \rightarrow \text{Jets})/\sigma(\pi\text{p} \rightarrow \text{Jets})$. The beam for this plot consisted of positive particles at 130 GeV/c, Cerenkov tagged. Jets with $P_T$ between 2.5 and 2.75 GeV/c were used to determine cms angles. Results are qualitatively very much as one expects they would be if jets are the result of parton scattering.

1.    E. Malmud, Proceedings of VIII International Symposium on Multiparticle Dynamics, Kayserberg, (1977).
2.    A. E. Brenner, D. C. Carey, R. Pordes, and J. H. Friedman.  NVERTX, A Computer Program for Monte Carlo Phase Space with Importance Sampling and Histogram Display.  Fermilab preprint.  We are indebted to D. Carey for his help with this program.
3.    D. Antreasyan et al., Phys. Rev. Letters $\underline{38}$, 112 (1977).
4.    We are indebted to Francis Halzen for discussions on this matter.
5.    D. Sivers, Brodsky, R. Blankenbecler, Phys. Reports $\underline{C23}$, (1976).

E395 JET EXPERIMENT

### WISCONSIN

M. CORCORAN
A. ERWIN
E. HARVEY
R. LOVELESS
M. THOMPSON

### PENNSYLVANIA

L. CORMELL
M. DRIS
B. ROBINSON
W. SELOVE
B. YOST
W. KONONENKO

### LEHIGH

A. KANOFSKY
G. LAZO

### FERMILAB

P. GOLLON

Figure 1

72

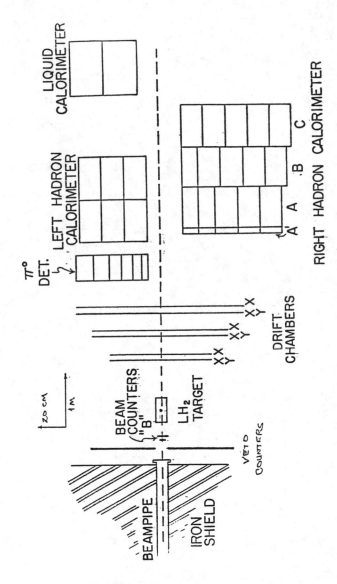

Fig. 2 Schematic floor plan of the experimental apparatus.

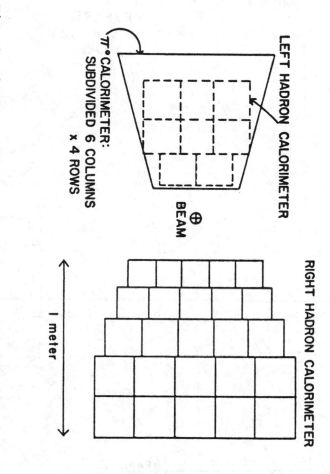

Fig. 3   The two calorimeter arms as seen from the target.

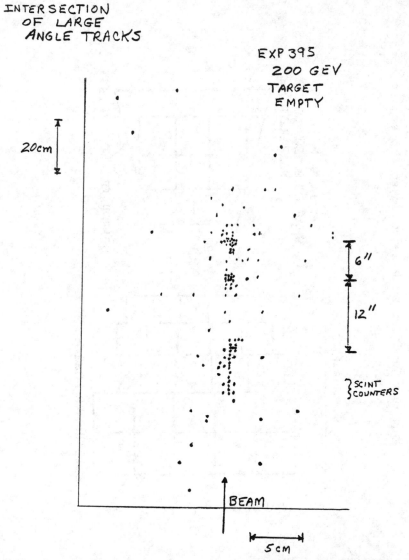

Fig. 4   Target empty run.  A plot of "interaction" vertices obtained by using the two largest angle
tracks in each event.  Clusters of vertices
occur at the superinsulation for the two targets and at scintillation counters just before
the target.

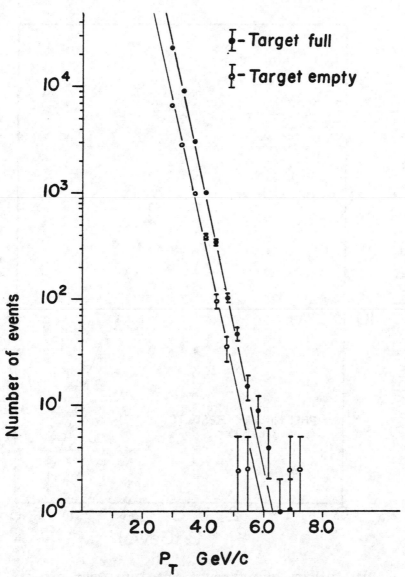

Fig. 5   Number of events versus $P_T$ at 400 GeV for double
arm (L + R) triggers. Target empty data has the
same slope as target full data and accumulates
at about 1/5 the rate of target full data.

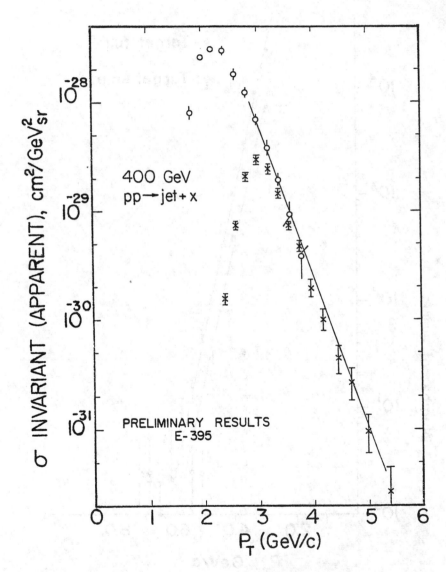

Fig. 6  Preliminary estimate of the invariant cross
section for 400 GeV pp events producing a jet
using right arm (R) triggers.

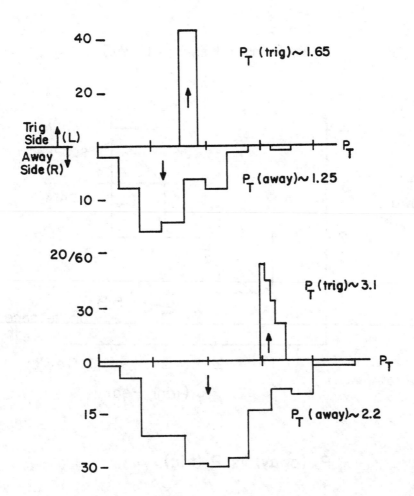

The Away-Side $P_T$ Spectrum for Two $P_T$(trig) Bands

Fig. 7  $P_T$ spectrum for the calorimeter on the away
side (inverted histogram) when the calorimeter
on the trigger side has $P_T$ in the bins shown
(erect histogram).

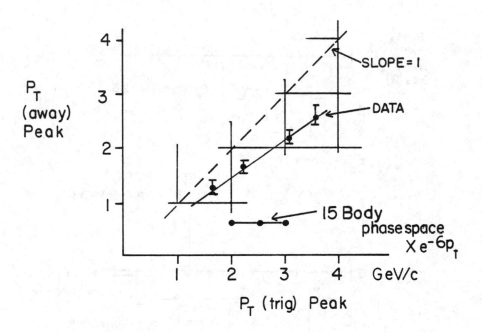

Fig. 8   Relation between peak $P_T$ on the away side to
trigger $P_T$ (solid curve). Three points generated by a 15 body phase space program
illustrate the small extent to which $P_T$ conservation forces particles into the away side at
these values of $x_T$.

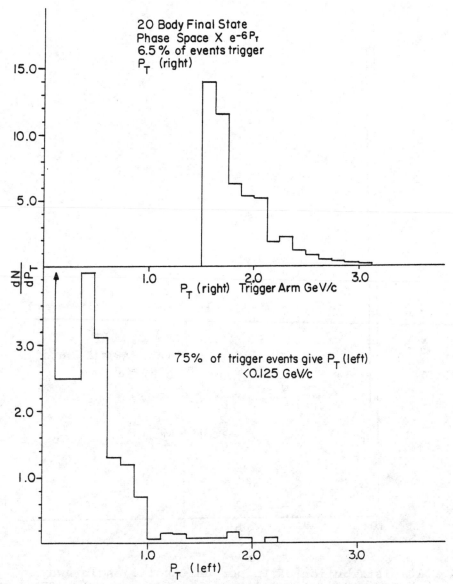

Fig. 9     Response of the calorimeters to 20 body events
generated by invariant phase space times
$\exp(-6P_T)$ when the right arm (R) threshold is
set at $P_T = 1.5$ GeV/c.

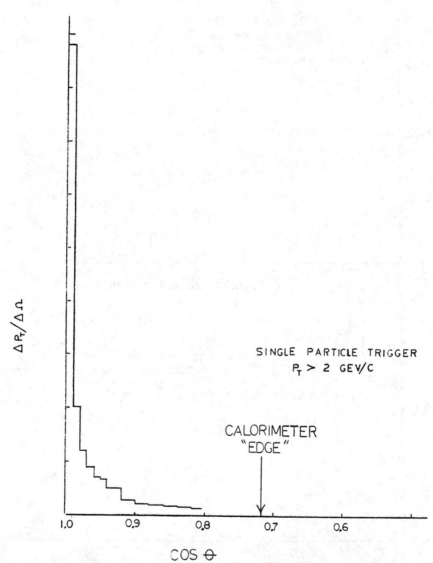

Fig. 10 Distribution of $P_T$ per unit solid angle away
from the jet axis when a single particle (SPR)
trigger is used. A random distribution of 2 or
more particles produces only a slight peaking on
such a plot.

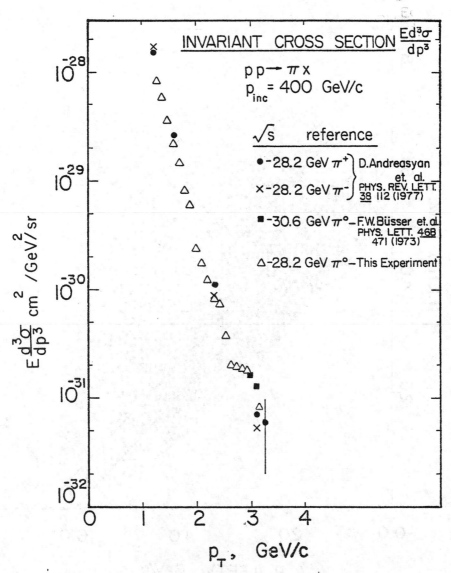

Fig. 11  Invariant cross section for single particle
(SPL) triggers obtained in this experiment compared to more precise  data of other experiments.

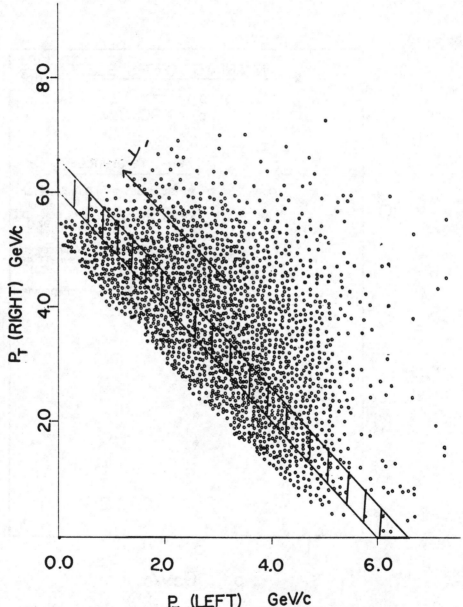

Fig. 12 Scatterplot of $P_T$ in the left and right arm for
double arm (L + R) triggers at 400 GeV/c. The
logarithmic intensity scale used for this plot
suppresses the fact that $P_T$ frequently balances.
For comparison with hard quark scattering
theories all events in the cross-hatched band
are assumed to originate from fundamental quark
scatters with the same $P_T$ (∿3.1 GeV/c in the figure).

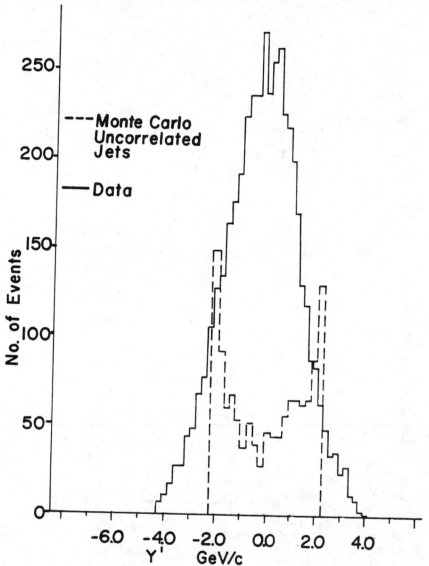

Fig. 13  Solid Curve:  True density of events in the
cross-hatched band of figure 11 as a function
of y', the distance from perfect $P_T$ balance
on the diagonal.  Dashed Curve:  Density of
Monte Carlo events when each arm has an independent $P_T$ spectrum satisfying previous single
arm spectrometer experiments.

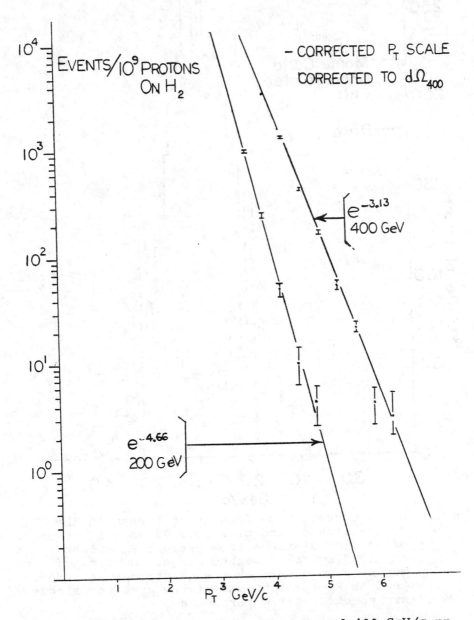

Fig. 14  Comparison of dσ/dP_T for 200 and 400 GeV/c pp
events giving double arm (L + R) triggers.  The
two sets of data would have the same slope if
plotted as a function of x_T.

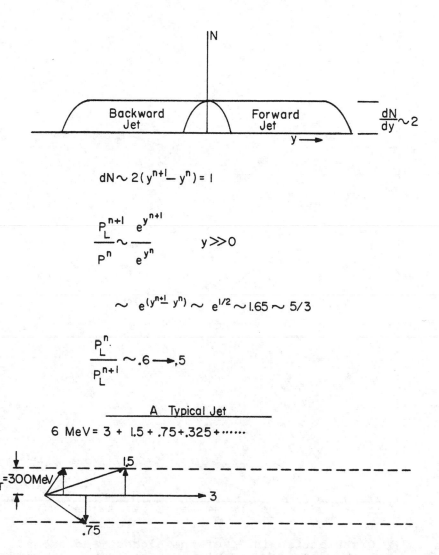

Fig. 15  Features of a jet model showing that average adjacent momenta are in the ratio 2:1 and that some fragments will always have a chance of being missed if internal transverse momentum in the jet is about 300 MeV/c.

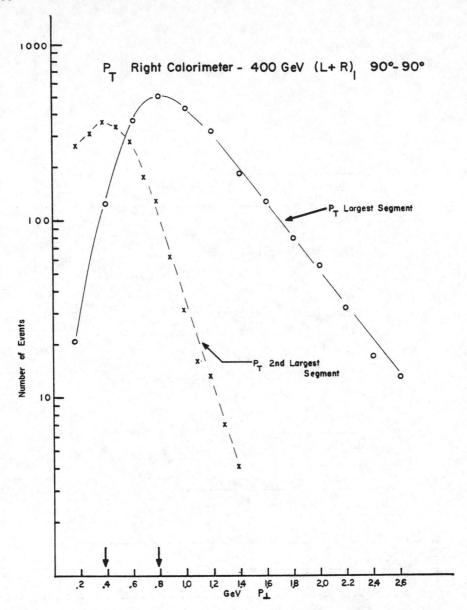

Fig. 16  Distribution in $P_T$ of the highest and second
highest $P_T$ segment of an event.  These should
be correlated with the $P_T$ of individual particles in the jet.  The ratio of $P_T$ for the
leading segments is $\sim 2{:}1$.

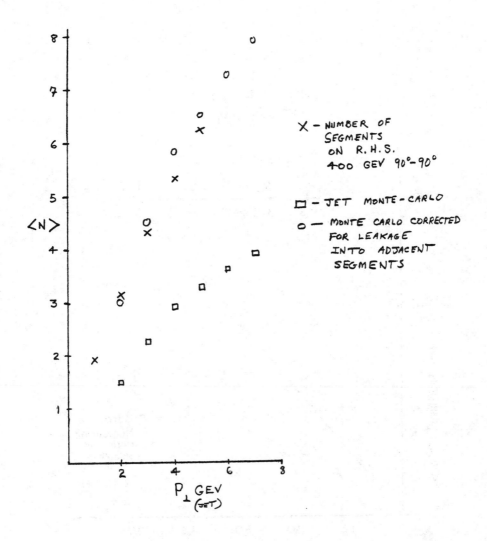

Fig. 17  Comparison of measured multiplicity based on
number of segment "hits" versus perdiction of
jet model.

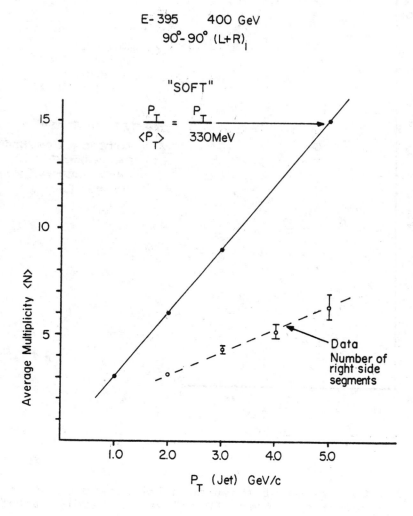

Fig. 18  Comparison of multiplicity versus $P_T$ expected
from soft pions and from data.

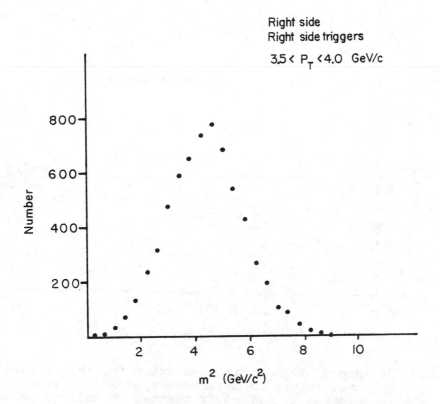

Fig. 19 Spectrum of measured jet masses for jets with
3.5 < $P_T$ < 4.0 GeV/c centered in a fiducial
area of the right calorimeter.

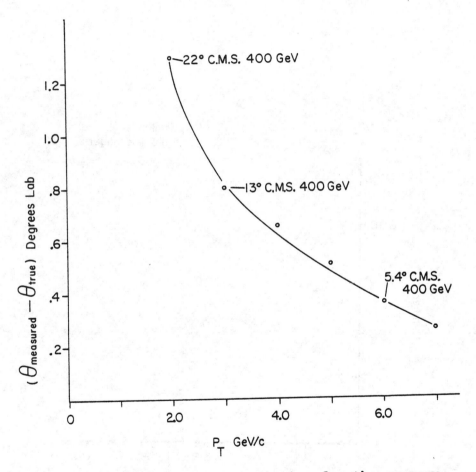

Fig. 20 Monte Carlo model prediction for the average
uncertainty in measurement of a jet direction
because of missed fragments. No allowance is
made for the possibility that a calorimeter
"prefers" jets without missed fragments.

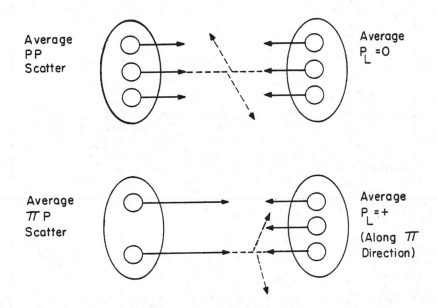

In Overall C.M.S.

Average PP Scatter

Average $P_L = 0$

Average $\pi$ P Scatter

Average $P_L = +$

(Along $\pi$ Direction)

In the Overall C.M.S. one should find both jets in the forward hemisphere more frequently in $\pi$P collisions than in the PP collisions.

So plot

$$R\left(\theta_L^*, \theta_R^*\right) = \frac{\sigma(PP \rightarrow Jets)}{\sigma(\pi P \rightarrow Jets)}$$

Use $P_T$ large enough to get angle definition in the C.M.S.

Fig. 21   Naive picture showing why parton structure functions tend to make both jets go forward in the $\pi$p center of mass system.

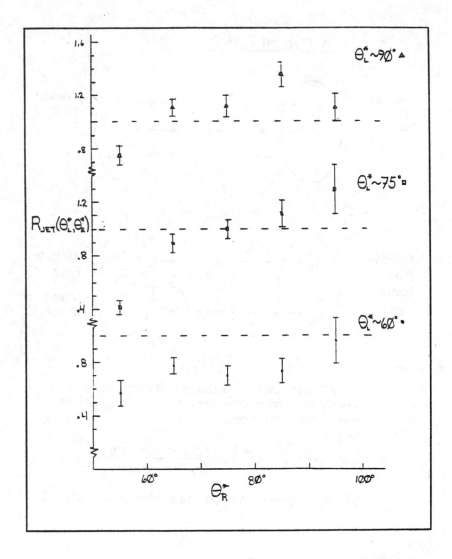

Fig. 22   Ratio of the number of pp to πp produced jets
as a function of cms angles in double arm
(L + R) triggers at 130 GeV/c.  The $p_T$ of a jet
was between 2.5 and 2.75 GeV/c as defined in
figure 11.

DIMUON PRODUCTION BY PROTONS IN TUNGSTEN[*]

W.P. Oliver
Tufts University, Medford, Massachusetts 02155
R. Gustafson, L. Jones, M. Longo, T. Roberts and M. Whalley
University of Michigan, Ann Arbor, Michigan 48104
D.A. Garelick, P.S. Gauthier, M.J. Glaubman,
H. Johnstad, M.L. Mallary, J. Moromisato, E. von Goeler
and R. Weinstein
Northeastern University, Boston, Massachusetts 02115
S. Childress, P. Mockett, J.P. Rutherfoord,
S.R. Smith and R.W. Williams
University of Washington, Seattle, Washington 98105

## ABSTRACT

The mass spectrum of dimuons produced by 400 GeV/c protons in
Tungsten has been measured for masses greater than 7 GeV. A clear
resonant signal of 2700 upsilons has been seen. The dependence of
dimuon production on $x_F$ and $p_T$ has been determined for dimuon
masses in the range 7.0 to 8.25 GeV as well as for masses in the
upsilon region 8.25 to 9.75 GeV.

## INTRODUCTION

This is a report on an experiment currently running at the
Fermi National Accelerator Laboratory. The experiment is primarily
a search for high-mass ($\gtrsim$ 7 GeV) muon pairs. The experiment was
designed to have very large acceptance (essentially the forward
hemisphere in the center-of-mass system) yet be capable of operating
at beam intensities in excess of $10^{11}$/sec. The design objectives
were achieved at the expense of resolution. The mass resolution,
$\sigma_m/m$, is 6% at the upsilon mass and worsens slowly as the mass increases.

Our large acceptance enables us to measure the dependence of
muon pair production on $x_F$ and $p_T$. The acceptance extends over the
range $-0.1 < x_F < 1$ and is essentially independent of $p_T$. Our large
acceptance also allows us to look for events with more than two
muons in the final state.

We report preliminary results based on data taken during a
three week period in January 1978. Only results from the analysis
of dimuon data are reported. We observed a clear signal for production of upsilons by 400 GeV/c protons incident upon a tungsten
target. Our upsilon signal consists of $\sim$2700 events. We are able
to accumulate upsilons at the rate of 100 events/hour under optimal
conditions. We report results for the $x_F$ and $p_T$ dependence of dimuon production for masses in the range 7.0 to 8.25 GeV as well as
for masses in the upsilon region 8.25 to 9.75 GeV.

[*]Work done under the auspices of the U.S. Department of Energy, and
the National Science Foundation.

## EXPERIMENTAL APPARATUS

The experiment is set up in the M2 beam line of the Meson Laboratory at Fermilab. The high intensity 400 GeV/c diffracted proton beam is dumped in a 0.30-m-long tungsten target. Immediately downstream of the target, a muon spectrometer is formed by a series of three gapless steel magnets. The magnets are operated at the saturation field of 21 kgauss. The field is horizontal. The total length of the three magnets is 5.5 meters.

Downstream of the magnets are the particle detectors. The detectors are arranged to form two arms separated by a gap of 0.33 m. The sensitive area of each arm is 0.71 x 1.02 m. Each arm consists of 5 Cerenkov counter hodoscopes and 9 planes of multiwire chambers. A plan view of the experiment is shown in Fig. 1.

The Cerenkov hodoscopes form the basis for the trigger. The two planes of horizontal Cerenkov counters are used to require stiff trajectories in the muon spectrometer. The three planes of vertical counters are used to require trajectories which originate in the tungsten target. The back plane of vertical counters is wedge-shaped to avoid muons from the $\rho$ and $\psi$ background.

The multiwire chambers provide fine spatial information used to reconstruct the muon trajectories. The chambers were operated with a sensitive time of 200 nsec. At a beam intensity of $2 \times 10^{11}$/sec, 40,000 protons interact in the tungsten target within this sensitive time. Nevertheless, clean muon trajectories are observed in the chambers.

## ACCEPTANCE

The combined requirement of the horizontal and vertical Cerenkov hodoscopes sets a threshold for dimuon mass at $\sim$7 GeV. The mass acceptance in the threshold region as calculated by Monte Carlo technique is shown in Fig. 2. The mass acceptance remains high for indefinitely large masses.

The acceptance in the Feynman-x variable, $x_F$, as calculated for 9.5-GeV mass, 0.0-GeV/c $p_T$ dimuons is shown in Fig. 3. The $x_F$ acceptance is similar for all masses above our threshold and for all $p_T$ within reason.

While the $x_F$-acceptance is not flat, it is still large over a wide range. For $x_F = 0$., well down from the peak, the acceptance is 1.7%.

## DIMUON RECONSTRUCTION

Events which satisfied the dimuon trigger requirement were reconstructed by the procedure:

1) Roads determined by the pattern in the Cerenkov hodoscopes were drawn through the multiwire chambers.

2) The roads were searched for wire hits. Straight line fits to the hit wires were made. At least 7 of the 9 chambers were re-

quired for a fit.

3)  Events with two good tracks were analyzed assuming both tracks originated at a point along the beam line and one absorption length (10.3 cm) into the tungsten target.  The muon momentum and vertical projection of the production angle were determined from the vertical projection of each track.  The single parameter, the vertical projection of the muon momentum, was varied until the track could be traced back through the spectrometer to pass through the assumed production point.

4)  The horizontal projection of each track was extrapolated back to the region of the assumed production point.  The horizontal deviation was required to be less than 3.5 times the expected deviation.  The calculation of the expected deviation was based on the spatial resolution of the chambers and the multiple Coulomb scattering at the measured momentum.

## RESOLUTION

Our resolution is primarily limited by the multiple Coulomb scattering the muons suffer in passing through the 5.5 meters of steel in the muon spectrometer.  The rms projected $p_T$ so acquired is 0.26 GeV/c.  The maximum $p_T$ which can be transferred in any single collision with the nuclear Coulomb field is only $\sim$0.04 GeV/c.  Consequently we expect the errors in projected $p_T$ to be Gaussian-distributed to an excellent approximation.  The magnetic field of 21 kgauss acting for 5.5 meters transfers 3.5 GeV/c of momentum to the muon.

Detailed analysis shows that our mass resolution is maximized for symmetric pairs in which each muon bends back toward the beam line.  In our treatment of the data, we enhance our mass resolution by requiring that the momentum ratio of the muons be less than 4 and that each muon bend back to the beam-line.  Since our acceptance is largest for bend-back pairs, the bend-back cut results in only a mild (25%) reduction of our data sample.  The present momentum ratio cut excludes virtually none of our data.

The results of a Monte Carlo calculation of the dimuon mass, $x_F$ and $p_T$ resolutions are shown in Table I.  The expected error in $p_T$ is approximately proportional to the dimuon mass.

## RESULTS

The distribution of reconstructed $\mu^+\mu^-$ mass is shown in Fig. 4. The distribution has been corrected for our acceptance.  A clear resonant signal of $\sim$2700 upsilons is seen.  At 9.5 GeV, the ratio of apparent resonant production to non-resonant production is 0.41. The ratio of the total $\Upsilon \to \mu^+\mu^-$ signal to the continuum signal at 9.5 GeV is 0.82 GeV.  Our resolution is insufficient to discern the individual members of the upsilon family.

A subsample, 16%, of our January 1978 data was analyzed to determine the $x_F$ and $p_T$ dependence of $\mu^+\mu^-$ pair production in two mass regions: 7.0 to 8.25 GeV and 8.25 - 9.75 GeV.  The upsilon region contained $\sim$20% upsilons.

The observed distributions in $p_T^2$ for the two regions are shown in Figs. 5 and 6. The fall-off with $p_T^2$ is strikingly different for the two regions. The fits shown in Figs. 5 and 6 were determined by eye. The fit shows that the average $p_T^2$ for the continuum region, 7.0 - 8.25 GeV, is 1.5 $(GeV/c)^2$. The average value of $p_T^2$ for the upsilon region is 2.2 $(GeV/c)^2$. These fits were inserted into the Monte Carlo program. The Monte Carlo-calculated distributions reproduced the data, verifying that apparent slopes in $p_T^2$ are very close to true slopes.

It is not possible to fit the $p_T^2$ distribution in the upsilon region with a two-component model which maintains the continuum slope at its value in the 7.0-to-8.25 GeV mass region. It seems that the continuum slope must be changing with mass. Thus, with our present analyzed data sample, we are unable to unravel the upsilon and continuum production. With our complete data sample, we expect to measure the slope in $p_T^2$ for continuum production above the upsilon region as well. We should then be able to draw conclusions about different production mechanisms.

Our data sample peaks in both mass regions at $x_F \simeq 0.25$. We assumed production cross sections of the form

$$\frac{d\sigma}{dx dp_T^2} = (1 - |x_F|)^N \, e^{-p_T^2/A}$$

and varied the parameter N in the Monte Carlo program until the observed $x_F$ distributions were reproduced. In both mass regions, good fits were obtained for N = 3.0.

TABLE I. RESOLUTION

DIMUON MASS RESOLUTION
(For $x_F$=.2 and $p_T$=0.)

| m(GeV) | $\sigma_m/m$ |
|--------|--------------|
| 9.5 | .059 |
| 18 | .077 |

DIMUON $x_F$ AND $p_T$ RESOLUTION
(For m=9.5 GeV and $p_T$=0.)

| $x_F$ | $\sigma_{x_F}$ | $\sigma_{PT}$ (GeV/c) |
|-------|----------------|------------------------|
| .2 | .03 | .62 |
| .6 | .06 | .76 |

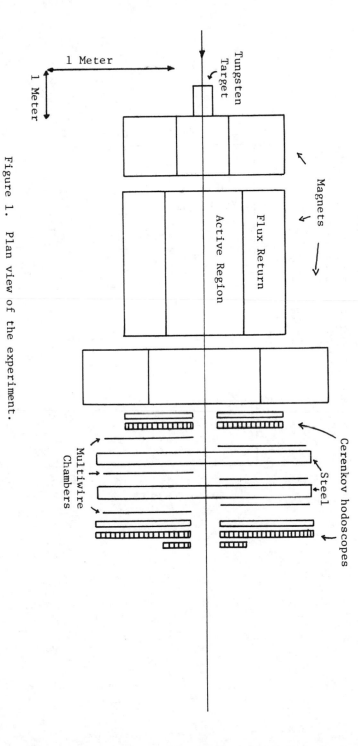

Figure 1. Plan view of the experiment.

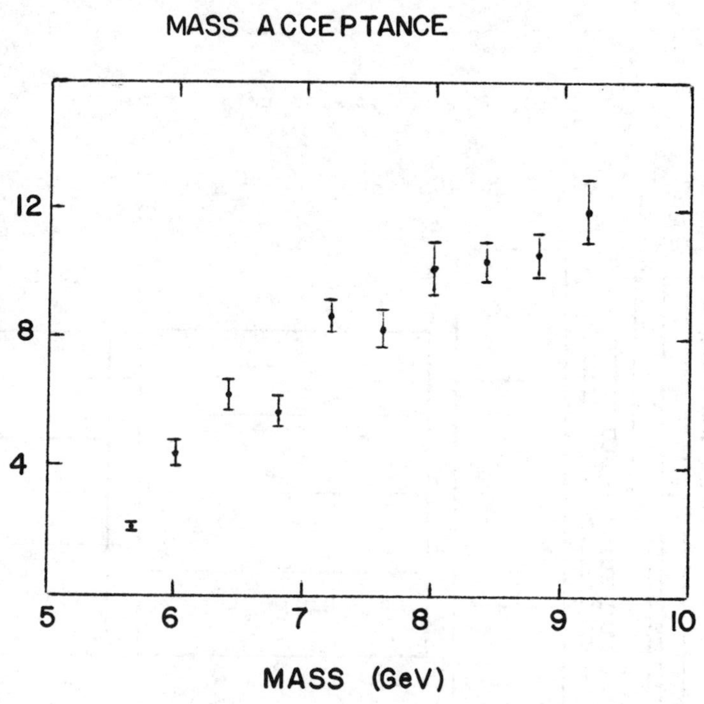

Figure 2. Mass acceptance as calculated by Monte Carlo program. The generated $x_F$ and $p_T$ distribution was $(1 - |x|)^2 \, e^{-p_T^2/1.8}$.

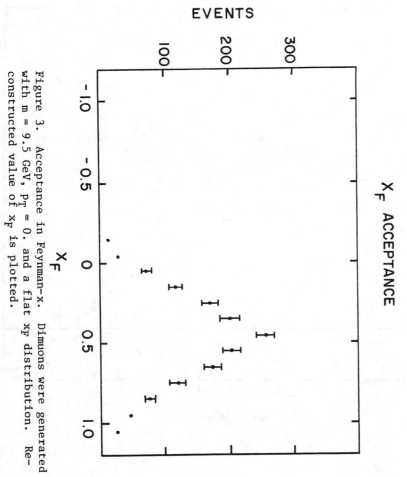

Figure 3. Acceptance in Feynman-x. Dimuons were generated with m = 9.5 GeV, $p_T$ = 0. and a flat $x_F$ distribution. Reconstructed value of $x_F$ is plotted.

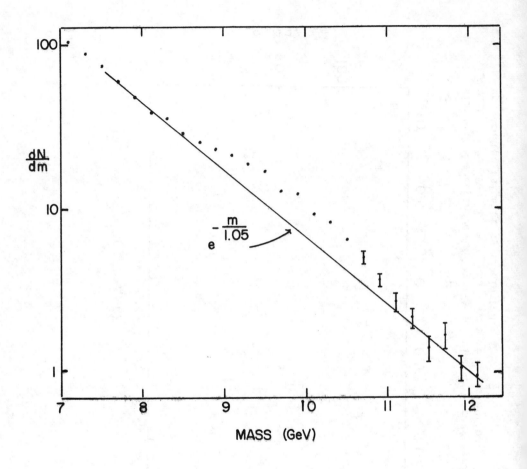

Figure 4. Dimuon mass spectrum after correction for
acceptance. There are ∿2700 events in the upsilon peak.

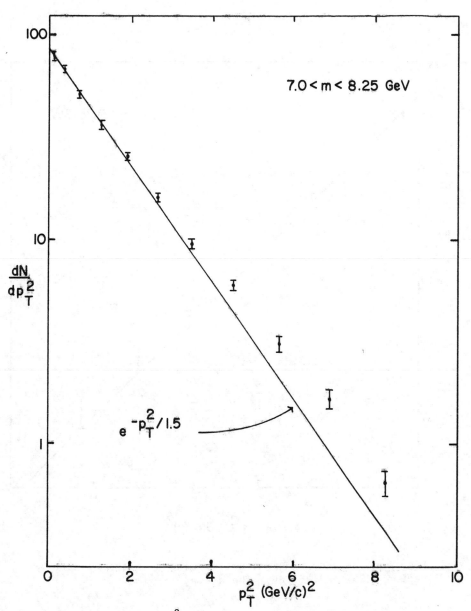

Figure 5. Dimuon $p_T^2$ spectrum for masses in the region 7.0 to 8.25 GeV. The average $x_F$ is 0.25.

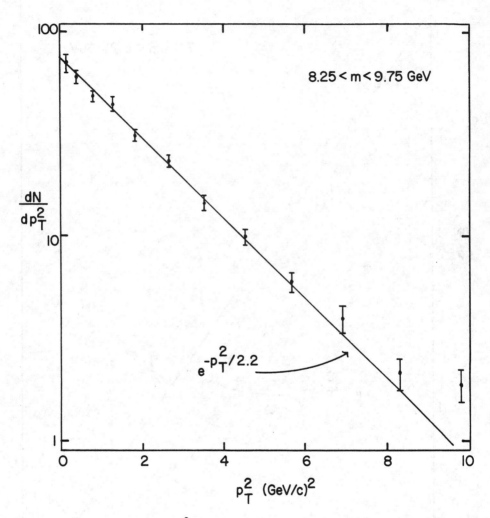

Figure 6. Dimuon $p_T^2$ spectrum for masses in the upsilon region 8.25 to 9.75 GeV. The average $x_F$ is 0.25.

# PRODUCTION OF LEPTONS AND LEPTON PAIRS IN $\pi^{\pm}p$ INTERACTIONS

R. F. Mozley
Stanford Linear Accelerator Center
Stanford University, Stanford, California 94305

## INTRODUCTION

Two low sensitivity but excellent mass resolution lepton production experiments are reported on, both performed at SLAC.

One is a study of $e^{\pm}$ production in $\pi^-$ interactions at 18 GeV/c performed using the SLAC hybrid bubble chamber. The other is a study of muon pair production in 15.5 GeV/c $\pi^-$ interactions.

## e PAIR EXPERIMENT

The $e^{\pm}$ experiment was performed by the following persons: (SLAC) J. Ballam, J. Bouchez, T. Carroll, G. Chadwick, V. Chaloupka, C. Field, D. Freytag, R. Lewis, K. Moffeit, and R. Stevens; (Duke University) H. Band, L. Fortney, T. Glanzman, J. S. Loos, and W. D. Walker; (Imperial College) P. A. Baker, P. J. Dornan, D. J. Gibbs, G. Hall, A. P. White, and T. S. Virdee. It utilized a large number of techniques for identification of electrons, in particular the introduction of three one-radiation-length tantalum plates into the hydrogen bubble chamber. Figure 1 shows a photograph of an electron pair event in the

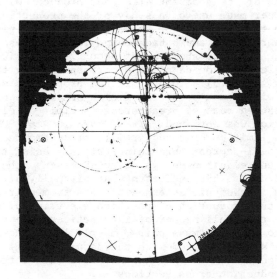

Fig. 1. Bubble chamber photograph

ISSN: 0094-243X/78/103/$1.50 Copyright 1978 American Institute of Physics

chamber. The following methods were used for electron identification;
1. Spiralling (for tracks down to 1 MeV/c).
2. Bremsstrahlung in hydrogen.
3. High energy delta rays.
4. Ionization (useful to $\sim$ 200 MeV/c).
5. Interactions in the plates.

Criteria for electron identification based on studies with an electron beam were:

    a. Electrons interact visibly before third plate.

    b. Transverse momentum of tracks in the electron shower does not exceed 30 MeV/c.

    c. No track with momentum > 10 MeV/c at $\theta > 90°$.

    d. Total $E_{vis}$ in shower behind at least one plate > 7% of incoming.

    e. No hadron signature (i.e., heavy ionization, stub tracks, etc.)

The $4\pi$ solid angle and selective criteria allowed a study of single electron production with $10^{-4}$ discrimination from hadrons.

An exposure of 250,000 $\pi p$ interactions was made and resulted in the following number of events:

A. Single $e^{\pm}$ candidates with all other tracks identified as hadrons. 22 events.

B. Candidates in pairs, from 55% of the data available so far. 2,000 pairs.

C. Single $e^{\pm}$ candidates with at least one unidentifiable track of opposite charge, from 55% of the data available so far. 400 pairs.

The analysis of the apparent electrons produced without an opposite charge partner made use of a study of the properties of pion induced showers in the plates. For this, pure beams of pions at 1.57 and 3.14 GeV/c were used, and a background level was established for pions to simulate electron showers (at about 1 in $2 \times 10^4$ pions).

The 22 single electron candidates corresponded by themselves to an $e/\pi$ ratio of $(4.3 \pm .9) \times 10^{-5}$. All the properties of the events were consistent with the hypothesis of pion breakthrough, and subtracting the measured rate of breakthrough a limit on direct, unpaired, electron production was obtained: $e/\pi < 2.4 \times 10^{-5}$ at 90% C.L.

From this a limit on charmed particle production could be derived in terms of the semi-leptonic branching ratios. Using the recent measurement for the electron decays of the D mesons, the limit is $\sigma < 13$ $\mu$b at 90% C.L. This indicates that virtually all of the reported electrons in hadronic interactions are from pair production.

A study of the production of positron electron pairs was also performed. Here the small number of interactions made the experiment sensitive only to very low mass pairs, mostly lower than the $\pi^0$ mass. The results were again compared with the detailed Monte Carlo calculation. Pair production from all known sources was introduced using the following production cross sections:

$$\begin{array}{lll} \eta & \sigma = 1.3 \text{ mb} & \\ \omega & \sigma = 3.4 \text{ mb} & \\ \rho & \sigma = 4.0 \text{ mb} & \text{(Ref. 1)} \\ \pi^{\pm} & \sigma = 85 \text{ mb} & \\ \pi^0 & \sigma = 36 \text{ mb} & \end{array}$$

The results are compared in Tables I and II.

Table I  A Comparison of the $p_t$ Distribution of $e^+$ and $e^-$
Tracks with That Expected from a Monte Carlo
Calculation Encompassing All Known Sources

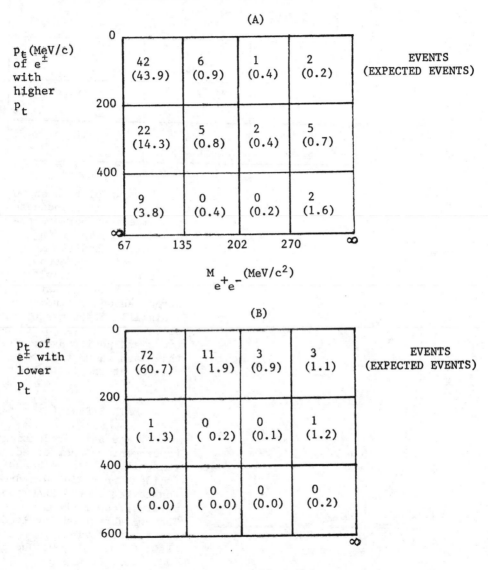

(A)

$p_t$ (MeV/c)
of $e^\pm$
with
higher
$p_t$

|       | 42 (43.9) | 6 (0.9)  | 1 (0.4) | 2 (0.2) | EVENTS (EXPECTED EVENTS) |
|       | 22 (14.3) | 5 (0.8)  | 2 (0.4) | 5 (0.7) |                          |
|       | 9 (3.8)   | 0 (0.4)  | 0 (0.2) | 2 (1.6) |                          |

0  200  400  ∞

67    135    202    270    ∞

$M_{e^+e^-}$ (MeV/c$^2$)

(B)

$p_t$ of
$e^\pm$ with
lower
$p_t$

|       | 72 (60.7) | 11 (1.9) | 3 (0.9) | 3 (1.1) | EVENTS (EXPECTED EVENTS) |
|       | 1 (1.3)   | 0 (0.2)  | 0 (0.1) | 1 (1.2) |                          |
|       | 0 (0.0)   | 0 (0.0)  | 0 (0.0) | 0 (0.2) |                          |

0  200  400  600  ∞

$M_{e^+e^-}$ (MeV/c$^2$)

Table II   The Excess of Measured $e^\pm$ Tracks Over Those Predicted by a
Monte Carlo Calculation Encompassing all Known Sources.
Both $e^+$ and $e^-$ are Counted

|  | | OBSERVED – MONTE CARLO | | Expected at $\pi^\pm$ rate $\times 10^{-4}$ |
|  |  | $M_{ee} > 67$ | $M_{ee} > 135$ |  |
| --- | --- | --- | --- | --- |
|  | 0 |  |  |  |
|  |  | 35.0 ±18.0 | 25.7 ±5.6 | 13.4 |
|  | 200 |  |  |  |
|  |  | 11.0 ±6.1 | 9.5 ±3.7 | 15.6 |
| $P_t$ (MeV/c) of $e^+$ or $e^-$ | 400 |  |  |  |
|  |  | 4.3 ±3.5 | -0.4 ±1.5 | 14.2 |

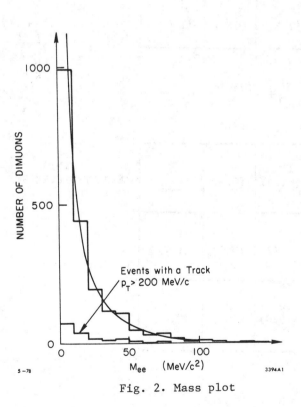

5 –78          3394A1

Fig. 2. Mass plot

A mass plot is shown in Fig. 2. This indicates the agreement between the Monte Carlo, including resolution, and the $\pi^0$ Dalitz pair dominated pair spectrum. Only above 135 MeV/c$^2$ is the data strong enough to indicate a significant excess of pairs. The excess of electrons and positrons in this mass range tend to cluster at low $p_T$ (Table II).

## $\mu$ PAIR EXPERIMENT

The $\mu$ pair production experiment was performed using 15.5 GeV/c $\pi^\pm$ beams incident on a liquid hydrogen target in the SLAC two-meter streamer chamber. The collaboration involved was the following:
SLAC: K. Bunnell, M. Duong-van, E. Kogan, B. Haber, R. Mozley, A. Odian, F. Villa, L. Wang.

Vanderbilt: R. Cassell, R. Panvini, S. Poucher, A. Rogers,
          S. Stone.
U.C.S.C.  : T. Schalk.
M.I.T.    : W. C. Barber

The results reported here are from an exposure of $10^8$ πp interactions which yielded 335 pairs with $X_F > 0.3$. This is half of the total exposure the rest of which is being analyzed.

The experimental layout is shown in Fig. 3. The pion beam is incident from the left into a 3cm diameter one-meter-long liquid hydrogen target mounted inside the visible region of the streamer chamber. The chamber is mounted in a 13 Kgauss magnet and viewed from above in three camera stereo. Directly downstream of the chamber a hadron absorber of copper and lead is erected with a hole allowing passage of the noninteracting pions in the beam. A scintillation counter hodoscope is interspersed in the wall with horizontal counters A, C, and E and vertical counters B and D. A trigger consisted of an interaction in the hydrogen target (i.e., a particle-incident and non-emergent) and the firing of a pair of appropriately lined up A and C counters and two nonadjacent B counters. The system would trigger on particles of momenta greater than 2 GeV/c. Hits on the D, E, and F scintillation counters were recorded as well as those in two wire chamber planes(W).

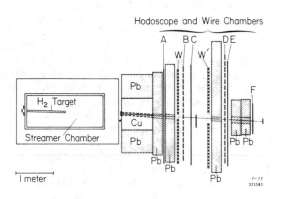

Fig. 3. Plan view of streamer chamber

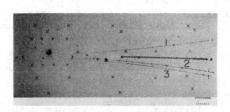

Fig. 4. Streamer chamber photograph with muon candidates (tracks 1, 2, and 3.)

Figure 4 shows a photograph of a muon pair candidate in the streamer chamber. Hadron punchthrough was reduced by the following procedure: Muon candidate tracks such as 1, 2, and 3 were measured in the streamer chamber and extrapolated positions calculated at the counter and wire chamber locations. A measured position over three standard deviations from the projected position was taken as evidence that the particle was in fact a hadron.

Figure 5a shows a plan view of three muon candidates visible in Fig. 4. All of the tracks satisfy muon criteria in this view. Fig. 5b shows a side view of the same three tracks. There is no appropriate hit for track 2 in the first plane of wire chambers, and as a result tracks 1 and 3 are identified as muons and track 2 as a hadron.

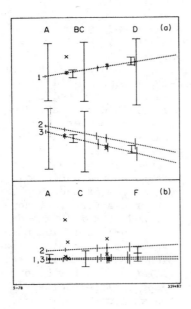

Fig. 5a. Plan view of arrangement of downstream scintillation
counters and wire chambers. The
location of counters hit is
shown by lines with horizontal
bars I while the predicted
track location is shown by
lines without horizontal bars
which extend ± 1 σ; wire chamber hits are shown by crosses.
5b. Side view of arrangement
of downstream counters and wire
chambers.

To calculate the punchthrough
expected with the use of these criteria, an exposure was made to events
resulting from beam interactions and
another exposure to events with a
single muon trigger. The particles
causing single muon triggers are primarily hadrons ∿ 30/1 and hence can
serve as an excellent model for
hadron punchthrough. As a result
the muon candidate tracks from the
single muon events were measured and
subjected to the same scrutiny as
that for the two muon trigger events.
These were then treated as examples
of hadron punchthrough and the hadron
punchthrough probability calculated
as a function of energy. The interaction trigger events were measured
and used as a sample of the hadrons
incident on the lead wall. The measured punchthrough probability was
applied to single muon events and the
probability of their simulating dimuons was then calculated.

It was found that excessive
punchthrough occurred if all events
were used which penetrated through
to the C counters (2 GeV/c particles),
and as a result the events reported
on here had muon candidates which
registered as passing through enough
lead to reach the D and E counters.
The minimum momentum required for
this was 2.4 GeV/c. This resulted in
a minimum $X_F$ of 0.3. Figure 6 shows
the acceptance for $X_F$ (c of m momentum in beam direction/maximum possible
c of m momentum) and dimuon mass.

Table III summarizes the data
obtained while Fig. 7 shows the dimuon mass spectrum. A clear ω signal can be seen, plus a ρ signal and a considerable enhancement below
the ρ. It was not possible to make a clean fit to the ρ signal without a better understanding of the nature of the background under it.
Allowing the background to vary could give an excellent fit for a very
large range of ρ cross sections and hence none is determined here. The
calculated punchthrough background is shown crosshatched.

Estimates of the maximum size of η → 2μ decay can be made because
of the good mass resolution. For this purpose the mass binning is adjusted to allow a single bin to be centered on the m mass. An insig-
nificant signal thus observed allows a cross section calculation to be
made.

Table III. Data Summary

| | $\pi^+$ | $\pi^-$ | Combined |
|---|---|---|---|
| Interactions | $45.3 \times 10^6$ | $51.1 \times 10^6$ | $96.3 \times 10^6$ |
| Sensitivity (events/$\mu$barn) | 1,880 | 1,990 | 3,870 |
| Dimuons found (background) $\mu^+\mu^-$ | 137 (17) | 198 (44) | 335 (61) |
| $\mu^+\mu^+$ | 15 (8) | 3 (9) | 18 (17) |
| $\mu^-\mu^-$ | 1 (1) | 22 (19) | 23 (20) |
| Cross section observed $x_F \to 0.3$ (nanobarn) | | | |
| $\pi p \to \mu^+\mu^- X$ | $250 \pm 70$ (a) | $320 \pm 90$ | $280 \pm 70$ |
| $\pi p \to \omega X \to \mu^+\mu^- X$ | | | $30 \pm 13$ |
| $\pi p \to \eta X \to \mu^+\mu^- X$ | | | $3 \pm 3$ |

$$\sigma(\pi^- p \to \mu^+\mu^- X)/\sigma(\pi^+ p \to \mu^+\mu^- X) = 1.28 \pm 0.23$$

$$\frac{\sigma(\mu^+\mu^-)}{\sigma(\pi^+\pi^-)} = (3 \pm 1) \times 10^{-5}$$

(a)  This number can be compared with the value of $340 \pm 70$ obtained at 150 GeV/c.[6]

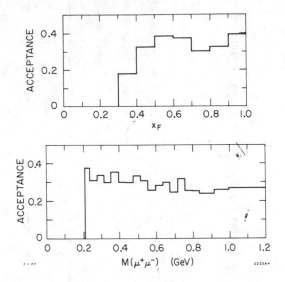

Fig. 6. Acceptance of the apparatus for dimuons.

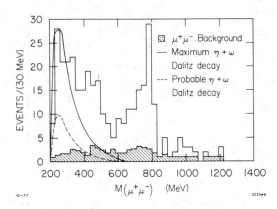

Fig. 7. Measured dimuon mass spectrum. Estimates of the hadron punchthrough background and contributions from η and ω Dalitz decay are indicated.

Figure 8 shows the mass spectrum with the background subtracted. The possible contribution from Dalitz decays of the η and ω can be estimated in two ways. One way is to use our measured values of η and ω → 2μ cross sections and to calculate their Dalitz decay contribution from this (Ref. 2). This is shown as the dotted curve in Fig. 7 and the mass spectrum with this and the background subtracted is shown in Fig. 9.

A more conservative way of estimating Dalitz decay backgrounds is to make use of the excellent mass resolution to introduce the maximum signal which could be present. This involves only knowing the shape of the Dalitz contribution, and here we use a shape with a very large ω contribution to emphasize the width of the distribution. The maximum possible contribution is shown in Fig. 7. If the contribution were larger than this it would cause the mass plot to peak at a higher value near threshold. The mass plot with the background and this maximum Dalitz contribution subtracted is shown in Fig. 10. Thus the good mass resolution allows this low statistics experiment to show for the first time model independent evidence for the existence of a dimuon mass continuum below the ρ which cannot be derived from Dalitz decay of known mesons.

There are an adequate number of theoretical papers predicting such behavior. A calculation by Bjorken and Weisberg (Ref. 3) produced a

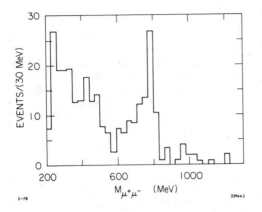

Fig. 8.  Dimuon mass spectrum after
subtraction of the calculated hadron
punchthrough.

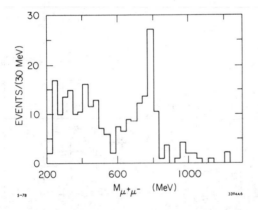

Fig. 9.  Dimuon mass spectrum after
subtraction of the calculated hadron
punchthrough and the most probable
contribution from $\eta$ and $\omega$ decays.

qualitative enhancement here
by quark antiquark annihilation similar to that pre-
dicted by the Drell-Yan model
but enhanced by the annihilation with the additional
quarks produced in pion production.  Calculations by
Blankenbecler and Duong-van
(Ref. 4) of pion pion annihilation produced a phenomenological more quantitative
estimate.  A similar en-
hancement would occur from
the pion bremsstrahlung prediction of Frautshi and
Farrar (Ref. 5) but here we
would expect a mass spectrum
peaking more at threshold.

It is interesting to
study the $X_F$ spectrum as a
function of the mass region.
The result is shown in Fig.
11.  The region of the $\rho$
shows a flatter spectrum as
would be expected from the
known enhancement of $\rho$ production at high $X_F$.  The
transverse momenta of the
observed pairs is shown in
Figs. 12 and 13.  Differences
between the mass regions are
not statistically significant.

The decay angle of the
muons with respect to the
direction of the muon pair
can be examined but unfortunately the acceptance of the
trigger lessens the value of
this.  Muons produced at almost all c of m angles are
incident on the lead absorber
but those produced at large angles have insufficient energy to penetrate the lead.  The acceptance is shown in Fig. 14.  Figure 15 shows
the resulting distribution.  The results don't clearly differentiate
between distributions although a $\sin^2\theta$ distribution is slightly preferred over $(1 + \cos^2\theta)$.  The same data is shown in Fig. 16 broken up
into different mass intervals.  No statistically significant differences are observed.

Since the entire final state can be observed in the streamer chamber, it is interesting to examine the hadrons in the events to see if
their characteristics differ from those in nondimuon events.  To make

112

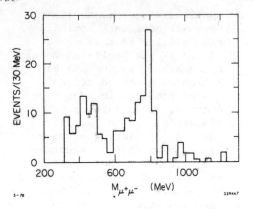

Fig. 10. Dimuon mass spectrum after subtraction of the calculated hadron punchthrough and the largest possible contribution from η and ω Dalitz decays.

Fig. 11a. $X_F$ distribution of muon pairs with a mass between 200 MeV and 380 MeV.
11b. $X_F$ distribution of muon pairs with a mass between 380 MeV and 600 MeV.
11c. $X_F$ distribution of muon pairs with a mass between 600 MeV and 900 MeV.

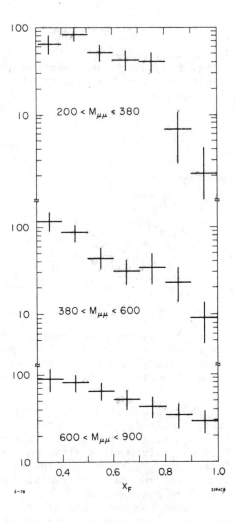

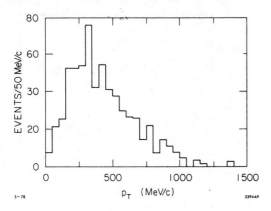

Fig. 12. Transverse momentum distribution of muon pairs <$p_T$> 420 MeV.

Fig. 13a. Transverse momentum distributions of muon pairs $200 < M_{\mu^+\mu^-} < 380$.
b. Transverse momentum distribution of muon pairs $380 < M_{\mu^+\mu^-} < 620$.
c. Transverse momentum distribution of muon pairs $620 < M_{\mu^+\mu^-}$

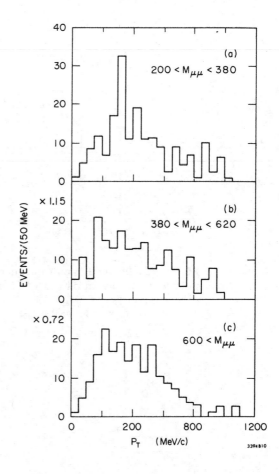

Fig. 14. Acceptance for muon
pairs as a function of the decay angle of the $\mu^+$ with respect to the pair direction in
its c of m system.

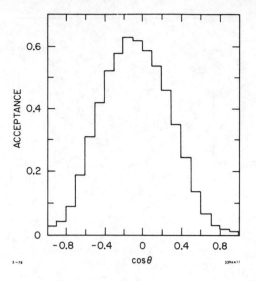

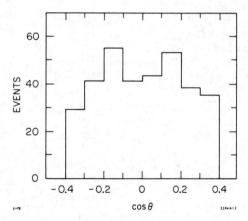

Fig. 15.  Decay Distribution of $\mu^+$ with respect to
pair direction in its c of m system.

$\chi^2$ for 4 Degrees of Freedom

| Flat | $\sin^2\theta$ | $(1 + \cos^2\theta)$ |
|------|------|------|
| 7.1 | 4.6 | 10.3 |

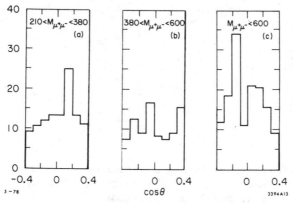

Fig. 16.  Decay distributions for different $M_{\mu^+\mu^-}$ intervals.

$\chi^2$ for 4 Degrees of Freedom

Flat  $\sin^2\theta$  $(1 + \cos^2\theta)$  Flat  $\sin^2\theta$  $(1 + \cos^2\theta)$  Flat  $\sin^2\theta$  $(1 + \cos^2\theta)$

3.5   2.5          4.6        1.5   1.9          1.6        10.6 8.5          13.0

this comparison we use the interaction trigger events mentioned earlier with respect to our background calculations. For studies using these events we identify pions in the nondimuon events which would be possible muon candidates if a trigger had in fact occurred. Fortunately very few events show more than one pair of candidates. Five percent of the sample had two candidate pairs and both of the pairs are included in the comparison. Figure 17 shows a comparison of the accompanying multiplicities for a $\pi^+$ beam and Fig. 18 for $\pi^-$ events. Muon pairs are subtracted from the dimuon event multiplicities while the appropriate two pions are subtracted from the interaction events with which a comparison is made. No statistically significant difference can be observed.

CONCLUSION

In conclusion it is clear that these two experiments utilize their large solid angle coverage and good resolution to produce unique results in spite of very low sensitivity. The electron study produced the first clear evidence that single electrons are not produced in anomalously large quantities but that those observed in other experiments are almost completely due to electron pairs. The streamer chamber experiment makes use of its good resolution to show the first model independent evidence for a mass enhancement below the $\rho$ which is not due to Dalitz decays of known mesons.

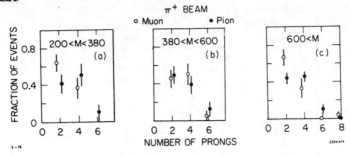

Fig. 17. Accompanying multiplicity distributions for muon pair events and for pairs of pions which are incident on the hadron absorber in an interaction experiment.

$\langle M_{\mu\mu} \rangle = 288$     $\langle M_{\mu\mu} \rangle = 470$     $\langle M_{\mu\mu} \rangle = 864$

$\langle M_{\pi\pi} \rangle = 316$     $\langle M_{\pi\pi} \rangle = 512$     $\langle M_{\pi\pi} \rangle = 882$

$\langle n_{\mu\mu} \rangle = 2.71 \pm .2$     $\langle n_{\mu\mu} \rangle = 3.2 \pm .28$     $\langle n_{\mu\mu} \rangle = 2.83 \pm .27$

$\langle n_{\pi\pi} \rangle = 3.30 \pm .3$     $\langle n_{\mu\mu} \rangle = 3.27 \pm .26$     $\langle n_{\mu\mu} \rangle = 3.31 \pm .14$

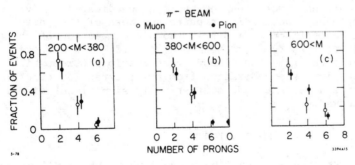

Fig. 18. Accompanying multiplicity distributions for muon pair events and for pairs of pions which are incident on the hadron absorber in an interaction experiment.

$\langle M_{\mu\mu} \rangle = 309$     $\langle M_{\mu\mu} \rangle = 473$     $\langle M_{\mu\mu} \rangle = 788$

$\langle M_{\pi\pi} \rangle = 309$     $\langle M_{\pi\pi} \rangle = 494$     $\langle M_{\pi\pi} \rangle = 895$

$\langle n_{\mu\mu} \rangle = 2.63 \pm .2$     $\langle n_{\mu\mu} \rangle = 2.69 \pm .2$     $\langle n_{\mu\mu} \rangle = 3.05 \pm .26$

$\langle n_{\pi\pi} \rangle = 2.88 \pm .22$     $\langle n_{\pi\pi} \rangle = 3.09 \pm .18$     $\langle n_{\pi\pi} \rangle = 3.12 \pm .10$

## REFERENCES

1. J. Bartke et al. Nucl. Phys., B118, 360 (1977).
2. C. H. Lai and C. Quigg, FN-296 (1976); C. Quigg and J. D. Jackson, (U.C.R.L.-18487 (1968).
3. J. D. Bjorken and H. Weisberg, Phys. Rev. D13, 1405 (1976).
4. M. Duong-van and R. Blankenbecler, to be published.
5. G. R. Farrar and S. C. Frautschi, Phys. Rev. Lett. 36, 1017 (1976).
6. K. J. Anderson et al., Phys. Rev. Lett. 36, 799 (1976).

# PRODUCTION OF $e^+e^-$ AND $\pi^\circ$ PAIRS AT THE ISR

Presented by M. Goldberg
for

J.H. Cobb, S. Iwata[1], R.B. Palmer, D. Rahm, P. Rehak and I. Stumer
Brookhaven National Laboratory[2], Upton, New York   11973, USA
C.W. Fabjan, E.D. Fowler[3], I. Mannelli[4], P. Mouzourakis,
K. Nakamura[5], A. Nappi[4], W. Struckzinski[6] and W.J. Willis
CERN, Geneva, Switzerland
M. Goldberg, N. Horwitz and G.C. Moneti
Syracuse University[7], Syracuse, New York   13210, USA
C. Kourkoumelis and L.K. Resvanis
University of Athens, Athens, Greece
T.A. Filippas
National Technical University, Athens, Greece
and
A.J. Lankford
Yale University, New Haven, Connecticut   06520, USA

## ABSTRACT

Recent results of an experiment studying large transverse momentum phenomena at the ISR are reviewed.  Data relevant to dielectron studies have yielded cross sections for the $T(9.5)$, possible association of the $J/\psi$ with production of $\chi(3.5)$, and a stringent upper limit for direct $\gamma$ production.  In addition, azimuthal correlations of high transverse momentum $\pi^\circ$ pairs have been investigated.

## INTRODUCTION

The central purpose of experiment R806 at the CERN ISR is an investigation of inclusive production of electron-positron pairs. The large azimuthal acceptance of our apparatus enables us to study a wide range of pair effective masses and measure their transverse momenta.[1]  After a brief review of the new techniques associated with the apparatus, the spectrum of pair invariant masses obtained will be explored with reference to $T(9.5)$, $\chi(3.5)$ and direct photon production.  In addition, correlations between high transverse momentum $\pi^\circ$ pairs, relevant to hard scattering models will be discussed.

---

[1] Permanent address: Nagoya University, Nagoya, Japan.
[2] Research under the auspices of U.S. Department of Energy
[3] Permanent address: Purdue University, Lafayette, In.  47907, USA.
[4] On leave of absence from the University of Pisa and INFN Sezione di Pisa, Italy.
[5] Permanent address: University of Tokyo, Tokyo, Japan.
[6] Now at: Phys. Inst. T.H. Aachen, Aachen, Germany.
[7] Work supported by the National Science Foundation.

ISSN:  0094-243X/78/118/$1.50  Copyright 1978 American Institute of Physics

# APPARATUS

The apparatus, composed of four subassemblies known as octants, is shown schematically in Fig. 1. Each octant covers 45° in azimuth and the polar angle interval 45° - 135°. An electron emerging from intersecting region and passing through an octant encounters

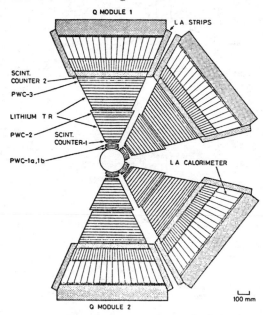

Fig. 1 Vertical section of the apparatus transverse to the proton beams.

(i) Two low mass PWC's utilizing charge division readout. These chamber's primary function is to provide tracking to the production vertex.

(ii) Two sets of scintillators. These are used in tracking and provide an important minimum ionization selection for electrons.

(iii) Two lithium foil transition radiators each followed by a Xe filled PWC with charge division read out. Also used for position measurement, the PWC pulse height contributes to hadron rejection when minimum ionizing electrons are selected.

(iv) The liquid argon calorimeter. This device provides position information ($\sigma \simeq 5$ mm), electromagnetic shower energy measurement ($\sigma/E \simeq 12\%/\sqrt{E}$) and additional hadron rejection, in addition to its utilization as a basic component of the various triggers. The two triggers important to the data under discussion require either

a) local energy deposit (> 1 GeV) characteristic of showers in calorimeters of 2 octants (Double High Trigger)

b) crude track segment correlations formed by the calorimeter hit, scintillation counters, and PWC's, with reduced calorimeter thresholds (Double Correlation Trigger).

All aspects of octant performance were determined by exposure to electron and hadron beams at the CERN P.S.

In Fig. 2 we show, for a small sample of data, a procedure for extracting di-electron signals. Starting from a triggered effective mass spectrum where two tracks are associated with two showers, offline scintillator cuts reject slow hadrons and small angle electron pairs. Next, selection of events with large pulse-heights in the Xenon PWC's results in an already clear J/$\psi$ signal. Further cuts involve combining electrons with other showers to reject Dalitz pairs from $\pi^o$'s, requiring good $\chi^2$ for fitted tracks to eliminate

120

near overlaps (e.g. $\pi^-, \pi^0$) and requiring normal electromagnetic
shower development. The end result is a J/$\psi$ signal with low background.

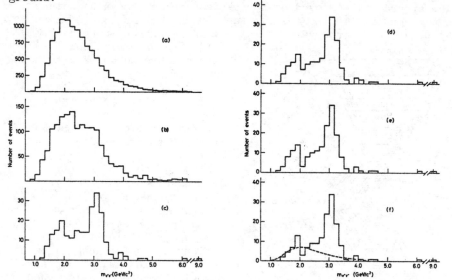

Distribution of effective masses of double-electron candidates:

a) before any off-line requirement is applied;
b) after scintillator cut is applied;
c) as (b) with the addition of TR cut;
d) as (c) with the addition of "$\pi^0$ removal" cut;
e) as (d) with the additional requirement of a good fit of the track in the calorimeter;
f) as (e) with the additional requirement of $E_{ass}/E_{tot}$, i.e. after all cuts applied (background is also shown as dashed curve).

Fig. 2

Since, in the absence of a magnetic field, we cannot select
$e^{\pm}e^{\pm}$ spectra from which to derive background levels, we rely on a
plateau and extrapolation method described below (see Fig. 3). From
exposures of the apparatus to electron beams, we calculate the
efficiency for electron pairs as a function of cut-strength, a cut
being a combination of the off-line selections shown in Fig. 2. A
true dielectron signal corrected for cut efficiency will not vary as
a function of cut-strength. The actual behavior of the data,
initially uncut, is a rapid falloff with increasingly tight cuts,
which plateaus at the "real" $e^+e^-$ level. The background estimate is
extrapolated from the initial falloff. This procedure has been
checked by applying it to our J/$\psi$ signal and using adjacent mass bins
as a background estimate.

The spectra we will discuss was derived from $\sim$ 800 hours of ISR
running, with integrated luminosity $5.8 \times 10^{36}$ cm$^{-2}$ at $\sqrt{s}$ = 52 GeV;
$2.6 \times 10^{36}$ cm$^{-2}$ at $\sqrt{s}$ = 62 GeV. From $\sim 10^{11}$ beam-beam interactions,
the double high and double correlation triggers selected $\sim 10^7$
triggers (events on 1000 tapes). Of these, the reconstruction
programs yielded $10^5$ ee candidates and after appropriate cuts, $\sim 10^3$
ee events remained. We are now processing 2000 additional tapes and
currently taking data at the ISR.

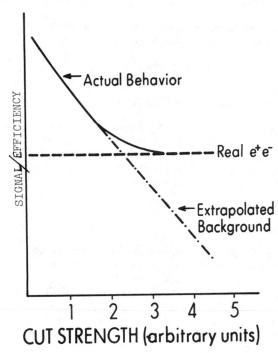

Fig. 3

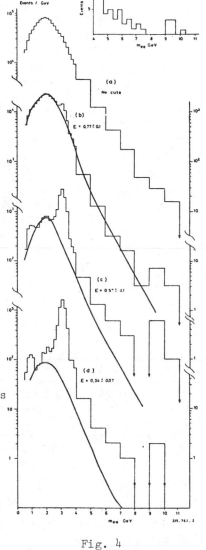

Fig. 4

## PRODUCTION OF T(9.5)

Figs. 4a-d show the pair effective mass spectra obtained as a function of cut strength as described above. The efficiency of each cut for real pairs is noted, and the plateau after initial falloff is clear. The most prominent signal is that of the $J/\psi$, and an enhancement in that of the $\rho$ region is also evident. We will return to the $\rho$ signal later, but now we focus on the signal above 9 GeV seen as a shelf in Fig. 4b and as a 7 event signal in Figs 4c and 4e. Although we are limited in the strength of our statements about these few events by their number, we note that if we associate them with the continum, then that cross section would be significantly higher than scaling and our own data at lower masses would indicate, and if we compare them with background in this mass region, we find that their average transverse momentum $\langle P_T \rangle = 1 \pm 0.4$ GeV is substantially

lower than that of background ($\langle P_T \rangle = 3.5$ GeV).

Identifying these events as members of the T(9.5) gives the following results for cross section times branching ratio

$$B_{ee}\left.\frac{d\sigma}{dy}\right|_{y=0} = 30 \pm 15 \text{ pb at } \sqrt{s} = 52 \text{ GeV}$$

$$B_{ee}\left.\frac{d\sigma}{dy}\right|_{y=0} = 60 \pm 35 \text{ pb at } \sqrt{s} = 62 \text{ GeV}.$$

We note that this is a two order of magnitude increase over what is observed at $\sqrt{s} = 28$ (see reports on high mass di-muons presented at this conference). Such a steep excitation function for the T is unexpected, considering the factor of 3 rise in the J/$\psi$ cross section over the same energy interval, or predictions of continum scaling.

## PRODUCTION OF χ(3.5)

The four steradian acceptance of our apparatus enables us to search, with good geometrical acceptance, for photons accompanying the triggering electron pair. A falling reconstruction efficiency requires us to accept only photons with energy greater than 0.4 GeV. When such photons are found accompanying pairs having J/$\psi$ effective mass, and photons which have $\pi^{\circ}$ mass when combined with another shower are excluded, we obtain the J/$\psi$ + $\gamma$ effective mass spectrum of Fig. 5. It is clear that with a falling intensity of photons as a function of energy, random combinations of J/$\psi$ + $\gamma$ will peak at low effective mass. The back- ground spectrum formed in this way and normalized to the mass spectrum above 3.7 GeV is shown. An intriguing peak at the mass of χ(3.5) stands above the background. Monte Carlo studies show that these 25 ± 10 χ events imply that

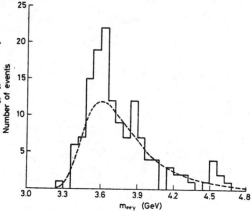

$$\frac{\text{J/}\psi \text{ from } \chi(3.5)}{\text{all J/}\psi} = 0.4 \pm 0.2.$$

It has been suggested that a substantial number of the J/$\psi$'s produced could come from χ's[2].

Fig. 5. Invariant mass distribution of J/$\psi$ plus photon with estimated background (dashed line)

## PRODUCTION OF DIRECT PHOTONS

As noted above, an enhancement in the ρ,ω region is seen in our electron pair mass spectrum. The spectrum for triggers in- volving adjacent octants only is shown in more detail in Fig. 6a. This well understood trigger enables us to check a special double correlation trigger which attempts to select pairs in a single octant. After appropriate cuts, the resulting mass spectrum is shown in Fig. 6b. A clear $\rho,\omega$ signal is seen, along with a continuum of less intensity in the $0.2 < m_{ee} < 0.5$ GeV region, where trigger acceptance is good above $P_T$ (pair) = 2 GeV. This single octant spectrum allows us to place a stringent upper limit on the ratio $\gamma$(direct)$/\pi^0$ in this transverse momentum range. Several models predict $\gamma$(direct)$/\pi^0 \approx 0.1$.

The method can be understood by referring to Fig. 7 produced by Cragie and Schildknecht[3]. If direct $\gamma/\pi \approx 0.1$ and $\sigma(\omega)=\sigma(\rho)=\sigma(\pi)$, then the spectrum of those direct $\gamma$'s internally converting to electron pairs can be predicted relative to the cross section times branching ratio of $\rho,\omega \to e^+e^-$(the internal conversion calculation is analagous to the Dalitz decay calculations). The resulting predicted spectrum is shown as $(\gamma/\pi = 0.1) + \rho+\omega$ in the figure. The spectrum for $\rho+\omega$ alone is also shown. A vector dominance model,in which the photon propagator is factored with the large $\rho$ width is responsible for the enhancement at $q^2_{ee} = 0$ in this case. The main

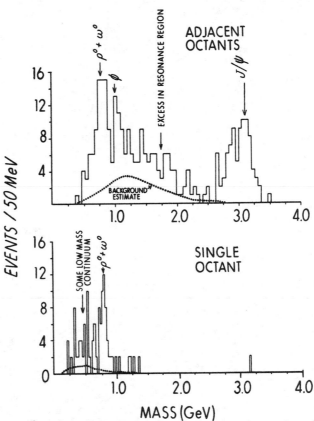

Fig. 6. $e^+e^-$ Pairs effective mass distributions from adjacent octant and single octant triggers.

point is that, when the resolution of our apparatus is taken into account, the $\rho,\omega$ peak should be a barely visible bump on a large continuum, if $\gamma/\pi = 0.1$. We have already seen that this is not the case.

To arrive at an estimate of an upper limit, we note from the figure that every postulated $\gamma/\pi$ ratio is in one-to-one correspondence with the ratio $\gamma \to e^+e^-$ $(0.2 < m^2_{ee} < 0.5)/\rho \to e^+e^-$. By taking this ratio for events with 2 GeV $< P_T$(pair $< 3$ GeV many

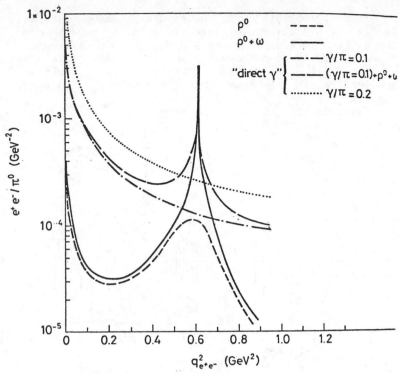

Fig. 7   Mass distribution of $e^+e^-$ pairs for $q_T \gg 1$ GeV. The
distribution from $\rho$-$\omega$ dominance is compared to that
expected for large $\gamma/\pi$ ratios (from Craige and Schildknecht).

systematic effects cancel out. The correspondence is shown in Fig.
8, together with the ($\gamma \to e^+e^-/\rho \to e^+e^-$) ratio observed for various
cuts. It can be seen that most of these pairs with masses less than
the $\rho$ mass are due to Dalitz decays of $\omega$, $\eta$ and $\eta'$, where measured
$\eta$ and $\eta'$ cross sections ($\sim 1/2$ the $\pi^0$ cross section) have been
folded in.

Our result, assuming isotropic $\rho,\omega,\gamma$ decay, allowing for variations in $E_T$ dependence for $\gamma/\pi$ up to $E_T^{\pm 2}$ and including variations in
possible polarization, and errors in measured cross sections and
branching ratios is: for $\sqrt{s} = 52$; $2 < P_T(\text{pair}) < 3$ GeV

$$\frac{\gamma(\text{direct})}{\pi^0} < .006 \pm .010.$$

This ratio is at the level of the vector dominance prediction of
Fig. 7.

## AZIMUTHAL CORRELATIONS OF $\pi^0$ PAIRS

The double high trigger enables us to investigate neutral pion
pair, as well as electron pair production at the ISR. The results
of the previous section gives us confidence that the high energy
photons we detect ($P_T > 1.2$ GeV/c) are due primarily to $\pi^0$'s. Most

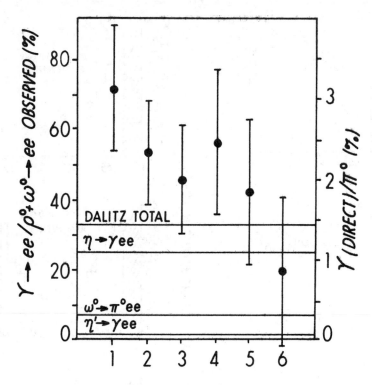

**CUT STRENGTH (arbitrary units)**

Fig. 8

studies of correlations between two particles of high transverse
momentum have been limited to nearly coplanar configurations
(azimuthal angle difference $\Delta\phi \sim 0$, $\Delta\phi \sim 180°$), while in this experiment we cover nearly the full angular range ($23° \leq \Delta\phi \leq 180°$)
for $\pi°$ pairs.

The motivation for this study is derived from hard scattering
model predictions, which suggest that a constituent of each proton
sometimes scatters at large angles and fragments into a "jet".
Some of the time each jet will give most of its energy to a single
$\pi°$. One way the underlying two body scattering is revealed is by a
"back to back" configuration of the transverse momenta of the $\pi°$'s,
i.e. a peak at $\Delta\phi = 180°$

In Fig. 9, we show the $\Delta\phi$ spectrum for several selections of
increasing $P_T(\pi°)$. We note a narrowing peak favoring back to back
configurations. However, there are reasons to believe that such a
structure can arise from effects having no relation to hard
scattering. As transverse momentum increases, momentum conservation
could contribute to such an effect. In addition, since increasing
transverse momentum is associated with a decreasing cross section,
it is expected that $\pi°$'s produced at $\Delta\phi \sim 90°$ and therefore
<u>requiring</u> additional balancing transverse momentum are less

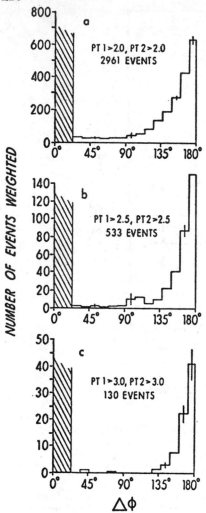

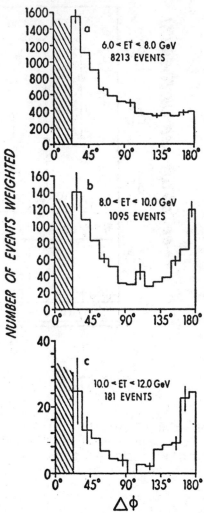

Fig. 9. Fully corrected distributions of azimuthal differences Δϕ between two π°'s emitted with polar angles 45° < θ < 135° and Δϕ > 23° Each of the transverse momentum greater than a) 2.0 GeV/c; b) 2.5 GeV/c; c) 3.0 GeV/c.

Fig. 10. Azimuthal differences of two π°'s with angular acceptance as in Fig. 9, but with total transverse energy (see text) in the interval a) 6 GeV < $E_T$ < 8 GeV; b) 8 < $E_T$ < 10 GeV; c) 10 GeV < $E_T$ < 12 GeV. The minimum π° transverse momentum required for inclusion is $P_T$ = 1.2 GeV/c.

copiously produced than those at 180°
    To eliminate these possible biases, we define the total transverse energy $E_T$ as follows: first we construct for each π° pair a missing, or "ghost" transverse momentum

$P_{TG} = -(\vec{P}_T(\pi^o{}_1) + \vec{P}_T(\pi^o{}_2))$ which must be non zero for $\Delta\phi < 180°$
Then $E_T = |P_T(\pi^o{}_1)| + |P_T(\pi^o{}_2)| + |P_{TG}|$. We see no obvious reason
reason for $\pi^o$ pairs having the same $E_T$ to prefer a back to back configuration, and Fig. 10a indicates they do not, for the $E_T < 8$ GeV
region where many "jet" phenomena have been analyzed. The observed
preferred configuration when $\pi^o$'s are in the same hemisphere may be
due to resonances or clusters. As the selected $E_T$ interval
increases, however, we note the reappearance of a prefrence toward
"back to back" pairs, together with the highly populated configuration at small $\Delta\phi$.

The three transverse momentum variables plus $E_T$ selection
define a "Dalitz type" plot in which all of the variables can be
displayed simultaneously. Fig. 11 shows a normalized plot with
coordinates $u_i = (2/E_T) P_T(\pi^o{}_i)$ for $10 < E_T < 12$ GeV. It is clear
that these events populate the plot boundary indicating their high
degree of alignment. Although this is qualitatively suggestive
of hard scattering models, we are investigating the possibility
that more subtle kinematic effects, or high mass resonant states,
may lead to such a distribution.

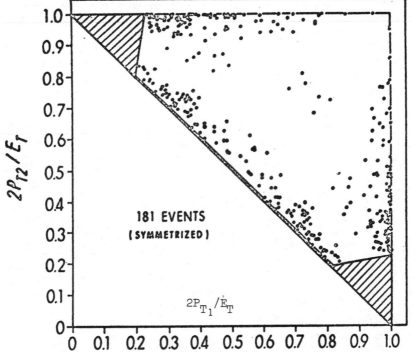

Fig. 11   Normalized symmetrized di-plot with horizontal and vertical
co-ordinates $u_i = 2P_{T_i}/E_T$ for each $\pi^o$ (uncorrected of
Fig. 10c). The excluded striped regions inside the tri-
angular boundary have $P_T < 1.2$ GeV/c. Also excluded (by
the requirement $\Delta\phi > 23°$) is a small region close to the
diagonal boundary.

REFERENCES

1. Further details regarding this experiment may be found in
   J. H. Cobb et al., PL 68B, 101 (1977); PL 72B, 273 (1977);
   PL 72B, 497 (1978); C. Kourkoumelis, CERN 77-06.

2. C. E. Carlson and R. Suaya, Phys. Rev. D15, 1416 (1977);
   S. D. Ellis, M. B. Einhorn and C. Quigg, PRL 36, 1263 (1976);
   A. Donnachie and P. V. Lansdorf (to be published).

3. N. S. Craige and D. Schildknecht, TH. 2193 CERN.

4. For π° correlation see:  K. Eggert et al., Nuc Phys B98, 49
   (1975).  F. W. Busser et al., Phys Lett 51B, 311 (1974).  For
   charged particle correlations see:  M. Della Negra et al.,
   (submitted to Nuc Phys B) CERN/EP/PHYS77-10; See also recent
   reviews by G.C. Fox and H.J. Frisch presented at Oct. APS-DPF
   meeting Brookhaven, 1976; by H. Bøggild at the VII Int. Symp on
   Multiparticle Dynamic's (Kayserberg June 1977); by
   M. Della Negra at the Europ. Cong. on Particle Physics
   at Budapest, (July 1977); Also see talk of H. Bøggild at
   this conference.

PROPERTIES OF PROMPT MUONS PRODUCED BY 28 GEV PROTON
INTERACTIONS[*]

W.M. Morse, K.-W. Lai, R.C. Larsen, Y.Y. Lee, L.B. Leipuner, and
D.I. Lowenstein
Brookhaven National Laboratory, Upton, New York 11973

D.M. Grannan, R.K. Adair, H. Kasha, R.G. Kellogg, M.J. Lauterbach
and M.P. Schmidt
Yale University, New Haven, Connecticut 06520

## ABSTRACT

We have measured prompt dimuon production from the interactions of 28.5 GeV protons with nuclear targets. The dimuon differential cross section $d\sigma/dx$ and the prompt muon to pion ratio are
equal within errors to that found at an incident proton beam energy
of 400 GeV. The atomic number dependence is found to be the same
as that of the total proton nucleon cross section. The dimuon
invariant mass distribution is presented.

## I. INTRODUCTION

Prompt single muon and dimuon production have been studied
extensively at Fermilab energies.[1] These measurements indicate that
prompt muons produced at different energies and angles are unpola-
rized[2,3,4] and derived largely from the production of pairs[5,6] and
that the intensity is roughly three times larger than expected from
electromagnetic decays of known mesons.[7,8] We present here the
results of a prompt muon experiment performed at 28.5 GeV. We compare the intensity and characteristics of muon production at this
energy with that at energies ten times higher. These measurements
support the view that a new electromagnetic process which we identify, tentatively, with the parton annihilation mechanism suggested
by Bjorken and Weisberg[9] contributes substantially to lepton pair
production.

## II. APPARATUS

The diagram of Fig. 1 presents a schematic view of the
essentials of the experimental apparatus. Protons from the Brookhaven AGS were focused onto a target assembly which was designed to
allow a wide variety of targets to be placed before the beam; variable density targets of wolfram (hevi-met), iron, and carbon were
used for the single muon intensity and polarization measurements
(the polarization measurements are reported elsewhere[4]), while

*Research supported by the U.S. Dept. of Energy under Contract
No. EY-76-C-02-0016.

130

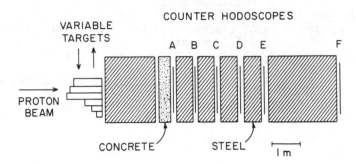

VARIABLE
TARGETS

COUNTER HODOSCOPES

A  B  C  D  E          F

PROTON
BEAM

CONCRETE          STEEL          I m

Fig. 1.  A schematic diagram of the experimental design.

different solid targets of wolfram, iron, and carbon were used for
the muon pair measurements.  The target assembly was followed by a
2 m thick steel hadron absorber and a 60 cm thickness of concrete
which helped to attenuate evaporation neutrons produced in the nuclear cascades in the steel.  This was followed by six scintillation
counter hodoscopes which are labeled A through F.  Each array
covered an area 1.8 m by 1.8 m and the arrays were separated by steel
absorbers; the counters A through F are separated by 60 cm steel
absorbers; hodoscope F is separated from the counter assembly E by a
further 2.4 m of steel.  The A array consisted of 32 counters arranged to define the path of the muon with a maximum uncertainty of
7 cm at the position of the array for those muons which passed within 50 cm from the center of the array and with a poorer spatial resolution for muons produced at larger angles.  The spatial resolution
of the downstream hodoscopes was less precise and they were used,
primarily, to differentiate between the passage of one or two muons
through the plane which they defined.

    The different trigger logic used for different aspects of the
experimental program set latches for each of the 98 hodoscope counters and these latches were read into a computer memory.  After a
200 nsec delay, the latches were read again so that accidental rates
could be monitored constantly.  While the information from each event
and from each machine pulse was written on magnetic tape for possible
off-line analysis, most of the results reported here were derived
from analyses of data written out in the course of the run.

III. MEASUREMENTS OF THE PRODUCTION OF SINGLE MUONS

    Measurements of the fluxes of single muons were made using tech-
niques described previously[8] which determine the ratio of single
prompt muons to muons from meson decay in a dense target.  Such measurements are especially useful, inasmuch as the measurements define
the ratio of muon pair production to the production of mesons in

a manner which is quite insensitive to systematic uncertainties.
Here we consider the production of muons produced by the interaction
of 28.5 GeV protons with wolfram targets of various densities. A
target of solid wolfram (hevi-met) was used as well as targets of
wolfram plates separated by air so as to give average densities of
1/2 and 1/3 that of the solid target.

The variation of the intensity with respect to target density,
$\rho$, can be written as

$$I(\rho_1/\rho) = I_p + I_m \cdot (\rho_1/\rho) \qquad (1)$$

where $\rho_1$ is the density of the solid wolfram; the values of $(\rho_1/\rho)$
are then 1, 2, and 3 for the various measurements. With this
parameterization, $I_p$ is the intensity of prompt muons and $I_m$ is the
intensity of muons from the decay of mesons in the solid target.
The variation of $d\sigma/dx$ for both meson production and dimuon production with x is very steep, the cross sections decrease sharply with
increasing x. As a consequence of the steepness of the spectrum,
both $I_p$ and $I_m$ take especially simple forms. In particular, to a
good approximation, we can write

$$I_p = 2(d\sigma/dx)_p \, A_p \, d\,\Omega p \quad \text{and} \quad I_m = (d\sigma/dx)_m A_m S_m d\Omega_m \qquad (2)$$

where A are longitudinal acceptances equal to $E_{max}/(E_{max}-E_{min})$ for
the decay of relativistic mesons and dimuons and $E_{max}$ and $E_{min}$ are
the maximum and minimum kinematically possible energies from the
decay of the parent state in the laboratory system; the acceptances
$d\Omega$ represent the effective solid angles subtended by the target and
$S_m$ is the probability of a meson decaying in the target material
before interacting and being effectively eliminated from considera-
tion. The value of $S_m$ will be

$$S_m = L/ \, (c^T E/m) \qquad (3)$$

where L is the mean-free path for the meson to interact strongly in
the target material, m is the meson mass, E the meson energy and $\tau$
the mean life. To a good approximation, the acceptance of the appar-
atus covers the whole forward production and $d\,\Omega_m = d\Omega_p$; $A_m \approx 2.5$
while $A_p \approx 1$, except for that small portion of the flux where the
invariant mass of the muon pair is very small; the factor of 2 in
$I_p$ accounts for the two muons produced by the dimuon decay.

There are corrections to the simple picture outlined above:
corrections for the production of muons from secondary mesons and
nucleons; corrections and adjustments to account for the contributions of K- mesons, as well as pions, to the muon flux and corrections which take into account the explicit variation of the dimuon

flux and the meson flux with x. But these corrections and adjustments are all small and the adequacy of the very simple picture outlined above produces confidence in the more complete calculations which we use to determine the ratio of $(d\sigma/dx)_m$ and $(d\sigma/dx)_p$. The ratios may then be in error by about 15%, but it is most unlikely that they are grossly in error.

The scaling properties of the single muon production may be important. If the dimuon differential cross sections obey a limited form of Feynman scaling and $(d\sigma/dx)_p$ is independent of energy, the known scaling of the meson production cross sections requires that

$$(I_p/I_m)_{28 \text{ GeV}} = (I_p/I_m)_{400 \text{ GeV}} \cdot (28.5/400) \cdot (184/63.6)^{1/3} \quad (4)$$

for any given value of x where the 400 GeV single muon intensities were measured using a copper target with A = 63.6 and the measurements reported here used wolfram with A = 184. The factor (28.5/400) represents the difference in time dilation and the two factors give the difference in meson survival for the two situations. The graph of Fig. 2 shows the ratios $(I_p/I_m)$ plotted as a function of x for

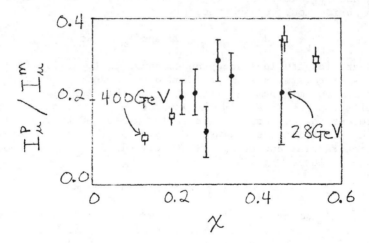

Fig. 2. The ratio of prompt single muons to muons produced in the decay of mesons in the solid target.

the 28 GeV data and for 400 GeV data[8] taken previously at Fermilab multiplied by the appropriate differential factor. The agreement of the two sets of data shows immediately that the single muon production scales and then that dimuon production also scales if we accept the evidence which suggests that the single inclusive muon flux is derived from muon pair production. We note that systematic uncertainties are minimized by this simple manner of measuring scaling.

## IV. MEASUREMENTS OF MUON PAIR PRODUCTION

Measurements of the production of muon pairs were made by requiring signals from two separate muons in the down stream hodoscope counters. Solid targets, one-mean-free-path thick, of wolfram, iron, and carbon were used. With these targets, muons with energies upon production greater than 2.9 GeV would pass through the A-plane while an initial energy of about 9.9 GeV is required for passage through the F-plane. The trigger, devised to detect muon pairs produced in the target, required a count in a 1.25 cm x 1.25 cm scintillation counter situated just upstream of the target together with two or more counts from both the A and B hodoscope arrays. About 200,000 protons were accepted per AGS pulse generating about 5 triggers per pulse.

The triggers were generated both by muon pairs and by hadron showers which penetrate the shielding. In an off-line analysis, both muons were required to penetrate to the C array; 4600 such events were recorded. Assuming all particles reaching the C-array are muons the trimuon to dimuon rate is 2%, while the requirement that one muon penetrate to the D array results in a ratio of 0.4%. The above rate of (nominal) trimuons is consistent with rates expected from the hadronic background. Thus the background in the dimuon sample due to hadron showers is negligible.

The intensity of dimuons plotted as a function of Feynman x is shown in Fig. 3. The errors shown are statistical only. The data

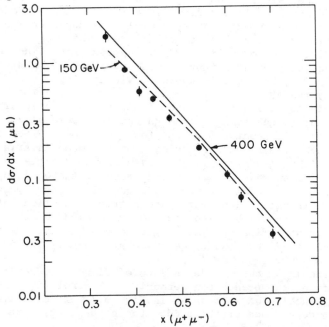

Fig. 3. The intensity of dimuons plotted as a function of x for pairs produced at 28 GeV, 150 GeV, and 400 GeV.

are well represented over this range of x by the functional form

$$d\sigma/dx = 40 \ \mu b \ e^{-10x} \tag{5}$$

Although corrections for muons from the decay of mesons are negligible, there are important modifications of the raw data to account for an acceptance which is a strong function of x. Since we are only sensitive to muons with energies greater than 4.5 GeV, a correction must be made for those muon pairs which decay such that one of the muons does not enter the apparatus. This correction will depend upon the angular distribution of the dimuons in their own center-of-mass system. These acceptance corrections were made assuming that the dimuons are emitted isotropically in their own center-of-mass system. The data did provide a measure of this decay distribution which was consistent with that hypothesis.

We also show on Fig. 3 curves which represent parameterizations of distributions at[7] 150 GeV and at[5] 400 GeV. Although the cross sections appear to increase slightly with energy, the differences of the 28 GeV data reported here and the 400 GeV results with the parameterization of the 150 GeV data is not outside of the systematic uncertainties which accompany these measurements and we do not consider that the set of results is inconsistent with the Feynman scaling hypothesis that the differential cross sections, $d\sigma/dx$, are independent of energy. This conclusion would be in accord with the analysis of the single muon data which is subject to different and, we believe , much smaller systematic errors. Dimuon production and hadron production appear to have a similar s dependence over the range of dimuon masses and x studied here. The cross section[10] for the inclusive reaction $pp \rightarrow \pi^- + X$ is approximately independent of s for beam energies greater than 28 GeV and x > 0.20. Large mass dimuon production, however, is known to violate Feynman scaling over this energy range.

The graphs of Fig. 4 show the distributions in invariant mass for different ranges of x. Although the mass resolution is poor -- the resolution is about 250 MeV/$c^2$ for a mass of 750 MeV/$c^2$ -- this does not much affect the conclusion that the mean invariant mass increases with increasing x. The graph shown in Fig. 5 shows the variation of mean mass as a function of x for these measurements together with the results given by Chicago-Princeton[7] parameterization of their data taken (with much better resolution) at 150 GeV. The small differences between the two expressions of the data are probably not significant. The result is also in agreement with an early conclusion[5] that, at x near 0.50 at 400 GeV, the mean invariant mass was 900 ± 200 MeV/$c^2$.

The atomic number (A) dependence of dimuon production yields information on the production mechanism. The relative dimuon intensity as a function of A is shown in Fig. 6. This experiment measures the A dependence of the production cross section relative to

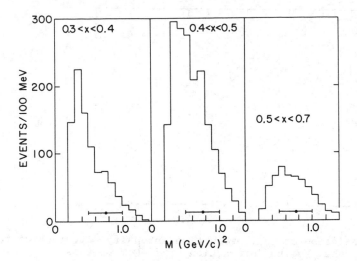

Fig. 4. Dimuon invariant mass distributions for various ranges of x. The horizontal bar represents the mass resolution.

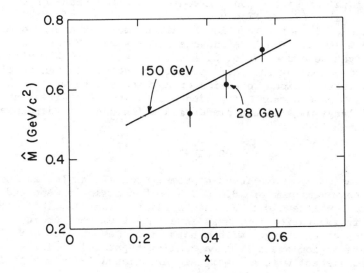

Fig. 5. The mean invariant mass of pairs produced by 28 GeV protons in this experiment. The solid line shows the mean invariant mass taken from data[7] at 150 GeV.

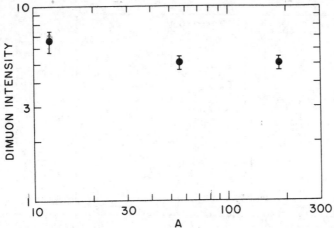

Fig. 6. The variation of pair production intensity with respect to A for pairs produced by the interaction of 28 GeV protons.

that of the total cross section. We note that the flux is independent of the choice of target indicating an A dependence similar to that of the total cross section. Assuming that the total cross-section varies as $A^{2/3}$, we find a dimuon A dependence of $A^{0.64\pm.03}$. This A dependence is not a function of dimuon mass or energy over the range we measure. The A dependence so measured is not quite that which might be expected for hadrons. At an x of 0.45 which is about the mean value for the pairs considered here, Eichten et al[11] find a dependence of pion production on A which can be written as $A^{0.54}$. The low mass dimuon A dependence is also quite different [12] from that of large mass dimuon production which has an A depdendence of approximately $A^1$.

## V. CONCLUSIONS

In conclusion, we have found a prompt muon signal for an incident proton beam momentum of 28.5 GeV. Both the single muon and dimuon data are consistent with Feynman scaling - that is, the intensity in the fragmentation region is independent of beam energy. The dimuon mass spectrum changes with x both at 28 GeV and 150 GeV - thus each mass region obeys Feynman scaling separately. The A dependence is consistent with a power law variation $A^{2/3}$ in a x region where pion production varies as $A^{.54}$.

REFERENCES

1. Pertinent reviews are presented by L. Lederman, Proc. Int. Symp. on Lepton and Photon Int. at High Energies (Stanford, 1975), and by B.G. Pope, Proc. E.P.S. Int. Conf. (Palermo, 1975).
2. L.B. Leipuner et al., Phys. Rev. Lett. 36, 1011 (1975).
3. M.J. Lauteraach et al., Phys. Rev. Lett. 37, 1436 (1976).
4. D.M. Grannan et al., Phys. Lett. 69B, 125 (1977).
5. H. Kasha et al., Phys. Rev. Lett. 36, 1007 (1976).
6. J.C. Branson et al., Phys. Rev. Lett. 38, 457 (1977).
7. K.J. Anderson et al., Phys. Rev. Lett. 37, 799 (1976).
8. L.B. Leipuner et al., Phys. Rev. Lett. 35, 1613 (1975).
9. J.D. Bjorken and H. Weisberg, Phys. Rev. D13, 1405 (1976).
10. W. Morse et al., Phys. Rev. D15, 66 (1977).
11. T. Eichten et al., Nucl. Phys. B44, 333 (1972).
12. M. Binkley et al., Phys. Rev. Lett. 37, 571 (1976).

# OBSERVATION OF PROMPT SINGLE MUON PRODUCTION
# BY HADRONS AT FERMILAB[*]

B. Barish, A. Bodek,[**] K. W. Brown, M. H. Shaevitz, and E. J. Siskind,
  California Institute of Technology, Pasadena, California  91125

A. Diamant-Berger,[†] J. P. Dishaw, M. Faessler,[††] J. K. Liu,[†††] F. S. Merritt,
  and S. G. Wojcicki
Stanford University, Stanford, California  94305

## ABSTRACT

Talk given by M. H. Shaevitz at the Third International Conference
at Vanderbilt University, March 3-6, 1978, on New Results in High Energy
Physics.  In an experiment at Fermilab, we have measured prompt single
and dimuon production by 400 GeV protons in a large acceptance detector.
A substantial prompt single muon rate has been observed in the region
$.7 < p_\perp < 1.5$ GeV/c and $x_F < .25$ indicating the production and weak decay
of heavy short-lived sources.  If interpreted as charmed particle
production, a model dependent calculation indicates a cross section for
$pp \rightarrow$ charm + anything of approximately 40 μb.

## INTRODUCTION

The discovery[1] in 1974 of large rates for prompt lepton production by hadrons started a great effort to determine the short lived
sources. Recent experiments[2] have indicated that most prompt leptons are
from lepton pair sources and are, therefore, electromagnetic in origin.
A question immediately arises - are there also weakly decaying prompt
sources?  Experimentally, a weak decay is characterized by a single
charged lepton and an undetected neutrino.  We have performed an experiment at Fermilab to isolate these weak sources, by separating single
muon from dimuon events and by measuring "missing energy" indicative of
undetected neutrinos.  This report is intended to be a brief statement
of our progress; the results are very preliminary and are obtained from
a very small part of the data sample (~15%).

* Work supported by the U.S. Department of Energy and the National
  Science Foundation.

** Present address:  University of Rochester, Rochester, N.Y.  14627.
 † Permanent address:  Department de Physique des Particules Elementaries,
     Saclay, France.
†† Present address:  CERN, Geneva, Switzerland.
††† Presently with American Asian Bank, San Francisco, Calif.

## APPARATUS

Basically, the experiment consists of a target-calorimeter followed by a muon spectrometer (Fig. 1). A 400 GeV proton beam is incident on the target-calorimeter; the momentum and direction of the beam are measured by an upstream PWC spectrometer. Beam halo and off-momentum particles are eliminated by counters upstream of the calorimeter. Good beam particles are required by the trigger to interact in the first 15 inches of Fe in the calorimeter and to be unaccompanied by additional beam within 60 nanoseconds.

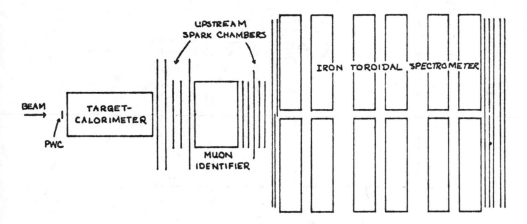

Fig. 1 - Plan view of the apparatus (not to scale).

The target-calorimeter consists of 1.5 and 2" Fe-scintillator sandwiches and measures the total energy in the hadron cascade. Phototube gains and rate effects are monitored by several devices allowing a measurement of total energy to ± 3.6% at 400 GeV. Downstream are spark chambers to measure the angle of muons exiting the calorimeter, a 1m muon identifier and a 40 foot long 12 foot diameter iron toroidal magnet system to measure muon momentum. The total $\int Bdz$ of the magnet system is 3200 Kg-in, corresponding to radial kick of 2.4 GeV/c and a momentum resolution of 9%.

## DIMUON RESULTS

The experiment ran for 12 weeks and recorded about four million triggers with a total integrated flux of $\sim 10^{11}$ protons. The large acceptance of the apparatus makes it ideally suited for measuring the production distributions of high mass dimuon sources. The dimuon invariant mass distribution (uncorrected for acceptance) for 25% of the data is shown in Fig. 2; a clear $\psi$ peak is in evidence with a width consistent with momentum resolution and multiple scattering in the calorimeter. Defining the $\psi$ region as $M_{\mu\mu}$ between 2.4 and 4.0 GeV

## 2 MUON INVARIANT MASS

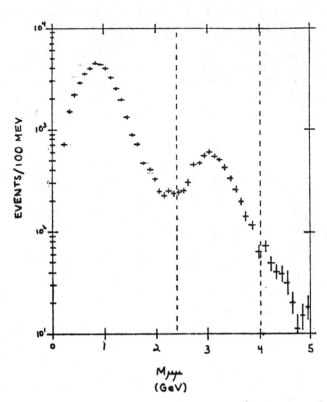

$M_{\mu\mu}$
(GeV)

Fig. 2 – Dimuon invariant mass (uncorrected
for acceptance). Approximately
5300 events in the $\psi$-region.

yields 5300 events covering the range $0 < p_{\perp} < 5$ GeV/c with good
acceptance; Fig. 3 shows the $p_{\perp}$ distribution for these events corrected
for trigger efficiency. (The efficiency was calculated via Monte Carlo
assuming a flat decay angular distribution in the $\psi$ rest frame and $p_{\perp}$
and $x_F$ distributions consistent with those observed). The best fit to
the form $dN/dp_{\perp}^2 \propto e^{-bp_{\perp}}$ for $p_{\perp} > 1.2$ gives $b = 2.04 \pm .05$ in good agreement with the $p_{\perp} > .4$ GeV data of Branson et al.[3] A similar fit gives
$\frac{dN}{dx_F} \propto (1 - x_F)^c$ with $c = 4.87 \pm .26$ for positive $x_F$. It should be noted
that contributions from secondary interactions have been determined to
be small in this region. Integrating the distributions (assuming a
linear A dependence for production and assuming symmetric distributions
around $x_F = 0$) yields a total cross section

$$B(\psi \to 2\mu) \cdot \sigma_{tot} (pN \to \psi + X) = 17.0 \pm 3.1 \text{ nb/nucleon.}$$

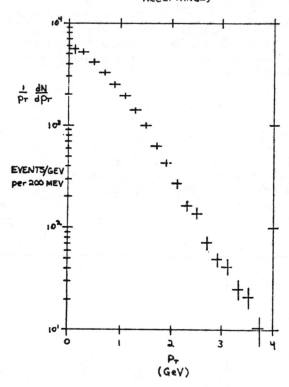

$P_T$ DISTRIBUTION FOR $\psi$'s

(CORRECTED FOR TRIGGER ACCEPTANCE)

$\frac{1}{P_T}\frac{dN}{dP_T}$

EVENTS/GEV per 200 MEV

$P_T$ (GeV)

Fig. 3

$p_\perp$ distribution for events in the $\psi$ region with 2.4 < $M_{\mu\mu}$ < 4 GeV, which have been corrected for acceptance but not resolution smearing. Best fit to the form $\frac{dN}{dp_\perp^2} \propto e^{-bp_\perp}$ yields b = 2.04 ± .05 for the region with $p_\perp$ > 1.2 GeV/c.

## SINGLE MUON RESULTS

A primary motivation of this experiment is to search for prompt single muons. Events with one observed muon can come from three sources: (1) decays of long lived particles (non-prompt), e.g. pions, kaons, and hyperons; (2) two muon events in which only one muon is observed (di-muon feed-down), and (3) weak decays of short lived particles (prompt single muons). Non-prompt decay contamination is empirically removed by varying the density of the target-calorimeter while keeping the mean interaction point fixed to minimize any differences in geometric acceptance, and extrapolating the observed rate to infinite density. In Fig. 4(a), an example of this extrapolation is shown for a region dominated by non-prompt decay ($p_\perp$ < .2 GeV/c); the data points are linear and consistent with no signal at the intercept.

Dimuon feed-down into single muon events is primarily eliminated by detecting both muons in the apparatus. Geometrically, the muon identifier and upstream spark chambers cover greater than 90% of $4\pi$ steradians in the center of mass; muons are detected with momentum as low as 3.5 GeV/c in the lab frame.

The small number of events in which one muon from a pair is unobserved has been calculated using a Monte Carlo program that generates
events according to previously measured dimuon distributions;[4] these events
typically come from moderate mass pairs ( $\overline{M}_{\mu\mu}$ ~ 700 MeV) where one muon
ranges out in the calorimeter. This correction is always small (less
than 12% of the single muons and averaging 6%) because second muons are
detected down to 3.5 GeV/c. (We plan, in the future, to use our measured
two muon distributions to do this correction empirically.)

Fig. 4 – Density extrapolation plots. Prompt
muons appear as a signal at 1/density = 0;
the slope of each plot is proportional to
the rate from muons from non-prompt decay.
(Rates are those observed and have not
been corrected for geometrical acceptance).
The 1μ signal has been corrected for 2μ
feed-down (typically 4-10%).

LOW $P_T$

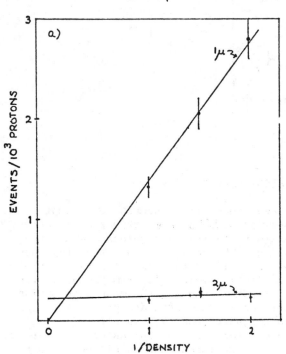

Fig. 4(a)

Very low $p_\perp$ region –
dominated by non-prompt
decay.

A combination of triggering configurations allowed the measurement
of single muon production in two regions: (1) $x_F$ ~ 0 and $p_\perp \geq .5$ GeV/c
and (2) $x_F \geq .05$ and $p_\perp \geq 0$. The single and dimuon rates for these
regions are shown in Fig. 4(b) and (c) versus inverse density; the extrapolations are linear with significant positive intercepts for the single
muons .

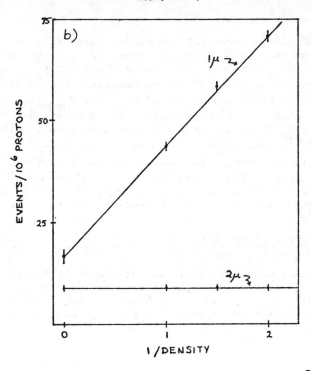

b)

EVENTS / $10^6$ PROTONS

75

50

25

1μ

2μ

0          1          2

1/DENSITY

Fig.4(b)

Region (1) –
$p_\perp$ > .5 GeV/c and
$x_F \sim 0.$

REGION 2

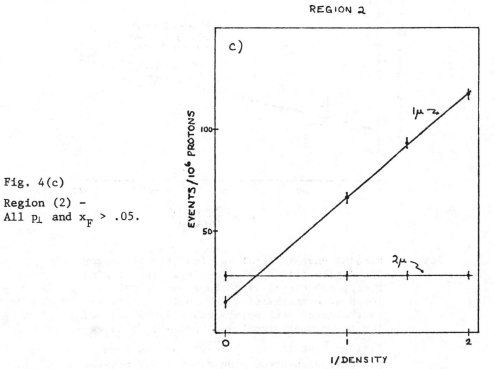

c)

EVENTS / $10^6$ PROTONS

100

50

1μ

2μ

0          1          2

1/DENSITY

Fig. 4(c)

Region (2) –
All $p_\perp$ and $x_F$ > .05.

Further evidence that single muons are produced is available by comparing the energy of the beam to the visible energy ($E_{vis}$); the energy sum of the hadron cascade and all muons. A difference between $E_{beam}$ and $E_{vis}$ could indicate missing energy carried off by neutrinos from a weakly decaying parent. An event by event comparison of $E_{vis}$ and $E_{beam}$ is precluded by the energy resolution of the calorimeter. However, a shift in the mean of the $E_{vis}$ distribution for single muon events relative to muonless events can be measured. (The error in the mean is 14 GeV/$\sqrt{N_{events}}$.) The missing energy, defined as the shift in the mean, at each density has contributions from non-prompt decay and prompt single muon sources. To separate the two contributions, the missing energy is plotted (Fig. 5) versus the fraction, $\alpha$, of the single muons that are prompt for each density, ($\rho$).

$$\alpha(\rho) = \frac{\text{Prompt } 1\mu \text{ Rate}}{\text{Total } 1\mu \text{ Rate}}$$

The missing energy associated with the signal is then found by extrapolating to $\alpha = 1$. Using this procedure, the missing energy for the prompt signal in region (1), ($p_\perp \geq .5$ GeV/c) is $8.0 \pm 4.2$ GeV, consistent with the energy expected for an undetected $\nu$ from a heavy particle decay.

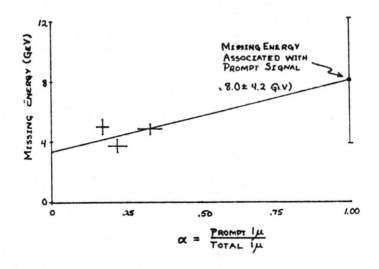

Fig. 5 – Missing energy signal vs. fraction of prompt 1μ for region (1) ($p_\perp > .5$ GeV/c). The missing energy associated with the prompt signla is found by extrapolating to $\alpha = 1$. A best fit to the three data points gives $8.0 \pm 4.2$ GeV for the prompt signal. A Monte-Carlo calculation using $(1 - |x_F|)^{10} e^{-4p_\perp}$ for the pion production spectrum predicts $\sim 4$ GeV missing energy from π-decay, qualitatively consistent with the intercept at $\alpha = 0$.

In order to compare with previous experiments, we have calculated from the density extrapolation plots the $\mu/\pi$ ratio for the sum of single and dimuons as a function of $x_F$ (Fig. 6). The ratio is about $10^{-4}$ at low $x_F$ and falls to $10^{-5}$ for $x_F$ above .3, in good agreement with previous measurements[2,6] at low $p_\perp$.

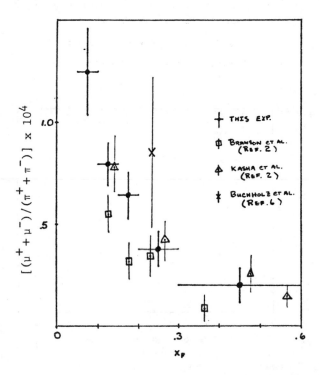

THIS EXP.

BRANSON ET AL.
(REF. 2)

KASHA ET AL.
(REF. 2)

BUCHHOLZ ET AL.
(REF.6)

Fig. 6

The $\mu/\pi$ ratio vs. $x_F$ for region (2) ($x_F > .05$). The different $x_F$ points are obtained from the density extrapolation for data in each $x_F$ bin. Basically, the ratio is found by dividing the total prompt signal (1$\mu$ and 2$\mu$) by the slope of the extrapolation curves multiplied by the probability for a pion to decay. The results of Ref. 2 and 6 are also shown.

In figures 7 and 8, the ratio of single prompt muons to total prompt is presented as a function of $x_F$ and $p_\perp$. The data show that the rate of single muons is small in regions explored by previous experiments at large $p_\perp$ and $P_{//}$. However, in the region of $x_F < .25$ or $p_\perp < 1.5$ GeV/c, the prompt single muon signal is sizeable and accounts for about 50% of the total prompt rate; i.e., half of the prompt muons in this region do not come from muon pairs.

One possible interpretation for these results is that the single muons are predominantly from charm production. The $p_\perp$ and $x_F$ distributions are qualitatively consistent with those expected from D(1865) production and subsequent three body decay ($D \rightarrow K\mu\nu$).[5] Assuming that the parent D meson is produced with the same distributions as we have measured for $\psi$'s ($d\bar{N}/dx_F dp_\perp^2 \propto e^{-2p_\perp} (1 - |x_F|)^5$ the observed prompt single muon signal corresponds to a charm cross section

$$\sigma \text{ (charm)} \sim 40 \text{ } \mu b$$

assuming BR(D → μ + ..) = .11 and a linear A dependence for production.
It should be noted that the above cross section is very dependent on
these assumptions. For example, with an assumed steeper dependence
$(dN/dx_F dp_\perp^2 = e^{-3p_\perp}(1 - |x_F|)^8)$ the cross section is a factor of 2.5
higher; for a flatter distribution the cross section would be correspondingly smaller. Also, for an $A^{2/3}$ dependence (assumed by the CERN
beam dump experiments) our result would be larger by a factor of 2.5;
including D → $\overset{*}{K}$ μν would increase the cross section, etc. With the above
caveats, we conclude that the prompt single muon signal, if interpreted
as charm, indicates a cross section of approximately 20-80 μb.

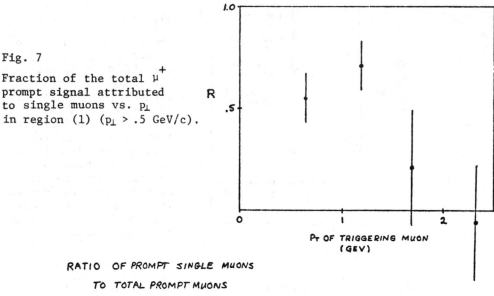

RATIO OF SINGLE PROMPT
MUONS TO TOTAL PROMPT MUONS

Fig. 7

Fraction of the total $\mu^+$
prompt signal attributed
to single muons vs. $p_\perp$
in region (1) ($p_\perp$ > .5 GeV/c).

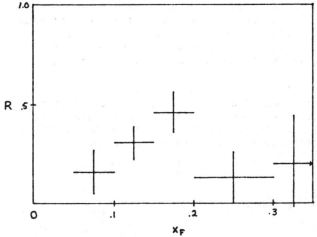

RATIO OF PROMPT SINGLE MUONS
TO TOTAL PROMPT MUONS

Fig. 8

Fraction of the total observed
prompt signal attributed to
single muons vs. $x_F$ in region
(2) ($x_F$ > .05). A correction
for difference in trigger
efficiencies between 1μ and
2μ, which will raise the
small $x_F$ points, has not been
applied to the data.

## CONCLUSIONS

We have presented evidence for a significant source of single prompt muons in high energy hadron-hadron collisions. The single muon signal accounts for about half of the prompt muons in the low $x_F$ region with $p_\perp \approx 1$ GeV/c and is associated with undetected energy (presumably associated with final state $\nu$'s); this behavior is consistent with production and weak decay of heavy short-lived particles, possibly charmed particles. As stated previously, the results are very preliminary and represent only a small fraction of the total data sample. In the near future, we plan to improve the statistics by a factor of five allowing a much better determination of the single muon signal at large $p_\perp$ and $x_F$.

Future experimentation will be needed to obtain the detailed dependence on $\sqrt{s}$, $p_\perp$, $x_F$, and incident particle type. We hope to carry out a program of such measurements in the future to help understand the sources and production mechanisms for the prompt single muons reported here.

## REFERENCES

1. J. P. Boymond et al., Phys. Rev. Lett. **33**, 112 (1974).
   J. A. Appel et al., Phys. Rev. Lett. **33**, 722 (1974).

2. J. G. Branson et al., Phys. Rev. Lett. **38**, 457 (1977).
   H. Kasha et al., Phys. Rev. Lett **36**, 1007 (1976).

3. J. C. Branson et al., Phys. Rev. Lett. **38**, 1331 (1977).

4. K. J. Anderson et al., Phys. Rev. Lett. **37**, 803 (1976).

5. M. Bourquin and J.-M. Gaillard, Nuclear Physics, **B114**, 334 (1976).

6. D. Buchholz et al., Phys. Rev. Lett. **36**, 932 (1976).

# THE OBSERVATION OF PROMPT NEUTRINOS
# FROM 400 GEV PROTON NUCLEUS INTERACTIONS

Aachen-Bonn-CERN-London-Oxford-Saclay Collaboration

presented by J.H. Mulvey

Department of Nuclear Physics, Oxford

## ABSTRACT

Events due to neutral primaries have been observed in BEBC during an experiment in which $3.5 \ 10^{17}$ protons of 400 GeV from the CERN-SPS were dumped into a large copper target. These events have the general characteristics of neutrino interactions however not all the 70 events of energy greater than 10 GeV can be attributed to conventional sources ($\pi$, K or hyperon decay); in particular the observation of 11 events with an emerging $e^-$ and 4 $e^+$ events requires a new, prompt source of neutrinos. The intensity of this source is estimated and possible origins such as charmed particle or heavy lepton decay are considered; an upper limit is also placed on the production of axions.

ISSN: 0094-243X/78/148./$1.50 Copyright 1978 American Institute of Physics

# 1. THE EXPERIMENT

In December 1977 a 'beam dump' experiment was performed at CERN. This report is based on observations made with BEBC in the beam configuration illustrated in figure 1(a). The 400 GeV beam of protons from the CERN-SPS, directed towards BEBC, interacted in a copper cylinder 2m long and 27cm in diameter; after a gap of 50cm any surviving hadron encountered a 1m cube of iron and a further 3m of iron placed 50m downstream of the target. The muon shield consisted of 180m of iron and 170m of rock and earth. The length of the copper target, about 10 absorption lengths, was sufficient to ensure that most $\pi$ and K-mesons would interact rather than decay so that the neutrino flux from these sources was expected to be about 2,000 times less than that in the normal wide band neutrino beam. On the other hand neutral, weakly interacting particles produced directly or in the decay of particles of lifetime $\ll 10^{-11}$ secs. would not have been attenuated.

BEBC was filled with a 72% (mole) neon-hydrogen mixture (radiation length 44cms) and a magnetic field of 35kg was used; under these conditions electron detection and identification is good (> 90% efficiency) including charge determination. Muons were identified using a two plane external muon identifier (EMI) arranged as shown in figure 1(b) and giving an efficiency > 98% for muon momenta > 5 GeV/c.

The fiducial mass was 13t and during the total exposure, of 70,000 pictures, $3.5 \ 10^{17}$ protons were incident on the target. The muon flux was monitored during the exposure using solid state counters placed in the iron muon shielding. In addition 14,000 pictures were taken while the proton beam was off to check cosmic ray induced background.

# 2. THE DATA

The film was double scanned (95% combined efficiency) for events with one or more prongs at least one of which had to be due to a forward going particle of momentum > 1 GeV/c. 180 events were found of which 93 had a total energy $E_{VIS} > 5$ GeV; no events of $E_{VIS} > 5$ GeV were found in the 14,000 beam-off pictures. The events were divided into five categories: those with identified leptons (C.C. events) $\mu^-$, $\mu^+$, $e^-$ or $e^+$ and those with no identified lepton which were classed as neutral current (N.C.) candidates. To be identified by the EMI muons were required to have momentum > 5 GeV/c. The energy distribution of these events is shown in figure 2(a) and figure 2(b) shows the distribution of $E_H/E_{VIS}$ (where $E_H$ is the hadron energy) for the CC events of $E_{VIS} > 10$ GeV.

Although the numbers are small the distributions in $E_H/E_{VIS}$ are consistent with the forms expected for $\nu$ and $\bar{\nu}$ ; the ratio of NC to CC events, 0.42 ± 0.17, is a little high but agrees with the expected ratio 0.35 for an equal mixture of $\nu$ and $\bar{\nu}$ ; the rate of V⁰ events was also similar to that expected. Overall, the character of the events strongly suggests that they are due to $\nu$ and $\bar{\nu}$ but there is one significant anomaly compared to a normal neutrino exposure and that is the high proportion of electron events.

## 3. EXPECTED RATES

The expected numbers and energy spectrum can be predicted on the assumption that the events are due mainly to neutrinos arising from the small proportion of $\pi$ and K-mesons decaying before interacting.

The relative number of electron and muon events can be estimated more readily than the absolute numbers since this depends largely on known branching ratios, absorption lengths, decay Q-values and the $(K/\pi)$ ratio at production (0.11). The results are: $e^-/\mu^- \sim 0.06$ and $e^+/\mu^+ \sim 0.10$, where in the last ratio $\bar{\nu}_e$ from $\Lambda$ and $\Sigma^-$ decay have been included. These ratios may be used to obtain predictions for the numbers of e⁻ and e⁺ events based on the observed numbers of muon events; these are (1.8 ± 0.4)e⁻ compared to 11 observed (of which 2 had $P_e$-< 5 GeV/c) and (0.5 ± 0.25)e⁺ compared to 4 observed. The observed numbers of electron events are about an order of magnitude greater than would be expected if $\pi$, K, $\Lambda$ and $\Sigma^-$ were the main sources.

The absolute numbers can also be predicted by assuming a model for the hadron production process. The angle subtended by BEBC at the target was ±1.8mr so that the neutrinos detected in BEBC were dominantly due to the decay of particles produced in the forward direction in the first generation of interactions and so of relatively high energy. A Monte Carlo calculation has been performed using the thermodynamic model [1] for the production of $\pi$ and K-mesons (corrected to agree with observed yields [2]). These are then allowed to decay within a distance of one absorption length (18cm for $\pi$, 21cm for K). This model gives the energy spectra shown as full curves in figure 2(a) and the numbers in Table I. The corresponding muon fluxes agree with those measured in the shielding to within ±30%, when allowance is made for prompt muon production, so this is used as an estimate of the uncertainty in prediction.

## TABLE I

Predicted and Observed Numbers of Events ($E_{VIS} > 10$ GeV)

|        | Predicted    | Observed |
|--------|--------------|----------|
| $\mu^-$ | $21 \pm 7$   | 30       |
| $\mu^+$ | $3.5 \pm 1.2$ | 5       |
| $e^-$  | $1.4 \pm 0.5$ | 11      |
| $e^+$  | $0.3 \pm 0.1$ | 4       |

Again the excess of e- and e+ events is evident; the numbers are also consistent with there being an equal excess of $\mu^-$ and $\mu^+$ events.

The excess of e- and e+ events must be attributed to some process of prompt production of neutrinos. The flux of prompt neutrinos required to explain the excess electron events is given in Table II and is compatible with the rate observed for prompt single muon production if this is about 20% of the total observed rate which is believed due mainly to muon pairs [3].

## TABLE II

Fluxes of Prompt $\nu_e$ and $\bar{\nu}_e$

| $E_{VIS}$(GeV) | 10-50 | 50-90 | 90-130 |
|----------------|-------|-------|--------|
| # e-           | 5     | 3     | 3      |
| # e+           | 0     | 2     | 2      |
| $10^{-11}$ N($\nu_e$)* | $1.33 \pm 0.6$ | $0.83 \pm 0.4$ | $0.53 \pm 0.2$ |
| $10^4$ N($\nu_e$)/N($\pi^+$) | $0.34 \pm 0.15$ | $0.17 \pm 0.08$ | $0.11 \pm 0.05$ |

-----------------------------------------------------------------

\*   N($\nu_e$) = ($\nu_e + \bar{\nu}_e$)/2 after subtraction of background.

In this table N($\pi^+$) is the number of $\pi^+$ emitted at the target into the solid angle subtended by BEBC and obtained from the thermodynamic model [1,2].

## 4. POSSIBLE SOURCES OF PROMPT NEUTRINOS

### 4.1 Short-lived hadrons

The decay of short lived hadrons ($\tau < 10^{-11}$secs.) such as charmed mesons produced in particle anti-particle pairs could contribute to the prompt production of neutrinos and give equal yields of $\nu_\mu$, $\bar{\nu}_\mu$, $\nu_e$ and $\bar{\nu}_e$. A first rough estimate of the production rate of these hadronic parents can be obtained as follows:

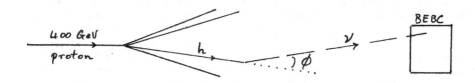

The angle in the laboratory frame between the parent particle, h, and the neutrino is $\sim 1/\gamma$ where $\gamma$ is $E_h/M_h$. Unless $M_h$ is very small this angle is much larger than $\alpha$, the angle subtended by BEBC at the target: $\alpha \sim 1.8$mr. Thus the fraction of produced neutrinos entering BEBC is $\sim (\alpha/\phi)^2 \approx \alpha^2 \gamma^2$; an order of magnitude estimate of the number of parents, $N_D$, is thus given by:

$$N_D \cdot B(e) \sim N_\nu /\alpha^2 \gamma^2$$

where B(e) is the branching ratio of h to electrons and $N_\nu$ the observed neutrino flux. With B(e)=0.10, $\gamma$= 60/2 and $N_\nu$ = 1.7. $10^{11}$ (Table II) the estimate for $N_D$ is:

$$N_D = 2.10^{15}$$

or

$$\sim 3.10^{-3} \ (D,\bar{D})$$
pairs per proton interaction.

A Monte-Carlo calculation has also been performed to estimate the possible yield from D,$\bar{D}$ mesons using various models to describe the production process including the form found for $\psi$-production:

$$d^2\sigma/dx.dP_T^2 \propto (1-x)^4\, e^{-2P_T} .$$

The production rates required to explain the yield of electron events lie in the range (3 to 12).$10^{-3}$ per incident proton or to a cross section:

$$(pp \rightarrow D + X) \sim 100\ \mu b.\text{to } 400\ \mu b. ,$$

taking 33 mb as the total inelastic cross section.

These values are one to two orders of magnitude greater than the currently accepted limits for charmed particle production by hadrons [4], especially the limit of 1.5 $\mu b$ for production by 300 GeV protons set in the emulsion experiment of Coremans-Bertrand et al [5]; however the search for charmed particle decays made in the latter experiment becomes inefficient for lifetimes in excess of a few $10^{-13}$sec or less than $10^{-14}$sec.

A production process peaked at x→1 (such as might be the case for $\Lambda_c$) would lead to a smaller required cross section ($\sim 20\ \mu b$) however this would appear to be inconsistent with the observation of an excess of $\nu_e$ events in the BEBC narrow-band anti-neutrino experiment; in that experiment the incident proton beam was directed 15mr away from the BEBC direction yet 8 e⁻ events were observed which cannot be explained unless they were due to prompt neutrinos generated in the berylium target by the same process as those found in this, $0^\circ$, beam-dump experiment.


4.2  Heavy leptons

In principle heavy leptons could be the source of the neutrinos however this seems unlikely for the following reasons.

(i)  The production cross section for the leptons,or their parents (say hadrons carrying $b$-type quarks), is required to be of the same order as that for charmed mesons.

(ii) In the case of the $\tau$-lepton the decay $\tau \rightarrow \nu_\tau +$ hadrons has a branching ratio $\sim 50\%$ [6]; as a consequence the interaction of $\nu_\tau$ through both the CC-mode:

$$\nu_\tau + N \rightarrow \tau^- + \text{hadrons}$$
$$\text{followed by } \tau \rightarrow \nu_\tau + \text{hadrons},$$

and the NC-mode:

$$\nu_\tau + N \rightarrow \nu_\tau + \text{hadrons} ,$$

would lead to an anomalously high apparent NC/CC ratio (> 1) and this is not the case.


## 4.3 Unknown particles of light mass

Low mass particles ($m < m_K$) of short lifetime and so far unknown could in principle be the source of the observed neutrinos and in this case the production cross section could, depending on the detailed mechanisms, be much smaller than the estimates made for D-production.


## 4.4 Axions

The existence of a light semi-weakly interacting scalar boson, called the axion, has been postulated [7] to avoid CP violation in QCD. This particle is expected to interact like a $\pi^0$-meson so leading to neutral current like events but with the important distinction that the net transverse momentum $P_T$, of the emitted hadrons would be small compared to the 2.5 GeV/c typical of neutrino induced N.C. events. The data show no evidence of an excess of N.C. candidates with $P_T < 0.5$ GeV/c and this observation can be used to set an upper limit to the product of production and interaction cross sections for axions (assuming the same x and $P_T$ distributions at production as for pions):

$$\sigma_{prod.} \times \sigma_{inter.} \text{ (axion)} < 2.10^{-67} \text{cm}^4 \; ; \; 90\% \text{ C.L.}$$

This is about 30 times lower than a recently predicted lower limit [8].

Ignoring the $P_T$ selection a limit can be set for any particle making N.C.-like events, based on the excess above expectation of ($4\pm7$) events:

$$\sigma_{prod.} \times \sigma_{inter.} < 2.10^{-66} \text{cm}^4 \; ; \; 90\% \text{ C.L.}$$


## 5. CONCLUSIONS

Prompt neutrino production has been observed from 400 GeV proton-nucleus collisions. The data are consistent with there being equal yields of $\nu_e$, $\bar{\nu}_e$, $\nu_\mu$ and $\bar{\nu}_\mu$ however the statistics are small and other ratios cannot be excluded. The average ratio $N(\nu_e)/N(\pi^+)$ is $(0.63\pm0.3)10^{-4}$ at production (for neutrinos and $\pi^+$-mesons at $0^0$ emitted into the solid angle subtended by BEBC).

No satisfactory explanation has been found for the source of these neutrinos; if they arise from charmed meson decay the production cross section for D's is required to be in the range 100 $\mu$b to 400 $\mu$b per nucleon at 400 GeV, which is in diagreement with the much lower limits set by other experiments on charmed particle production by hadrons. A smaller cross section would apply for as yet unknown light parents (m < $m_K$) of short lifetime. Heavy leptons (e.g. $\tau$) cannot be a main surce of the neutrinos and the axion seems excluded if its properties are as predicted.

## 6. ACKNOWLEDGMENTS

I wish to acknowledge the work done by my many colleagues in the ABCLOS collaboration, the operating staff of BEBC and the SPS and all the other staff at CERN and in the Universities who contributed to this experiment.

## 7. REFERENCES

1.   J.Ranft, Leipzig University Report KMU-HEP 75-03 (1975).

2.   W.F.Baker et al.  NAL-PUB 74/13 EXP;
     A.S.Carroll et al.  Phys.  Rev.  Lett.  33 (1974) 928.

3.   L.P.Leipuner et al., Phys.  Rev.  Lett.  35 (1975) 1613;
     D.Buchholz et al., Phys.  Rev.  Lett.  36 (1976) 932.

4.   W.R.Ditzler et al., Phys.  Lett.  71B (1977) 451;
     J.C. Adler et al., Phys.  Lett.  66B (1977) 401.

5.   G.Coremans-Bertrand et al., Phys.  Lett.  65B (1976) 480.

6.   A.Barbaro-Galtieri et al., Phys.  Rev.  Lett.  39 (1977) 1058.

7.   S.Weinberg, Phys.  Rev.  Lett.  40 (1978) 223;
     F.Wilezek, Phys.  Rev.  Lett.  40 (1978) 279.

8.   J.Ellis and M.K.Gaillard, Cambridge preprint 78/6 (1978).

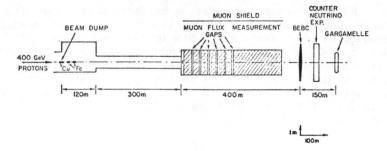

Figure 1(a)     Layout of the CERN-SPS proton beam dump experiments.

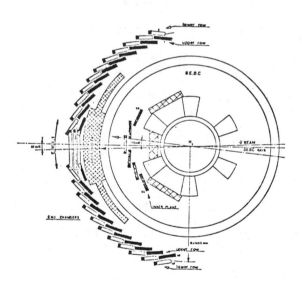

Figure 1(b)     Arrangement of proportional wire chambers forming
EMI for BEBC.

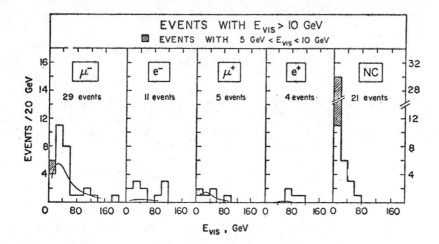

Figure 2(a)   Energy distribution for the five categories
of events:  $\mu^-$, $\mu^+$, $e^-$, $e^+$ and NC.  The curves show the
spectra expected if the neutrinos arise from conventional
sources, i.e. $\pi$, K and hyperon decays.

Figure 2(b)   The distri-
bution of $E_H/E_{VIS}$ for the
$\mu^-$, $\mu^+$, $e^-$ and $e^+$ events
with $E_{VIS} \geq 10$ GeV.

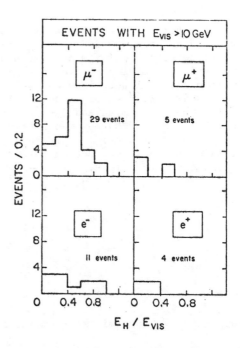

# RESULTS OF A BEAM DUMP EXPERIMENT AT THE
## CERN SPS NEUTRINO FACILITY

T. Hansl, M. Holder, J. Knobloch, J. May, H. P. Paar,
P. Palazzi, F. Ranjard, D. Schlatter, J. Steinberger, H. Suter,
W. Von Rüden, H. Wahl, S. Whitaker and E. G. H. Williams
CERN, Geneva, Switzerland

F. Eisele, K. Kleinknecht, H. Lierl, G. Spahn
and H-J. Willutzki
Institut für Physik[*] der Universität, Dortmund, Germany

W. Dorth, F. Dydak, C. Geweniger, V. Hepp, K. Tittel
and J. Wotschack
Institut für Hochenergiephysik[*] der Universität, Heidelberg, Germany

P. Bloch, B. Devaux, S. Loucatos, J. Maillard, B. Peyaud,
J. Rander, A. Savoy-Navarro and R. Turlay
D.Ph.P.E., CEN-Saclay, France

F. L. Navarria
Istituto di Fisica dell'Università, Bologna, Italy

## ABSTRACT

We report results from a beam dump experiment that has been
performed at the CERN SPS neutrino facility using the CDHS neutrino
counter detector.  Limits on dimuon and trimuon production by new
penetrating neutral particles are given.  A new source of prompt
electron and muon neutrinos has been observed giving
$(1.2 \pm 0.4) \cdot 10^{-7}$ $\nu_e$ or $\nu_\mu$ per incident proton with neutrino angle
smaller than 1.85 mrad and $E_\nu > 20$ GeV.  If these prompt neutrinos
are attributed to charmed meson pair production, the inclusive $D\bar{D}$
production cross section could be of the order of 30 μb.  If axions
are existing their production rate relative to $\pi^0$ mesons is found to
be less than 0.5 x $10^{-8}$.

[*]Supported by Bundesministerium für Forschung und Technologie.

## INTRODUCTION

The origin of trimuon events, that have been detected in high
energy neutrino experiments[1-4] is not yet definitively established.
It is conceivable that a fraction of these events is not due to the
interaction of ordinary neutrinos but due to a new type of penetrating neutral long-lived particle, produced either directly in the
hadronic collision at the $\nu$-target or in the prompt decay of a new
heavy short-lived particle. Interactions of new particles in the
detector might further give rise to abnormal charged current event
distributions or an abnormal ratio of muonless to charged current
events. This would for instance be expected for the interaction of
$\tau$-neutrinos or axions[5]. A beam dump experiment allows sensitive
searches for such particles.

The experiment was performed at the CERN SPS neutrino facility
in December 1977. The 400 GeV extracted proton beam was dumped at
zero degrees into a solid copper block (2 m long, 27 cm diameter
that replaced the ordinary wide band beam neutrino target $\sim$890 m
upstream of the detector. In copper the secondaries of the primary
proton interaction are moderated in one interaction length, approximately 15 cm, so that the neutrino flux from $\pi$ and K decays through
the detector is reduced by a factor of $\sim$1/3000 relative to the normal neutrino beam. On the other hand any short-lived particles
produced at the dump would decay promptly and their leptonic and
semileptonic decays would contribute to the neutrino flux through
the detector. In addition, new long-lived neutral weakly interacting particles produced at the dump might reach the detector
before decay. The CDHS neutrino counter detector[6] recorded every
event during the 1 msec spill time, provided the total energy
deposited was larger than $\sim$7 GeV. In addition events due to cosmic
ray particles were recorded within a 4 msec gate. The final event
selection required a total visible energy $E_{vis} > 20$ GeV for all
event types and used a fiducial volume with a radius of 1.65 m and
a total length of 9.3 m of iron for charged current events (1.6 m
radius and 7.1 m length of iron for muonless events). The recorded
event numbers are summarized in Table I.

## RESULTS ON MULTIMUON EVENTS

The raw measured rate of dimuon events with $E_{vis} > 30$ GeV in
the beam dump experiment is

$$\frac{N(\mu^-\mu^+)}{N(1\mu)} = (0.82 \pm 0.34)\%$$

in agreement with the observed rate of 0.5% for the interactions of
ordinary muon neutrinos ($\nu_\mu + \bar{\nu}_\mu$) as measured in the narrow band
beam experiment[7]. Also the kinematical properties of the observed
dimuon events show no qualitative differences with respect to the
events recorded during narrow band beam and wide band beam data

taking. Thus we see no evidence for multimuon production by a new type of neutral penetrating particles and the observed event numbers can be used to give upper limits for such a process. In a 400 GeV wide band beam run with $2.3 \times 10^{17}$ protons on target we have observed 13 trimuon events and ∿1500 dimuon events above 30 GeV visible energy. Normalizing to the same number of incident protons and comparing with the six dimuon and zero trimuon events observed in this experiment we can conclude immediately that with 90% confidence level less than 10% of the trimuon and less than 0.3% of the dimuon events observed in wide band beam running can be due to the interaction of new particles.


## RESULTS ON SINGLE MUON EVENTS

The observed energy distributions of single muon events are shown in Fig. 1 together with the calculated expectations from $\pi$ and K decay[8]), normalized to the total number of negative muons. The predicted $N_\mu+$ /$N_\mu-$ ratio for $E_\nu > 20$ GeV, 0.146 ± 0.015, falls significantly below the measured ratio of 0.22 ± 0.02. The excess of $\mu^+$ events indicates a new source of muon antineutrinos. We will come back to this point in Section 5.

The events have been analysed for possible differences in their scaling distributions $d\sigma/dx$ and $d\sigma/dy$ compared to charged current events observed in ordinary neutrino beams. No differences were found and we conclude that the single muon events observed in this experiment are in agreement with those expected from charged current interactions of muon neutrinos and antineutrinos.


## RESULTS ON MUONLESS EVENTS

Events without a muon are separated from events with muons by the technique described in Reference 9. We find 261 ± 20 muonless events and 311 ± 19 events with muons giving a ratio for muonless events to charged currents events of

$$R = \frac{N(\text{no muon})}{N(\geq 1 \text{ muon})} = 0.84 \pm 0.08$$

for shower energies above 20 GeV (the energy of muons is disregarded). This value of R is inconsistent with the expected neutral current over charged current ratio 0.30 ± 0.01 from muon neutrino ($\nu_\mu + \bar{\nu}_\mu$) interactions[9]). If we subtract the expected number of these neutral current events we are left with 168 ± 21 muonless events which are not due to muon neutrinos or antineutrinos. We expect 38 events from the CC and NC interactions of electron neutrinos due to $K^\pm_{e3}$, $K^o_{e3}$, and hyperon leptonic decays leaving 130 ± 22 events that must be attributed to a new source.

The origin of these excess events can be further explored by studying the longitudinal shower development to isolate a possible electromagnetic contribution. In Fig. 2 the shower development of the excess events is compared to that observed in the $\nu_\mu(\bar{\nu}_\mu)$ produced showers of charged current events. The early shower development is significantly different. We find that only (57 ± 10)% of the total pulseheight can be attributed to hadron showers (including their $\pi^0$ component). The remaining 43% can be attributed to an electromagnetic origin. This is expected for the interaction of electron neutrinos where the electrons deposit on average one-half of the pulseheight, but not for $\tau$ neutrinos or axions which would give essentially hadronic showers.

The prompt muonless signal of 130 events can now be attributed to $\nu_e$ and $\bar{\nu}_e$ interactions. Approximately 18 of these events should then be $\nu_e / \bar{\nu}_e$ neutral current events leaving 112 $\nu_e/\bar{\nu}_e$ charged current events. Their energy spectrum is plotted in Fig. 3.

## SUMMARY AND INTERPRETATION

As was shown in the foregoing, the existence of prompt muon antineutrinos is observed in the excess of $\mu^+$ events, and the existence of prompt electron neutrinos and/or antineutrinos is observed in both the magnitude and shower development distribution of the muonless events.

Keeping in mind the differences in the fiducial cuts and selection of the two data samples, the magnitudes of the two signals are consistent with equal fluxes of prompt muon- and electron-neutrinos and antineutrinos. 180 ± 74 of the single $\mu^-$ events and 78 ± 34 of the single $\mu^+$ events observed can be attributed to the new source. The corresponding numbers of charged current events due to prompt electron neutrinos, adjusted to the selection criteria of single muon events, are 104 ± 18 for $\nu_e$ and 50 ± 9 for $\bar{\nu}_e$. These event numbers for the new source may be compared to 314 ± 33 single $\mu^-$ events from kaon neutrinos, leading to a ratio of prompt $\nu_e$, $\bar{\nu}_e$, $\nu_\mu$ or $\bar{\nu}_\mu$ fluxes relative to the $\nu_\mu$ flux from $K^+$ decay of 0.33 ± 0.07. The absolute number of prompt neutrinos per incident 400 GeV proton going in the forward cone covered by our detector may be calculated on the basis of the neutrin cross section and the total number of protons on the dump target. We find $N_{\nu_e} = N_{\nu_\mu}$

= (1.2 ± 0.4) · $10^{-7}$/proton for neutrino angles smaller than 1.85 mrad and $E_\nu > 20$ GeV.

The only known particles which might be the origin of this prompt neutrino flux are the charmed particles. If we assume a charm origin, the neutrino flux given above can be used for cross section estimates, but only with additional assumptions about the production dynamics. We have used a simple model of $D\bar{D}$ production with an invariant cross section proportional to $(1 - x_F)^3$ for the D-meson, an average transverse momentum of 0.7 GeV/c and a semi-leptonic branching ratio of 10%. The predicted charged current event spectrum is in good agreement with the measured spectrum as shown in Fig. 3. The total inclusive cross section for $D\bar{D}$ production in this model is about 30 µb. This number may easily change by large factors if different production properties or meson-baryon instead of meson-pair production are assumed.

The data of this experiment can finally be used to set upper limits on axion production and interaction cross sections. In axion models the expected production rate relative to $\pi^0$ mesons $n_{a^0}/n_{\pi^0}$ is between $10^{-7}$ and $10^{-8}$ [5]. If they interact in the detector they would produce mainly hadronic showers, but in some cases also $\mu^+\mu^-$ pairs with very small invariant mass and no other visible energy. The estimated cross section for the latter process is $\sim$3pb for our energy range[10]. Up to 65 out of 130 excess muonless events could be purely hadronic showers and therefore due to axion interactions. From this number and the geometry acceptance of our detector for $\pi^0$ we obtain

$$\sigma(pN \to a^0) \times \sigma(a^0 N \to X) = \frac{n_{a^0}}{n_{\pi^0}} \, \sigma(pN \to \pi^0) \cdot \frac{n_{a^0}}{n_{\pi^0}} \, \sigma(\pi^0 N \to X) < 10^{-67} cm^4$$

(90% C.L.)

We see no dimuon event that could be due to axions such that

$$\sigma(pN \to a^0) \times \sigma(a^0 + Fe \to \mu^+\mu^-) = \frac{n_{a^0}}{n_{\pi^0}} \, \sigma(pN \to \pi^0) \cdot 3pb < 2.9 \times 10^{-69} \, cm^4.$$

Both measurements lead to the same upper limit for the ratio of axions to $\pi^0$ mesons produced. We find $n_{a^0}/n_{\pi^0} < 0.5 \times 10^{-8}$ with 90% confidence.

## REFERENCES

1) S. Mori et al., Phys. Rev. Lett. 40 (1978) 432.
   A. Benvenuti et al., Phys. Rev. Lett. 38 (1977) 1110.
2) B. C. Barish et al., Phys. Rev. Lett. 38 (1977) 577.
3) M. Holder et al., Phys. Letters 70B (1977) 393.
4) M. Holder et al., Characteristics of trimuon events in neutrino interactions, Phys. Letters, to be published.
5) S. Weinberg, Phys. Rev. Lett. 40 (1978) 223.
   F. Wilczek, Phys. Rev. Lett. 40 (1978) 279.
6) M. Holder et al., Nucl. Instr. Methods 148 (1978) 235.
7) M. Holder et al., Phys. Letters 69B (1977) 377.
8) We thank H. Wachsmuth for the calculation of the K and $\pi$ $\nu$-fluxes.
9) M. Holder et al., Phys. Letters 71B (1977) 222.
10) W. A. Bardeen et al., FERMILAB-Pub-78/20-THY.

TABLE I

Summary Of Event Numbers For $E_{vis} > 20$ GeV

| Number of protons on dump | $4.3 \cdot 10^{17}$ |
|---|---|
| Single muon events  $\mu^-$ | 727 |
| $\mu^+$ | 160 |
| Dimuon events  $\mu^-\mu^+$ | 6 |
| Trimuon events | 0 |
| Muonless events | 261 |

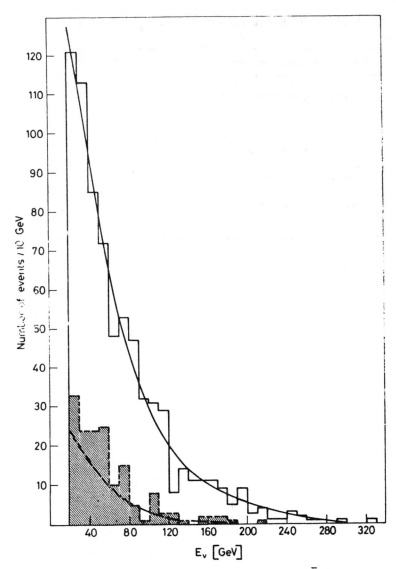

Fig. 1. Total visible energy spectra for single μ⁻ (upper histogram) and single μ⁺ events (shaded histogram) compared to the expectations from π- and K-decay neutrinos normalized to the total number of negative muons.

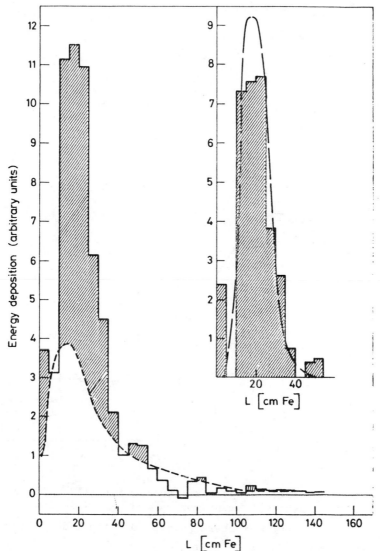

Fig. 2.   Longitudinal shower development in the iron calorimeter
          for excess muonless events compared to the shape of hadro-
          nic showers (dashed curve) as measured by single muon
          events.   The total hadronic pulseheight is determined from
          L > 40 cm.

          The insert shows the pulseheight not due to hadronic
          showers compared to the shape of purely electromagnetic
          showers (dashed curve).

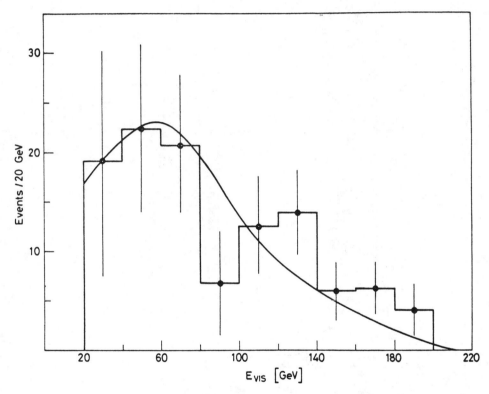

Fig. 3. Total visible energy distribution for (112 ± 22) charged current events from prompt $\nu_e$ and $\bar{\nu}_e$. The line shows the prediction of our charm pair production model ($N(D \rightarrow K^* e \nu_e)$ : $N(D \rightarrow K^0 e \nu_e) = 1 : 1$).

ELECTRON-MUON COINCIDENCES IN PROTON-PROTON COLLISIONS
AT THE CERN INTERSECTING STORAGE RINGS

A.G. Clark, P. Darriulat, K. Eggert[*], V. Hungerbühler,
H. Renshall, B. Richter[**], J. Strauss[***] and A. Zallo
CERN, Geneva, Switzerland

M. Banner, C. Lapuyade, T. Modis[†], P. Perez, G. Smadja,
J. Teiger, C. Tur, J.P. Vialle[††], H. Zaccone and A. Zylberstejn
CEN Saclay, Gif-sur-Yvette, France

P. Jenni[†††], P. Strolin and G.J. Tarnopolsky
Eidgenössische Technische Hochschule[×], Zürich, Switzerland

(Presented by G.J. Tarnopolsky)

## ABSTRACT

In an experiment carried out at the CERN Intersecting Storage Rings with a highly selective electron spectrometer system and a magnetized iron filter to detect muons, we have observed $32 \pm 16$ dilepton events of the type $p + p \rightarrow \mu^{\pm} + e^{\mp} + \ldots$ . The integrated luminosity of the experiment was $(2.0 \pm 0.1) \times 10^{37}$ cm$^{-2}$, and the over-all detection efficiency $0.14 \pm 0.07$. Interpreting this signal as due to charmed meson-pair production, we estimate a model-dependent acceptance of $6.5 \times 10^{-5}$ per event, and a cross-section $\sigma(p + p \rightarrow D + \bar{D} + \ldots) = (18 \pm 9)$ µb, with a scale uncertainty of 50% due to the detection efficiency.

## INTRODUCTION

In this talk I will present results on electron-muon coincidences observed at the CERN Intersecting Storage Rings, by the CERN-Saclay-ETH Zürich Collaboration.

---

[*] III. Physikalisches Institut der Technischen Hochschule, Aachen, Germany.

[**] On sabbatical leave from SLAC, Stanford, California, USA.

[***] Institut für Hochenergiephysik der Österreichische Akademie der Wissenschaften, Vienna, Austria.

[†] Now at Université de Genève, Geneva, Switzerland.

[††] LAL, Orsay, France.

[†††] Now at SLAC, Stanford, California, USA.

[×] Supported in part by the Swiss National Science Foundation.

ISSN: 0094-243X/78/167 /$1.50 Copyright 1978 American Institute of Physics

The idea behind this experiment is that proton–proton collisions might produce charmed states in the form of charm–anticharm pairs,

$$\text{proton} + \text{proton} \rightarrow C + \bar{C} + \text{anything} , \tag{1}$$

where C is generic for a charmed state. If this charmed object decays, fulfilling the $\Delta C = \Delta Q$ (C = charm, Q = charge) transition rule of weak semileptonic decay, then at some level we should observe $\mu^{\pm} e^{\mp}$ coincidences,

$$\text{proton} + \text{proton} \rightarrow \mu^{\pm} e^{\mp} + \text{anything} , \tag{2}$$

with a rate determined by the square of the weak semileptonic branching ratio and the charm production cross-section in proton–proton reactions.

We have measured the $\mu^{\pm} e^{\mp}$ rate, which potentially contains a charm signal, as well as the background channels $\mu^{\pm} e^{\pm}$. We assume here that the non-charm $\mu e$ rate is dominated by "charge-symmetric" sources, that is, background sources that populate equally, say, the $\mu^+ e^-$ and $\mu^+ e^+$ states. Typically, we have in mind events in which a muon of either sign is associated with a neutral pion that, via a photon conversion or via a Dalitz decay, produces electrons and positrons with the same frequency. The equal-sign dilepton events therefore measure the charge-symmetric background and should be subtracted from the experimentally observed opposite-sign $\mu e$ events. Thus, the signal is given by

$$S = (\mu^+ e^- - \mu^+ e^+) + (\mu^- e^+ - \mu^- e^-) . \tag{3}$$

In contrast to the charge-symmetric background, the hadron background of pion pairs, which (owing to imperfect particle identification) mimic genuine $\mu e$ coincidences, carries an apparent signal. This is due to the fact that in high-energy proton–proton collisions charged pion pairs of opposite sign occur more often that doubly-charged pairs[1]. This background was measured, and it contributes only a small number of events to the sample.

I will describe first the electron spectrometers, and discuss their capability of selecting single electrons from the much more abundant hadrons and electron pairs. I will then describe the characteristics and performance of the muon spectrometer. This will be followed by a description of the $\mu e$ event sample, the detection efficiency and the acceptance calculation, and the final results.

## ELECTRON SPECTROMETERS

Electrons were detected in a double-arm spectrometer, symmetrically arranged around 90° scattering angle, and with an acceptance of 0.5 sr each. An elevation view of one of the spectrometers is shown in Fig. 1.

Starting from the source, a particle penetrating the electron arm encountered the vacuum pipe of 0.3 mm corrugated titanium, two drift chamber planes, a double layer of sixteen ionization counters

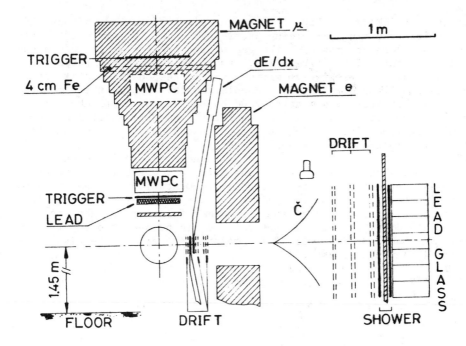

Fig. 1  View of the apparatus transverse to the beams.  A second
complete electron spectrometer (not shown) is symmetrically placed
to the left.

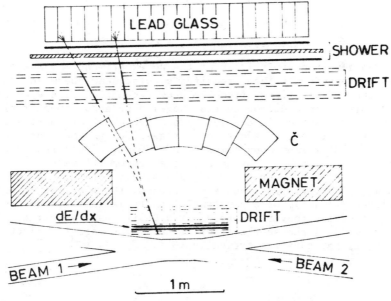

Fig. 2  Plan view showing a photon conversion in the pipe

(indicated by "dE/dx" in Fig. 1) used to reject converted photons and Dalitz pairs from singly-ionizing particles, and five more drift chamber planes before the magnet. The gap of the bending magnet contained the gas enclosure of and air-filled, atmospheric pressure, threshold Čerenkov counter, whose twelve mirrors and corresponding phototubes each viewed a slice of about 15° in polar angle. Behind the Č counter the outgoing trajectory was determined by additional drift-chamber planes. A shower counter, which consisted of a 22-counter hodoscope, a 2 cm thick iron plate, and a 20-counter hodoscope, all provided with pulse-height read-out, was placed immediately before a lead-glass wall of 138 blocks, each $15 \times 15$ cm$^2$ $\times 14.8$ X$_0$ and arranged in seven rows.

Figure 2 shows a plan view of the detector, where the components are labelled as before and the Č counter segmentation is clearly seen. In Fig. 2 I have drawn an e$^+$e$^-$ pair resulting from a $\gamma$ conversion in the beam pipe; the e$^+$ and e$^-$ share a common front track, and their trajectories diverge in the magnetic field, to produce two separate showers in the lead-glass wall. The main parameters of the electron arms are summarized in Table I.

Table I   Electron arm parameters

$$\int B \, d\ell = 0.34 \text{ T} \cdot \text{m}$$

$$\Delta p/p = \sqrt{[5.6p \text{ (GeV/c)}]^2 + 3^2} \%$$

$$55° \lesssim \theta \lesssim 125°$$

$$\Delta\phi \simeq \pm 15°$$

$$\Delta\Omega \simeq 0.5 \text{ sr}$$

The electron arm trigger of the μe experiment was one of two types. The hadron trigger required a coincidence between the dE/dx hodoscope and any signal from the shower counter. The electron trigger required, in addition to the above, a Čerenkov counter signal and an energy deposition in the lead-glass blocks of at least 400 MeV. As described later, an additional trigger for the muon arm, corresponding to a minimum muon momentum of 970 MeV/c, was required in coincidence with each of the hadron or electron triggers.

The hadron trigger was used extensively for the hadron background evaluation. The events selected by the electron trigger consisted primarily of a penetrating particle in the muon arm, in combination with a converted photon or a Dalitz pair in either electron arm, as well as the sought-after signal. The ratio of electron to hadron trigger rate was typically 10$^{-2}$.

# HADRON REJECTION

How did this detector fare in hadron rejection?  We had three electron-identifying elements:  the Čerenkov counters, the shower counter, and the lead-glass wall.  The Čerenkov rejection for hadrons was measured to be $P(\check{C})$ = (3.0 ± 0.4) × $10^{-3}$.  The combined shower-counter and lead-glass analysis proceeded as follows.  From a study of electrons and positrons from fully reconstructed photon conversions the energy resolution of the lead-glass array was measured to be

$$\Delta E/E = 0.12/\sqrt{E \text{ (GeV)}} \qquad (4)$$

in the energy range $0.5 \leq E \leq 1.5$ GeV, and over the entire life of the experiment.  For electrons, we compared the measured momentum p with the deposited energy E.  The correlated distribution appears in Fig. 3a.  A clear clustering along the diagonal is evident, the energy and momentum matching well.  The electron-identifying criteria used were a two standard deviation cut in (E − p), and both the energy and momentum above 600 MeV.  We compared the response to electrons with a similar scatter plot for hadrons, Fig. 3b, where the events cluster at relatively low measured energies for any value of the momentum. The distributions of Figs. 3a and 3b are nearly disjoint above p, $E \geq 600$ MeV.  The shower-counter pulse-height distributions (not shown) are dramatically different for electrons and hadrons, and the combined hadron rejection factor obtained is P(shower + lead-glass) = = (8.2 ± 0.6) × $10^{-3}$.  The over-all hadron rejection factor of the electron arms is P = (2.5 ± 0.4) × $10^{-5}$.

# PAIR REJECTION

The overwhelming majority of electron triggers originated from electron pairs, either from $\pi^0$ photon conversions in the vacuum pipe or from both $\pi^0$ and η Dalitz decays.  The contribution from these background sources was reduced by means of the dE/dx hodoscopes, and by studying the event pattern.

The response of the dE/dx hodoscope to single electrons was established from a sample of ≃ 1000 J/ψ events[2].  These electromagnetic J/ψ decays produced a single electron in each spectrometer arm. Fully reconstructed photon conversions in the vacuum pipe, such as that indicated in Fig. 2, were used to determine the double ionization lines. After establishing pulse-height cuts for the dE/dx counters we obtained a pair rejection P(dE/dx) = (1.0 ± 0.25) × $10^{-2}$.

Symmetric pairs, well centred in the spectrometer acceptance, resulted in readily-identified opening-track patterns (Fig. 2).  When the pair was produced either close to the acceptance edge, or sharing very asymmetrically the photon energy, it could result in a single track penetrating the spectrometer.  The front drift-chamber hit distributions from the tracks of each electron candidate were inspected in order to reject:  i) electron tracks associated with a track in the immediate neighbourhood, or pointing to the same Čerenkov mirror, ii) electron tracks (mainly from photon conversions in the counters) for which no drift chamber hits were recorded before the dE/dx

172

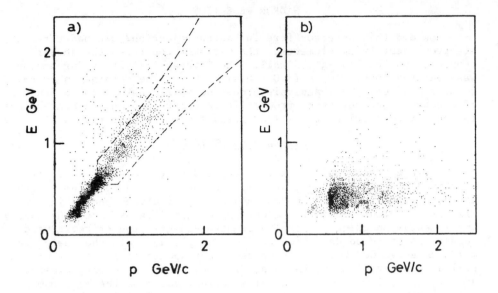

Fig. 3 Energy and momentum correlation plots; a) for electrons from converted photons, and b) for hadrons. The boundary in (a) indicates the energy and momentum cut for electrons.

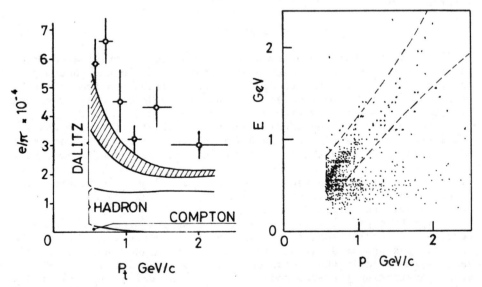

Fig. 4 The data points give the measured electron to pion ratio. The lines give the expected background level from the indicated source.

Fig. 5 Energy and momentum correlation plot for electrons of the final μe sample. There are 269 events in the accepted region.

counters, and iii) electron tracks with additional nearby hits not associated with any identifiable track. (For asymmetric pairs, the hits belonging to the lowest momentum track might have been deflected in the fringe field of the magnet to an extent that precluded fitting.)

This powerful strategy for the rejection of pairs involved a heavy price in terms of the detection efficiency, the latter being reduced by a factor of 0.5 ± 0.03 following the above pair cuts.

## CHECK ON SINGLE ELECTRON IDENTIFICATION

How can we check that this series of cuts indeed select single electrons? We have measured in a short 10 h run the inclusive e/π ratio. The result of this measurement is shown as a function of centre-of-mass transverse momentum in Fig. 4.

The expected background contributions are superimposed in Fig. 4. The Compton background, due to photon scattering in the vacuum chamber wall, is the result of a Monte Carlo calculation. The hadron background was evaluated, using the hadron trigger, from the number of hadrons that mimic in all respects electrons except for an absent Čerenkov signal. This number is then downgraded by the Čerenkov hadron rejection of $(3.0 \pm 0.4) \times 10^{-3}$, and normalized to the luminosity of the electron sample under discussion. The background from Dalitz decays and photon conversions could not be directly measured, and we relied on a calculation that included both $\pi^0$ and $\eta$ Dalitz decays[3]. The result of this calculation is indicated by the hatched band in Fig. 4. The width of the band reflects the uncertainty of the calculation resulting from such subtle effects as the absorption of soft electrons from asymmetric pairs in the corrugated vacuum pipe. What is left above background, with large experimental errors, is an e/π ratio that agrees well with other determinations[4] of this quantity. From this data we learned that the dominant background is indeed charge symmetric, populating equally same-sign and opposite-sign μe states.

## MUON SPECTROMETER

A particle originating at the source and penetrating the muon spectrometer, Fig. 1, passed in turn through a magnetic beam shield, a 10 radiation length shower absorber, a bottom trigger counter, and a six-plane MWPC (2 mm wire spacing), to determine the incoming particle trajectory. Subsequently, the particle penetrated a 60 cm thick iron filter magnetized to saturation with its field oriented along the bisector of the ISR beams. Emerging from the iron, a second (4 mm wire spacing) MWPC determined the outgoing trajectory. After penetrating an additional 4 cm thick iron absorber[5], the particle finally traversed the top trigger hodoscope. A muon trigger required the coincidence of the top and bottom hodoscopes.

For each particle, we measured the angular deflection and the spatial shift of the outgoing track compared with the extrapolated incoming track. Although the strong correlation of shift and deflection is weakened by multiple scattering in the iron, the bending angle due to the magnetic field was always ≥ 3 times the mean multiple-scattering deflection. This resulted in an unambiguous charge identi-fication. A Monte Carlo simulation, which included the ionization loss and the multiple scattering of muons, was used to determine the con-tour in the deflection-shift space that accepted 90% of the muons above threshold.

How good is this device in rejecting hadrons, that is, what frac-tion of hadrons fulfil the muon shift-deflection cuts? To assess this rejection we measured the inclusive cross-section $\Delta\sigma$ seen by the muon spectrometer, with a trigger that required the top and bottom muon arm counters only. We compared this with the known inclusive cross-section measured at the ISR[6], integrated over the momentum and geo-metric acceptance of the spectrometer. The cross-section seen by the muon arm is $\Delta\sigma = (2.2 \pm 0.1)$ μb. $\Delta\sigma$ includes an estimated 12% contri-bution from $\pi \to \mu\nu$ and $K \to \mu\nu$ decays. Subtracting it, the hadron re-jection given by the ratio of the known to the measured cross-section is 38 ± 2, corresponding to an effective attenuation length in iron of 15.5 ± 0.5 cm. The main parameters of the muon spectrometer appear in Table II.

Table II   Muon arm parameters

| |
|---|
| 60 cm thick magnetized iron |
| B = 16 kG |
| $P_{t(min)}$ = 970 MeV/c |
| $\lambda_{eff}$ = 15.5 ± 0.5 cm |
| 60° ≤ θ ≤ 120° |
| φ = ±15° |
| $\Delta\Omega$ = 0.55 sr |

RESULTS

We come now to the final electron-muon sample. Relaxing the E versus p cut described earlier, the (E, p) scatter plot of the sample is shown in Fig. 5. The band along the diagonal corresponds to elec-trons, and that parallel to the momentum axis to hadrons. The accep-ted region, indicated by the broken line, contains 269 events of all sign combinations. The distribution according to charged states ap-pears in Table III.

Table III    Final event sample

| State | Observed events | Hadron background | Subtracted sample |
|-------|-----------------|-------------------|-------------------|
| $\mu^+ e^-$ | 85 | 8.6 ± 1.3 | 76.4 ± 9.3 |
| $\mu^+ e^+$ | 64 | 8.0 ± 1.3 | 56.0 ± 8.1 |
| $\mu^- e^+$ | 67 | 8.3 ± 1.2 | 58.7 ± 8.3 |
| $\mu^- e^-$ | 53 | 6.3 ± 1.0 | 46.7 ± 7.4 |

From each charge combination we removed the hadron contamination (column 3, Table III) evaluated from the background runs, and obtained the subtracted number of events. The doubly-charged channels now give the background from photon conversions and Dalitz pairs.

The signal is

$$(\mu^+ e^- - \mu^+ e^+) + (\mu^+ e^- + \mu^- e^-) = 32 \pm 16 \text{ events} \qquad (5)$$

measured with the restrictions $E \gtrsim 600$ MeV, $p_e \gtrsim 600$ MeV/c, and $p_t \gtrsim 1$ GeV/c. The corresponding integrated luminosity is $(2.0 \pm 0.1) \times 10^{37}$ cm$^{-2}$, of which 1/3 was taken at 31 GeV/c and 2/3 at 26 GeV/c per beam.

To convert this rate into a cross-section we have to know the detection and reconstruction efficiency, and, within certain assumptions, the acceptance of our detector.

In the case of detection and reconstruction efficiencies, we have estimated the losses due to the many cuts applied to the data by using independent standard samples, and by studying the effect of each cut on these samples. The samples used were: i) converted photons, ii) hadrons, and iii) J/$\psi$ events. In this fashion we assessed the loss of genuine events induced by each separate cut.

Quite independently, we further checked all cuts not involving the Čerenkov and lead-glass analysis by measuring the inclusive hadronic cross-section seen by the electron spectrometers, and comparing it with the well-known result[6]. The ratio of our measurement to the known cross-section gives yet another estimate of the efficiency. The efficiency obtained from the product of each separate cut loss, and that obtained from the inclusive cross-section ratio agreed remarkably well. The over-all muon and electron detection and reconstruction efficiency was found to be

$$\varepsilon = 0.14 \pm 0.07 . \qquad (6)$$

This 50% uncertainty on the efficiency will reflect itself as an overall scale factor on the quoted cross-section. Notice, however, that this uncertainty will not affect the statistical significance of the result.

To compute the acceptance of our detector, we have ascribed the
μe coincidences to the process

$$p + p \rightarrow D + \bar{D} + \text{anything} , \qquad (7)$$

with the D mesons decaying semileptonically into either $\ell K\nu$ or
$\ell K*(892)\nu$ [7]. (We assumed that the leptonic decays proceed half of the
time through each of these modes.) We have used charm production
parametrization of Bourquin and Gaillard[8]. This parametrization re-
produces the desired features of hadron production, that is, that the
$p_t$ distribution flattens for low $p_t$ as the mass increases, and that
the rapidity distribution is flat in the central region, falling rapidly as the rapidity approaches its kinematic limit. The result of
this calculation gives an acceptance $A = 6.5 \times 10^{-5}$ per $D\bar{D}$ event.
Combining all these numbers we obtain

$$BR^2 (D \rightarrow \text{leptonic}) \cdot \sigma(p + p \rightarrow D + D + X) = (180 \pm 90) \text{ nb} \cdot \begin{cases} 1.5 \\ 0.5 \end{cases} \qquad (8)$$

where the last scale factor reflects the 50% uncertainty in detection
efficiency. Introducing branching ratios[9] of 10%, we get a charm
production cross-section of $(18 \pm 9)$ μb, which again should be multiplied by a factor between 0.5 and 1.5.

## CONCLUSIONS

We have observed a significant $\mu^{\pm}e^{\mp}$ signal, above known sources
of background, in proton-proton collisions at the highest ISR energies.
These events could be ascribed to $D\bar{D}$ production, followed by weak semi-
leptonic decays obeying the $\Delta C = \Delta Q$ rule. A model-dependent estimate
of the acceptance results in cross-section for charm production of the
order of $(18 \pm 9)$ μb. In a related experiment[2] we have also measured
the J/$\psi$ production cross-section at the ISR, and get for the ratio of
the assumed charm production to J/$\psi$ production a value

$$\frac{\sigma(p + p \rightarrow D + \bar{D} + \text{anything})}{\sigma(p + p \rightarrow J/\psi + \text{anything})} \sim 40 . \qquad (9)$$

## ACKNOWLEDGEMENTS

We acknowledge useful discussions with Drs. A. Ali, M. Bourquin
and J.-M. Gaillard. We wish to thank Drs. H. Hoffmann and H. Verweij,
and their collaborators, for their help in the initial stages of the
experiment. The technical assistance of Mr. A. Mottet and
Mr. M. Lemoine, as well as that of Messrs. H. Acounis, G. Bertalmio,
L. Bonnafoy, J.M. Chappuis, C. Engster, L. McCulloch and J.C. Michaud
is gratefully acknowledged.

FOOTNOTES AND REFERENCES

1. We found $\pi^\pm\pi^\mp$ pairs to appear $\sim 20\%$ more often than $\pi^\pm\pi^\pm$ pairs, with one of the pions detected with $p_t \gtrsim 1$ GeV/c in the muon spectrometer, and the second pion with $p \gtrsim 600$ MeV/c in the electron spectrometers.

2. The J/$\psi$ results are discussed by A. Zylberstejn at this Conference.

3. G.J. Donaldson et al., Phys. Rev. Letters <u>40</u>, 684 (1978). The $\eta$ transverse momentum spectrum of this reference was extrapolated below $p_t = 1.5$ GeV/c using the Bourquin-Gaillard formulae.

4. F.W. Büsser et al., Nuclear Phys. <u>B113</u>, 189 (1976) and references therein.

5. This absorber before the top trigger counter assured that a particle penetrating the apparatus had a momentum greater than 150 MeV/c at the top MWPC. Multiple scattering increases rapidly below this momentum with a corresponding deterioration of the sign identification.

6. B. Alper et al., Nuclear Phys. <u>B87</u>, 19 (1975).

7. A. Ali and T.C. Yang, Phys. Letters <u>65B</u>, 275 (1976).

8. M. Bourquin and J.-M. Gaillard, Nuclear Phys. <u>B114</u>, 334 (1976).

9. W. Bacino et al., Phys. Rev. Letters <u>40</u>, 671 (1978).
   J.M. Feller et al., Phys. Rev. Letters <u>40</u>, 274 (1978).

# MASSIVE LEPTON PAIR PRODUCTION IN HADRONIC COLLISIONS

Edmond L. Berger
High Energy Physics Division
Argonne National Laboratory, Argonne, Illinois 60439

## ABSTRACT

A review is presented of theoretical attempts to describe the massive lepton pair continuum produced in hadronic collisions. I begin with the classical Drell-Yan quark-antiquark annihilation model. Its expectations are compared with data. Scaling violations and corrections to the Drell-Yan picture are discussed. Next, I present original work bearing on the understanding of the transverse momentum distributions of the dileptons. In a parton model, the quark-gluon scattering graphs which provide scaling violations also endow the dileptons with relatively large $\langle p_T \rangle$. Comparisons with available data support these QCD expectations; further tests of the QCD predictions are proposed. Implications are drawn for W production experiments, and predictions are presented for reactions initiated by antiprotons.

## TABLE OF CONTENTS

ISSN: 0094-243X/78/178/$1.50

# I.   INTRODUCTION

My charge is to summarize our theoretical understanding of massive lepton-pair production in high energy collisions of hadrons:

$$h_1 + h_2 \to \ell^+ \ell^- \quad \text{plus anything,}$$

where $\ell$ is an electron, muon, or heavy lepton ($\tau$).  The theory is evolving rapidly, in response, in part, to the ever increasing quality of the data.[1,2] Recently good tests have been made of several theoretical expectations, including the scaling hypothesis.  A relatively new feature of the data is the observation that dileptons emerge with larger mean transverse momenta than previously supposed. The mean transverse momentum appears also to be independent of dilepton mass for $5 < M_{\mu\mu} < 10$ GeV.  These $\langle p_T \rangle$ properties may be interpreted in a QCD framework in terms of the same types of quark-gluon diagrams which provide scaling violations.

In this report I focus for the most part on the dilepton continuum, the "background" above which one observes the resonances of the J/$\Psi$ and T families.  I begin with a short review of the traditional Drell-Yan annihilation model[3] and its experimental successes.  Next I describe the reinterpretations of, or "corrections" to this model which are required in light of recent theoretical and experimental developments, including scaling violations in deep-inelastic electron and muon scattering.  Finally, I report on some very recent work which others and I have done in an attempt to interpret the transverse momentum distribution of lepton pairs.

There are various reasons for investigating the production of lepton pairs in hadronic reactions.  The J/$\Psi$ and T states were found this way, and it takes no imagination to suppose that the discovery of other new hadronic degrees of freedom (viz. heavier quarks) may be only a question of securing the greater acceptance, resolution, luminosity, and energy needed to probe even higher values of the dilepton mass.  If there are new quarks, the $J^P = 1^-$ $Q\bar{Q}$ state will couple through a virtual photon to lepton pairs.  Higgs mesons and the neutral vector boson mediator of weak interactions, the $Z^\circ$, should be observed as states in the $\ell^+\ell^-$ mass spectrum.  Measurements of dilepton yields at present energies, along with the conserved vector current and scaling hypotheses, permit estimates to be made for charged weak vector boson $W^\pm$ yields at the much higher energies which should be available soon at CERN, Fermilab, and Brookhaven.

In this report, I concentrate on yet another aspect of the physics of massive lepton pair production in hadron collisions.  The data provide a good opportunity to test various concepts regarding quarks and other partons, including:

a.   Scaling and scaling violations.
b.   The connection between $h_1 h_2 \to \ell^+\ell^- X$ and the "crossed" processes, in which one or two leptons are in the initial state:

$$ep \to e'X \, ;$$

$$\mu p \to \mu' X \; ;$$

$$\nu p \to \mu X \; ;$$

$$e^+ e^- \to \quad X \; .$$

c. QCD quark-gluon dynamical predictions for the $p_T$ distribution of high mass lepton pairs (including the M, s, and $x_F$ dependences of $\langle p_T \rangle$).

d. The data also provide an independent determination of anti-quark and gluon momentum distributions. In the case of $\pi$ induced reactions, e.g. $\pi N \to \mu\bar\mu X$, the quark structure functions for the pion may be deduced for the first time.

## II. THE CLASSICAL DRELL-YAN ANNIHILATION MODEL

I eschew the usual warnings regarding the inapplicability of the "rigorous" operator product expansion to $h_1 h_2 \to \ell^+\ell^- X$ and, instead, adopt the phenomenological Drell-Yan model[3] as a point of departure. In this model, it is supposed that when two hadrons collide, a quark constituent from one miraculously enough finds an anti-quark constituent in the other hadron, with which it then annihilates through a single virtual photon of mass $M = \sqrt{Q^2}$. This process is sketched in Fig.1. The probability that a quark [antiquark] is present with longitudinal momentum fraction x of the parent hadron's momentum is expressed by a function $q(x)[\bar q(x)]$. In this <u>classical</u> Drell-Yan model the functions $q(x)$ and $\bar q(x)$ are independent of $Q^2$— i.e. are scaling functions—and the quarks and antiquarks are assumed to carry negligible transverse momentum. Both assertions require modification, as I'll describe later. The longitudinal fractions x are positive, and $M^2 = Q^2 = s x_1 x_2$, where here $x_1$ refers to a quark (antiquark) from hadron 1, and $x_2$ labels an antiquark (quark) from hadron 2. It is supposed that $q(x)$ is a function derived from data on deep-inelastic electron and deep-inelastic muon scattering experiments, and that $\bar q(x)$ is likewise known as a result of studies of $\nu p \to \mu X$ and $\bar\nu p \to \mu X$. For large $M^2$, the oft-quoted Drell-Yan prediction for $h_1 h_2 \to \mu\bar\mu X$ or $h_1 h_2 \to e\bar e X$ is

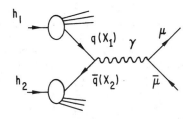

Fig.1.    Basic Drell-Yan quark-antiquark annihilation mechanism for lepton pair production in hadronic collisions, illustrated here for $h_1 h_2 \to \mu\bar\mu X$; q and $\bar q$ denote respectively a quark and an antiquark constituent.

$$M^4 \frac{d\sigma}{dM^2} = \frac{4\pi\alpha^2}{9} \sum_i e_i^2 \int\int dx_1 dx_2 \left[ q_i(x_1)\bar{q}_i(x_2) \right.$$

$$\left. + \bar{q}_i(x_1)q_i(x_2)\right] \delta\left(x_1 x_2 - \frac{M^2}{s}\right). \tag{1}$$

The sum is taken over the different quark flavors, usually restricted for practical purposes to $i = u, d$, and s. The factor 9 in the denominator is a product of two factors of 3, one derived from an angular integration in the final $\ell\bar{\ell}$ rest frame. The second is the famous "color factor"; $e_i$ is the fractional quark charge. An integration may be "undone" easily in Eq. (1) to obtain expressions for $d\sigma/dM^2 dy$ and $d\sigma/dM^2 dx_F$, where y and $x_F$ are the rapidity and scaled longitudinal momentum (Feynman x) of the lepton pair. I do not repeat those expressions here. For $h_1 h_2 \to \tau^+\tau^- X$, the lepton mass ($m_\tau \sim 1.8$ GeV) is no longer negligible with respect to typical values of M, and a threshold factor

$$\left(1 - \frac{4m_\tau^2}{M^2}\right)^{\frac{1}{2}} \frac{3}{8}\left[\left(1 + \frac{4m_\tau^2}{M^2}\right) + \left(1 - \frac{4m_\tau^2}{M^2}\right)\cos^2\theta^*\right] \tag{2}$$

must be inserted on the right hand side of Eq. (1). The $\tau$ is assumed to have spin 1/2. Displayed in Eq. (2) is the explicit angular dependence expected in the classical Drell-Yan model. In the $\tau^+\tau^-$ rest-frame, $\theta^*$ is the polar angle of a $\tau$ with respect to the axis defined by the collinear $q\bar{q}$ system.

My plan is to list some of the predictions of the classical Drell-Yan model and to compare them with data. This discussion leads naturally to questions of scaling violations, "corrections" to the classical model, and to $p_T$ spectra, which I take up in subsequent sections.

## III. PREDICTIONS OF THE CLASSICAL MODEL

1. <u>Scaling</u>. An immediate consequence of the form of Eq. (1), regardless of the explicit values of $q(x)$ and $\bar{q}(x)$, is the assertion of scaling, whereby an appropriately defined quantity depends only on the ratio $M/\sqrt{s}$, at fixed y or $x_F$. Specifically, we should find that

$$M^4 \frac{d\sigma}{dM^2 dy} = f_s\left(\frac{M}{\sqrt{s}}, y\right). \tag{3}$$

This equation may be rewritten in a variety of ways, including, for example,

$$\frac{M^2 d\sigma}{d\sqrt{\tau}dy} = g\left(\frac{M}{\sqrt{s}}, y\right). \tag{4}$$

Here $\tau = M^2/s$.

To test Eq. (3) [or Eq. (4)], we need precise data at several different energies $\sqrt{s}$, for values of M which exclude the resonances of the J/Ψ and Υ families; i.e., the acceptable regions of dilepton mass are $4.5 \lesssim M \lesssim 9$ GeV, or $M \gtrsim 11$ GeV. For practical reasons, only the mass range $4.5 \lesssim M \lesssim 9$ GeV has been explored thus far at several different energies. Experimental acceptance also limits the values of y at which data are available at different energies.

In the reaction $pN \to \mu\bar{\mu}X$ at $y \approx 0.2$, the Columbia-Fermilab-Stony Brook collaboration[1] has demonstrated that scaling holds to within

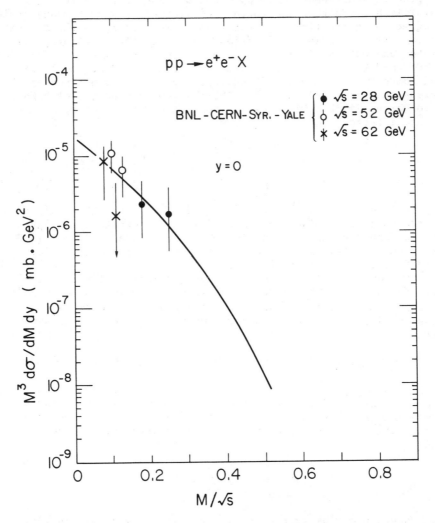

Fig. 2. ISR data from Ref.5 are compared with a scaling curve obtained from the hybrid model described in Section III.6. The model is constructed to fit lepton pair data at $p_{lab} = 400$ GeV/c, $y = 0$, and $0.2 < M/\sqrt{s} < 0.5$.

20% for $M/\sqrt{s}$ values between 0.2 and 0.4, for $p_{lab} = 200$, 300, and 400 GeV/c. Their target N is Platinum, 60% neutrons and 40% protons. This is the most precise test of scaling thus far. A previous investigation was made by a Chicago-Princeton group.[4] Obviously tests are desirable over a wider range of s, and for more values of y. In Fig.2, I compare some recent ISR data[5] with a scaling curve derived from the 400 GeV/c pN data of the Columbia-FNAL-SUNY group.[6] The comparison does not test scaling inasmuch as the ISR data are by and large limited to $M/\sqrt{s} < 0.2$, whereas the FNAL data are in the different range $0.2 < M/\sqrt{s} < 0.5$. Within the rather large errors of the present ISR data, the comparison in Fig.2 suggests that the function $f(M/\sqrt{s})$ derived from the FNAL data can be extrapolated into a lower region of $M/\sqrt{s}$ without gross error.

Scaling violations, owing to the fact that the functions $q(x)$ and $\bar{q}(x)$ are expected to be functions also of $Q^2$, suggest that Eq.(3) is to be replaced by the form

$$M^4 \frac{d\sigma}{dM^2 dy} = f_{NS}\left(\frac{M}{\sqrt{s}}, M, y\right) . \tag{5}$$

The explicit M dependence in $f_{NS}$ represents the scaling violation. How large should the deviations be from the perfect scaling predicted by the classical Drell-Yan model? Theorists are now trying to answer this question[7-9], and I will return to it below. To answer the question one must first devise a set of $Q^2$ dependent structure functions, which are consistent with the high energy deep-inelastic $\mu p \to \mu' X$ and $\nu p \to \mu X$ data, and which fit the (latest) data from $pN \to \mu\bar{\mu}X$ at one energy, say, 400 GeV/c. Then expectations can be calculated for, say, 200 GeV/c and ISR energies. Because the $Q^2 (=M^2)$ values are relatively large in dilepton production ($Q^2 \gtrsim 25$ GeV$^2$), whereas scaling violations are most pronounced for smaller $|Q^2|$ in deep inelastic processes, the interval in s over which data are compared may have to be very substantial before scaling violations are measurable in dilepton production by hadrons.

2. Quantum Number Effects. There are many tests of this nature, some more model dependent than others. I will mention only one. Consider the ratio of cross-sections $\sigma(\pi^+ T_0)/\sigma(\pi^- T_0)$ for producing high mass lepton pairs when $\pi^\pm$ beams impinge on an isoscalar target $T_0$. The sea component of the quark and antiquark distribution functions, $q(x)$ and $\bar{q}(x)$, dies off much more rapidly as x increases than does the valence part. Thus, at large enough lepton pair masses (recall, $M^2 = sx_1 x_2$), the cross section is dominated by the annihilation of a valence anti-quark in the pion beam with a valence quark in the target. The $\pi^+$ is a $(\bar{d}u)$ system and the $\pi^-$ is a $(\bar{u}d)$. Therefore, we expect $\sigma(\pi^+)/\sigma(\pi^-) = (e_d/e_u)^2 = 1/4$ for large dilepton masses. The recent Chicago-Princeton data support this expectation nicely.[2]

3. Linear Dependence on A. In the model, the quarks are assumed to act incoherently in a nucleon. Thus, they should also be incoherent in a nucleus. It is evident, therefore, that the cross-section for massive lepton pair production should be proportional to A, the total number of nucleons in a nucleus. More than a predic-

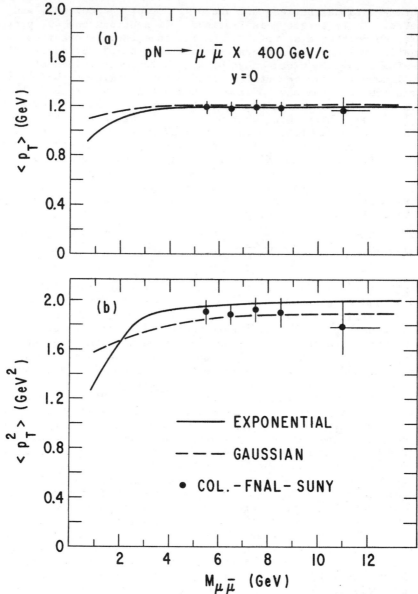

Fig. 3. The first two moments (a) $\langle p_T \rangle$ and (b) $\langle p_T^2 \rangle$ of the transverse momentum distribution of muon pairs produced in $pp \to (\mu\bar{\mu})X$ at 400 GeV/c and $y = 0$ are shown as a function of the mass of the muon pair. The data are from Ref. 6. The theoretical curves are calculated from simple models in which the initial quarks and antiquarks are assigned a distribution in their transverse momenta. The two models (Gaussian and exponential) are described in the text.

tion of the classical Drell-Yan model, this condition is a minimum requisite for applicability of the model. It seems to hold[10] for $M \gtrsim 4.0$ GeV.

4.   Transverse Momentum Distributions.   As remarked above, the quarks and antiquarks are assumed to carry "little" transverse momentum, and therefore the dilepton pairs should emerge with "small" $p_T$. Experimentally,[6] on the other hand, it is observed that $\langle p_T \rangle \simeq 1.2$ GeV, and $\langle p_T^2 \rangle \simeq 1.9$ GeV$^2$ for $5 < M < 10$ GeV in pN collisions at $p_{lab} = 400$ GeV/c.   The data are shown in Fig. 3.   Are these statements consistent? It seems obvious that quarks and antiquarks in the initial hadrons carry some non-zero $\langle k_T \rangle$ associated simply with the fact that they (several together) are confined in a region of transverse dimensions of order 1 fermi.   The uncertainty principle suggests $\langle k_T \rangle \simeq 300$ MeV. This is perhaps a lower limit.   More relevant for a determination of $\langle k_T \rangle$ is the fermi motion of quarks within a hadron, which is in turn associated with the inter-quark spacing.   Thus, $\langle k_T \rangle \simeq 600$ MeV, or more, is not obviously an unreasonable figure to assign for the mean transverse momentum of a quark or antiquark in the intial hadron wave function.   Specific (bag) models of quark confinement can be exploited[11] to refine this estimate, to suggest whether $\langle k_T \rangle$ should vary with x of the quark, and to provide predictions for possible differ-

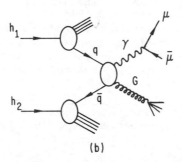

(a)

(b)

Fig. 4.   (a) Diagram which illustrates the scattering of a quark and a gluon constituent to produce a lepton pair and a quark (which, in turn, fragments into a jet of hadrons). (b) Quark-antiquark scattering to produce a lepton pair and a gluon jet.

ences between $\langle k_T \rangle_q$ and $\langle k_T \rangle_{\bar q}$. Turning to the data, now, if we assume that the entire experimental figure of $\langle p_T^2 \rangle = 1.9$ GeV$^2$ is to be associated with transverse momenta of the quarks and antiquarks in the wave functions of the incident hadrons, then we would conclude that $\langle k_T^2 \rangle_q \simeq 1$ GeV$^2$. [Here I have assumed $\langle k_T^2 \rangle_q \equiv \langle k_T^2 \rangle_{\bar q}$]. Adopting a Gaussian distribution in $k_T$, I deduce $\langle k_T \rangle_q \simeq 900$ MeV ($\sim$800 MeV if exponential). Although 50% larger than the "fermi motion" figure I quoted above, the value of 900 MeV is not outrageously large. Nevertheless, it is the judgment of many theorists[12,13] that a dynamical explanation should be sought for the "large" experimental $\langle p_T \rangle \simeq 1.2$ GeV in terms of hard-scattering models, rather than the bag or confinement explanation I sketched above. In the hard-scattering approach a substantial part of $\langle p_T \rangle$ derives from the scattering of quark and gluon (or meson) constituents, such as sketched in Fig.4. In this view, the transverse momentum of the dilepton is balanced by a quark (Fig.4(a)) or gluon (4(b)) jet in the final state. I describe specific models of this type in Sec.V.

The hard-scattering and confinement explanations differ in their predictions for the s dependence of transverse momentum effects. The observed growth[1] of $\langle p_T \rangle$, at fixed $M/\sqrt{s}$, when $p_{lab}$ is increased from 200 to 400 GeV/c is suggestive that the hard-scattering approach is important even at relatively small values of $p_T$. No doubt both confinement and hard-scattering components are present. In any case, the classical Drell-Yan model needs modification. I return to this issue below in Secs. IV and V.

In Fig.5, I compare the experimental distribution[6] in $p_T$ with calculations of $Ed\sigma/d^3p$ in which I replace $q(x)$ in Eq.(1) by the factorized form[14]

$$q(x, \vec{k}_T) = x_R^{-1} x\, q(x)\, f(|\vec{k}_T|), \qquad (6)$$

with $x_R^2 = [x^2 + 4k_T^2/s]$. An identical substitution is made for $\bar q(x)$. The forms I use for $q(x)$ and $\bar q(x)$ are described later, in Eqs.(9) and (10). For $f(|\vec{k}_T|)$ I tried both Gaussian and exponential forms, with $\langle k_T^2 \rangle_q \equiv \langle k_T^2 \rangle_{\bar q} = 0.5$ x (1.9) GeV$^2$. The description of the 400 GeV/c data is adequate with these naive models[14] of the confinement type, but energy dependent effects are not reproduced unless $\langle k_T \rangle_q$ is chosen to be a function of s. In Fig.3, I show how $\langle p_T \rangle$ and $\langle p_T^2 \rangle$ are expected to vary with M at 400 GeV/c in these naive models. Note the kinematic rise at small M before $\langle p_T \rangle$ becomes roughly independent of M for $M \gtrsim 4$ GeV, as in the data.

5. Angular Distributions in the Classical Drell-Yan Model.[14,15,16]
In the quark-antiquark rest frame (which is also the dilepton rest frame), the angular distribution of a final lepton is predicted to have the form

$$\frac{d\sigma}{d\Omega^*} = [1 + \alpha \cos^2 \theta^*], \qquad (7)$$

with

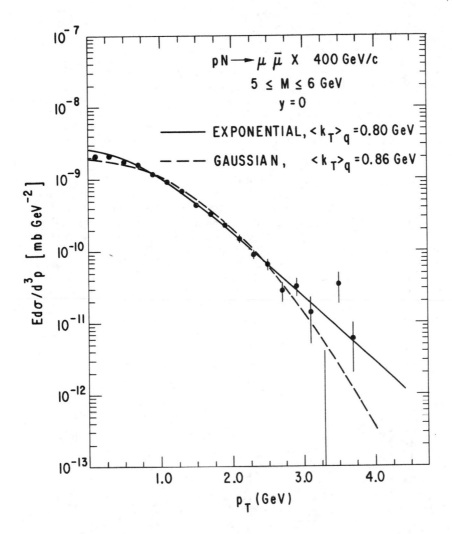

Fig.5. The inclusive yield of muon pairs $E d\sigma/d^3 p$ shown as a
function of $p_T$ for $5 < M_{\mu\bar{\mu}} < 6$ GeV, at $y = 0$ and 400 GeV/c.
The data are from Ref.6. The theoretical curves are
obtained from simple models in which the initial quarks
and antiquarks in the hadrons are assigned a distribution in their transverse momenta $k_T$. In one model, this
distribution is chosen to be an exponential in $|k_T|$ with
$\langle|k_T|\rangle = 0.8$ GeV; in the other, a Gaussian is chosen with
$\langle|k_T|\rangle = 0.86$ GeV. The $p_T$ distribution in the data at
other values of mass (not shown) is also described
equally well with these models.

$$\alpha = \left\langle \frac{M^2 - 4m_q^2}{M^2 + 4m_q^2} \right\rangle . \tag{8}$$

The average is taken over the different quark masses $m_q$. In the (usual) limit $M \gg m_q$, $\alpha \equiv 1$. The longitudinal direction ($\theta^* = 0$) is defined by the quark-antiquark collinear axis. If the quark and antiquark carry no transverse momentum, as in the classical Drell-Yan model, then Eq.(7) is true also in the "t-channel" dilepton rest frame, in which the $\theta^* = 0$ axis is specified by the (longitudinal) direction of the initial hadrons.

Owing to the fact that $\langle p_T \rangle \neq 0$, Eq.(7) with $\alpha \simeq 1$, should no longer hold in the t-channel frame (and even less so in the s-channel helicity frame). In general, both $\theta^*$ and $\phi^*$ dependences are expected. If the effective $\theta^*$ dependence is parametrized as $[1 + \alpha_t \cos^2 \theta_t^*]$, the value of $\alpha_t$ in the t-channel frame is expected to change with $p_T$, M, $x_F$ and s of the reaction. In Ref.14, specific forms are chosen for the distributions $f(|\vec{k}_T|)$ in Eq.(6), and explicit results are presented for the variation of $\alpha_t$ with $x_F$ and M. For $M > 5$ GeV, it is found that $\langle \alpha_t \rangle \gtrsim 0.8$ for all $x_F$; here the average is taken over $p_T$. The modification of $\langle \alpha_t \rangle$ due to $k_T$ smearing is not great for large enough dilepton masses. In Fig.6 I show the expected variation of $\alpha_t$ with $p_T$, for $M = 5.5$ GeV and $x_F = 0$. These

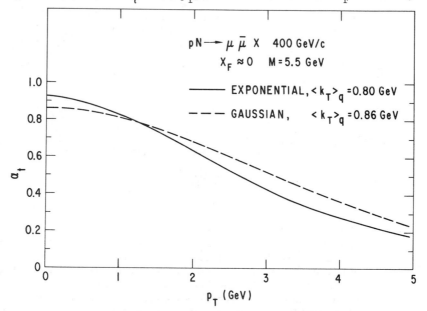

Fig.6. Predicted $p_T$ dependence of the coefficient $\alpha_t$ in the t channel angular distribution $d\sigma/d\Omega \propto [1 + \alpha_t \cos \theta_t]$ for lepton pairs of mass $M \simeq 5.5$ GeV produced in $pN \rightarrow \ell^+\ell^-X$ at 400 GeV/c and $y = 0$; $p_T$ is the transverse momentum of the $\ell^+\ell^-$ pair. The two models used are described in the text.

results are of both theoretical and practical interest. When sufficient data are available, they will permit another non-trivial check
of the Drell-Yan mechanism. At the moment the experimental acceptance
is restricted to a small region in $\theta_t^*$ about $\cos\theta_t^*=0$. The theoretical
results may therefore be useful now in estimating corrections to the
data for the limited acceptance. In Ref.6, the assumption is made
that $\alpha_t = 1$ for all M and $p_T$. The curves in Ref.14 suggest that this
assumption overestimates the cross-section $d\sigma/dMdy$ at $y = 0$ and $M = 5$
GeV by about a factor of ~1.07 relative to that at $M > 10$ GeV. Likewise, Fig.6 suggests that the assumption leads to an overestimate of
the experimental cross-section at $M = 5.5$ GeV and $p_T \simeq 4$ GeV, relative
to that at small $p_T$, by a factor of $\simeq 1.2$. While this latter error
leads to slight overestimates of $\langle p_T \rangle$ and $\langle p_T^2 \rangle$ in the data, the effect
is not substantial.

　　　　6.　Absolute Normalization. While predictions for the absolute
dilepton yields are perhaps the most interesting for experimental comparisons, and for estimates of $W^{\pm}$ rates, they are very sensitive to
model dependent assumptions about, for example, the function $\bar{q}(x)$ in
Eq.(1). To be specific, suppose we consider the observable $d\sigma/dMdy$
at $y = 0$ for $pN \rightarrow \mu\bar{\mu}X$ at 400 GeV/c. At $y = 0$, $x_1 = x_2 \equiv x = M/\sqrt{s}$, and $\sigma \propto$
$\langle q(x)\bar{q}(x) \rangle$. (For the moment I continue to ignore possible $Q^2$ depend-
ence). Above the $J/\Psi$ region, where $M > 5$ GeV, the Columbia-FNAL-SUNY
group[6] provide data in the range $0.2 < \frac{M}{\sqrt{s}} \lesssim 0.5$. Unfortunately, $\bar{q}(x >$
$0.2)$ is essentially unknown. Gargamelle neutrino data[17] at low ener-
gies provide $\bar{q}(x)$ for $x \lesssim 0.2$, and theoretical extrapolations are necessary for estimates of $\bar{q}(x > 0.2)$. Various such extrapolations have
been made.[18,19] I think it is fair to say that none was successful in
predicting the dimuon rate observed by the Columbia-FNAL-SUNY group[6].
One example, due to Field and Feynman[18], is compared with the data in
Fig.7. The Field-Feynman expectation falls below the data by about a
factor of 3 at $M \sim 5$ GeV, but appears to meet the data for $M \gtrsim 10$ GeV.
A second ingredient in the prediction of the absolute cross-section
is the color factor of 3 in Eq.(1). If it were removed, the Field-
Feynman curve would agree with the data at $M \simeq 5$ GeV and exceed experiment at larger values of M. However, few theorists would seriously
suggest dropping the color factor. Owing to present uncertainty in
our knowledge of $\bar{q}(x)$ in the relevant x range, it is impossible to
"test" whether the color factor is correct in Eq.(1). This situation
may change in the next year or so when good determinations are available of $\bar{q}(x)$ from neutrino counter experiments. For the time being
it is more sensible to retain the color factor and to await improvements in $\bar{q}(x)$. It may be remarked that the classical Drell-Yan model
has done astonishingly well in coming within a mere factor of 3 of
the experimental rate. Indeed, the data themselves[6] are assigned a
systematic uncertainty of 15% and a separate overall normalization
uncertainty of 25%.

　　　　Given the sensitivity of the data to $\bar{q}(x)$, the problem can be
inverted, and the data on $pN \rightarrow \mu\bar{\mu}X$ used to determine an average $\bar{q}(x)$.
(The average here is over quark flavors and $Q^2$). This procedure re-
quires an independent determination of $q(x)$, from some other source,
since $\sigma_{\mu\bar{\mu}} \propto \langle q(x)\bar{q}(x) \rangle$. One method was chosen by the authors of Ref.6,
who find $\bar{q}(x) \simeq 0.6(1-x)^{10}$. I use a different procedure. I adopt
Field and Feynman's parametrization of the valence part of $q(x)$,

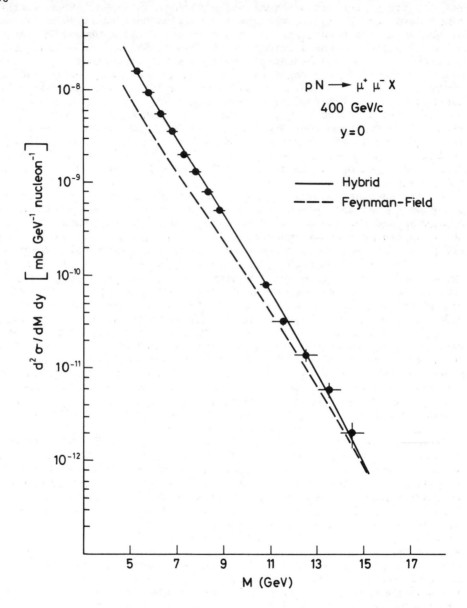

Fig. 7. Cross-section $d^2\sigma/dMdy$ for $pN \to \mu\bar{\mu}X$ at 400 GeV/c and y = 0 shown as a function of the lepton pair mass M. The data are from Ref. 6. The dashed curve is obtained from the Field-Feynman structure functions, Ref. 18. In the calculation, N = 60% neutrons and 40% protons. The solid curve is a fit to the data obtained from the hybrid model described in the text.

which they determine from deep-inelastic electron data. For each quark flavor, I write

$$q(x) = q_V^{FF}(x) + S(x) \qquad (9)$$

$$x\bar{q}(x) = xS(x) \equiv c_1(1-x)^{c_2} \qquad (10)$$

I determine that a good fit to the dimuon data is obtained if the average sea is parametrized as

$$xS(x) = 0.42(1-x)^9 \qquad (11)$$

This function is determined by data in the range $0.2 < x < 0.5$. My fit is shown in Fig. 7. I attribute no virtues to this "hybrid" model, but it does provide a set of quark and antiquark distribution functions which I need for calculations reported in Sec. V. The complete details of my parametrization are provided in Table 1. The

TABLE 1

A. Field and Feynman Model

$$xf(x) = g(x) \sum_{k=0}^{N} (a_k + \sqrt{x}\,b_k) C_k(x)$$

$$C_k(x) = \cos\left(k \cos^{-1}(2x-1)\right)$$

| $xf(x)$ | $xu(x)$ | $xd(x)$ | $xs(x)$ | $x\bar{u}(x)$ | $x\bar{d}(x)$ | $x\bar{s}(x)$ |
|---|---|---|---|---|---|---|
| $g(x)$ | $(1-x)^3$ | $(1-x)^4$ | $(1-x)^8$ | $(1-x)^{10}$ | $(1-x)^7$ | $(1-x)^8$ |
| $a_0$ | 161.579 | -3.175 | 0.10 | 0.17 | 0.17 | 0.10 |
| $a_1$ | 225.327 | -2.937 | 0.0 | 0.0 | 0.0 | 0.0 |
| $a_2$ | 70.699 | 1.082 | 0.0 | 0.0 | 0.0 | 0.0 |
| $a_3$ | 6.761 | 0.674 | 0.0 | 0.0 | 0.0 | 0.0 |
| $b_0$ | -177.909 | 5.607 | 0.0 | 0.0 | 0.0 | 0.0 |
| $b_1$ | -230.510 | 2.6340 | 0.0 | 0.0 | 0.0 | 0.0 |
| $b_2$ | -52.427 | -2.288 | 0.0 | 0.0 | 0.0 | 0.0 |
| $b_3$ | -1.371 | -0.247 | 0.0 | 0.0 | 0.0 | 0.0 |

B. Hybrid Model

Derived from the Field-Feynman Model by the substitutions (for each species: u, d, s)

$$x\bar{q}(x) = 0.42(1-x)^9$$

$$xq(x) = xq_{FF}(x) - x\bar{q}_{FF}(x) + 0.42(1-x)^9$$

parametrization satisfies various desirable sum rules.

Several reasons may be advanced for the differences between my sea distribution and that of Field and Feynman who use $x\bar{u}(x) = 0.17$ $(1-x)^{10}$ and $x\bar{d}(x) = 0.17(1-x)^7$. First, as remarked above, the Field-Feynman choices are pinned to data at $x < 0.2$, whereas my expression fits (different) data for $x > 0.2$. It is easy to concoct a form for $xS(x)$ which has the "Gargamelle value" 0.17 at $x = 0$, chosen by Field and Feynman, but which yields my expression in the range $x > 0.2$. This procedure is tantamount to suggesting that the Gargamelle and Drell-Yan sea distributions are not really different, but that they can be made to merge into one another if sufficient flexibility is adopted in parametrizing the function $xS(x)$. On the other hand, the difference can be viewed instead as a real physics difference associated with $Q^2$ dependence. This is the more popular theoretical interpretation. The Gargamelle data are confined to values of $|Q^2| <$ 2 GeV$^2$, whereas in the dimuon data $25 < Q^2 < 150$ GeV$^2$. If $\bar{q}(x)$ is replaced by the scaling violating form $\bar{q}(x, Q^2)$, then there is no reason to suppose that the Gargamelle $\bar{q}(x, Q^2)$ should apply for values of $Q^2$ more than an order of magnitude larger. It is more relevant then to compare the average $\bar{q}(x)$ extracted from the dimuon data with anti-quark distributions deduced from very high energy neutrino experiments. Recent BEBC data[20] are in fact consistent with my Eq.(11). I will return to scaling violations in Sec. IV.

7. Distributions in $x_F$ and y. The expected rapidity y and Feynman $x_F$ dependences of $d\sigma/dMdy$ and $d\sigma/dMdx_F$ are straightforward predictions of the classical Drell-Yan model once specific forms are chosen for $q(x)$ and $\bar{q}(x)$. I've discussed uncertainties associated with $\bar{q}(x)$ above; they are reflected in expectations for the y and $x_F$ variations of the dilepton yield.

## IV. CRITIQUE OF THE CLASSICAL MODEL

In Sec. III, I surveyed some of the successes of the classical Drell-Yan model. Rather than continue in that vein, I think it is appropriate to discuss the justification for the model and to ask what modifications or reinterpretations are necessary in the light of other recent experimental and theoretical developments. I mentioned that the "large" values observed for $\langle p_T \rangle$ in $pN \rightarrow \mu\bar{\mu}X$, and the observed energy dependence of $\langle p_T \rangle$, require modifications of the classical model. I also referred to the scaling violations which appear[21] to have been observed in high energy deep inelastic muon scattering $\mu p \rightarrow \mu'X$, and by the BEBC collaboration[20] in $\nu p \rightarrow \mu X$. These data suggest that the structure functions $q(x)$ and $\bar{q}(x)$ in Eq.(1) may have to be replaced by functions with explicit $Q^2$ dependence, which may or may not be identical to those determined in deep inelastic reactions at $Q^2 < 0$. Moreover, we may also ask what is special about the $q\bar{q}$ annihilation diagram in Fig.1. Why not calculate and include other contributions, for example, from the graphs shown in Fig.4?

To first order in the strong coupling constant $\alpha_s$, the constituent scattering portions of Fig.4 are provided by the (two body final state) "quark exchange" Compton and annihilation amplitudes shown in Fig.8. Higher order graphs may also be drawn. Those in

COMPTON

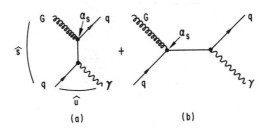

(a)          (b)

ANNIHILATION

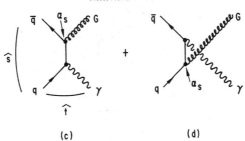

(c)          (d)

Fig.8.   First order processes in the strong coupling constant
$\alpha_s$.   Diagrams (a) and (b) represent quark-gluon Compton
scattering to yield a quark and a virtual photon.   Diagrams (c) and (d) represent quark-antiquark annihilation
into a gluon and a virtual photon.

Fig.9, with three particle final states (q,q,$\gamma$), are typical of
graphs of order $(\alpha_s)^2$.   [In Figs.8 and 9, the dilepton (not drawn)
emerges, as always, from the decay of the virtual photon].   Although
$\alpha_s \propto 1/\log Q^2$,   the cross section corresponding to the graphs in Fig.8
is proportional to $\alpha_s \log Q^2$, and the graphs in Fig.9 provide contributions proportional to $(\alpha_s \log Q^2)^2$.   Therefore, the higher order
terms are not necessarily small with respect to the simple annihilation graph in Fig.1.   They may also provide different $x_F$ and M dependences.   Noting that the graph in Fig.9(a) involves the scattering
of valence quark constituents, we may wonder why it is not the dominant contribution, especially in pN collisions where the annihilation
diagram in Fig.1 feeds on the relatively small antiquark sea.   The
process sketched in Fig.9(a), where the final photon may be joined
to any of the four quark lines, was in fact proposed to explain high
mass dilepton production by Berman, Levy, and Neff[22] at about the
same time as the original Drell and Yan proposal.

   If the various diagrams sketched and suggested in Figs.8 and 9
must be computed separately then, at the very least, the "model"
becomes cumbersome and may lose considerable predictive power.   The
amplitudes of different orders of $\alpha_s$ should be added coherently, be-

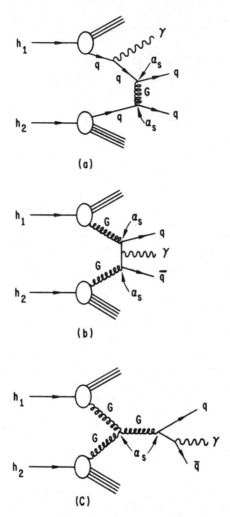

Fig.9.   Some second order processes which contribute to the
         production of massive virtual photons.

fore cross-sections are computed.   Otherwise, errors of "double-
counting" are committed.   The latter difficulty can be avoided only
at the price of a different or additional ambiguity; one may try to
compare different constituent scattering terms with data in different regions of phase space where the amplitudes have negligible overlap (e.g. try to separate the two jet, three jet, four jet, etc. contributions).[12]

Fortunately, there is growing support among theorists for the
conjecture[23] that the three problems mentioned above (scaling violations, large $\langle p_T \rangle$, and additional diagrams) are all part of the same
story, and that they may be resolved together.   Stated in oversim-
plified fashion, the idea is that the full series of constituent

scattering diagrams, to all orders in $\alpha_s$, generates $Q^2$ dependence of the structure functions, as in deep inelastic scattering. Thus, the cross-section for lepton pair production in hadron collisions, $d\sigma/dMdy$, may be computed from the simple annihilation graph in Fig.1(a), provided that $Q^2$ dependent structure functions are used. Moreover, these $Q^2$ dependent structure functions are identically those determined in deep inelastic electron, muon, and neutrino scattering at the same $|Q^2|$. In this fashion, for $d\sigma/dMdy$, effects of the higher order graphs are automatically included. When the experimentally extracted $Q^2$ dependent quark and antiquark distribution functions are used in Eq.(1), constituent subprocesses such as those sketched in Figs.8 and 9 should not be calculated independently, as they are already included.

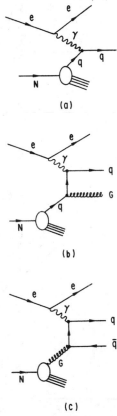

(a)

(b)

(c)

Fig.10.    Diagrams in deep inelastic electron scattering.
a) Classical quark model diagram in which an exchanged photon scatters from a quark constituent in the hadron N.   b) The first order contribution in which $\gamma q \rightarrow qG$, where G is a gluon.   c) The first order contribution in which the exchanged photon scatters from a gluon in the target to yield quark and antiquark systems in the final state.

The original papers[24] should be consulted for a full discussion
of scaling violations in deep inelastic processes and their interpretation in terms of asymptotically free QCD. I limit myself to a
few qualitative comments. In deep inelastic scattering, the quark
distribution functions represent not only the naive quark model contribution sketched in Fig.10(a), but include also other effects to
all orders in the strong coupling constant $\alpha_s$. The "extra" contributions to first order in $\alpha_s$ are illustrated in Figs.10(b) and 10(c).

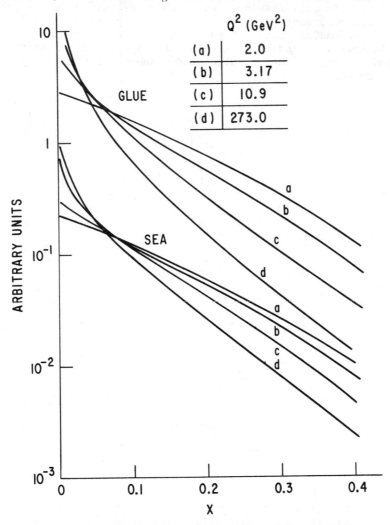

| | $Q^2$ (GeV$^2$) |
|-----|-----|
| (a) | 2.0 |
| (b) | 3.17 |
| (c) | 10.9 |
| (d) | 273.0 |

Fig. 11.   The x dependence of the gluon and sea parton distributions in a proton expected for different values of
$Q^2$ according to QCD. This figure is adapted from
Hinchliffe and Llewellyn Smith, Ref.7, with the scale
parameter $\Lambda^2 = 0.5$ GeV$^2$.

In Fig.10(b), the quark "first" radiates a gluon before scattering from the $Q^2 < 0$ exchanged photon and in 10(c), a gluon constituent dissociates into a quark-antiquark pair, one of which then scatters from the photon. The first order graphs in Figs.10(b) and 10(c), and those in higher order in $\alpha_s$, are understood to generate $Q^2$ dependence of the structure functions $q(x)$. If the quark, antiquark, and gluon distributions are supplied (as functions of x) at one initial starting value $Q^2 = Q_0^2$, then the renormalization group equations of the theory provide the x distributions at higher values of $|Q^2|$. These x distributions generally change with $Q^2$. For example, as $|Q^2|$ grows, the valence quark distributions are predicted to become more sharply peaked toward $x = 0$. The sea distributions are expected to increase in magnitude at $x = 0$, but to fall off more sharply with increasing x. This behavior of the sea is illustrated in Fig.11. The pattern and size of the predicted $Q^2$ dependence (scaling violations) agree qualitatively with experiment.[24]

The obvious similarity between the graphs in Figs.4 and 8, and those in Figs.10(b) and (c), encourages the conjecture mentioned above that the QCD $Q^2$ dependent corrections to the structure functions are the same in both deep inelastic processes and in lepton pair production reactions. However, the mathematical techniques available in the $Q^2 < 0$ deep-inelastic regime are inapplicable in the $Q^2 > 0$ domain of lepton pair production. A check of the conjecture must be made in perturbation theory, order by order. It is not obvious that the necessary factorization can be demonstrated whereby a Drell-Yan type annihilation formula will result in each order of $\alpha_s$. After a computation is made of the QCD diagrams, such as shown in Figs.8 and 9, it is necessary that the answer have the appearance of a product of terms, each associated with one of the initial hadrons, and that no leading terms appear involving the sum $(p_1 + p_2)^2$ of the initial hadron momenta.

The quark-gluon diagrams of Fig.8(a) and (b) provide a leading contribution to the lepton pair cross-section having the form[23]

$$\sigma_0 \left(1 - 2x_1(1-x_1)\right) \alpha_s \log (Q^2/p_1^2)$$

where $p_1^2$ is the four momentum of the initial gluon, and $\sigma_0$ is the classical Drell-Yan cross-section. The term multiplying $\sigma_0$ above is exactly the first term of the series for $P_{\bar{q}/G}(x,Q^2)$ representing the antiquark content of the gluon, as measured in the deep inelastic process in Fig.10(c). Thus, to this order in $\alpha_s$, the contributions of Fig.8(a) and (b) are already included in the annihilation process of Fig.1 if in Eq.(1) we make the replacement

$$\bar{q}(x) \rightarrow \bar{q}(x) + P_{\bar{q}/G}(x,Q^2), \tag{12}$$

or, more generally,

$$\bar{q}(x) \rightarrow \bar{q}(x,Q^2) . \tag{13}$$

The gluon-gluon graphs shown in Figs. 9(b) and 9(c) provide a leading contribution to the lepton pair cross-section having the form

$$\sigma_0 \left(1 - 2x_1(1-x_1)\right)\left(1 - 2x_2(1-x_2)\right)\alpha_s^2 \log(Q^2/p_1^2) \log(Q^2/p_2^2) . \qquad (14)$$

This expression manifests the necessary factorization, and is exactly the order $\alpha_s^2$ term in the product

$$P_{q/G}(x_1, Q^2) P_{\bar{q}/G}(x_2, Q^2) . \qquad (15)$$

In summary, the conjecture is that the Drell-Yan quark-antiquark annihilation formula is fully justified in a QCD framework, and that it includes in principle the sum of QCD graphs to all orders in $\alpha_s$, provided that $Q^2$ dependent structure functions are used in Eq. (1). Moreover, these structure functions are identically those extracted from deep-inelastic processes (with a trivial change of the sign of $Q^2$). Thus far, this important conjecture has been verified in perturbation theory only to order $(\alpha_s)^2$ and, then, only for the leading logarithmically divergent contribution in each order. [Conceivably the "non-leading" contributions in each order in $\alpha_s$ are different in deep-inelastic and in massive lepton pair production processes. However, estimates given below in Sec. V suggest that the non-leading terms in order $\alpha_s$ are negligible].

It is of substantial interest to check the above conjecture experimentally. This requires data from deep inelastic processes of sufficient precision to allow extraction of the structure functions $q(x, Q^2)$ and $\bar{q}(x, Q^2)$ at the same $|Q^2| > 25$ GeV$^2$ and $x = M/\sqrt{s}$ values at which data are taken in lepton pair experiments. For the time being, the conjecture instructs us to regard structure functions extracted from lepton pair data, as in Sec. III.6, as effectively as $Q^2$ dependent. Thus, the average sea distribution $0.42(1-x)^9$ in Sec. III.6 is one appropriate in the range $5 \lesssim Q \lesssim 12$ GeV. That this sea distribution has a greater intensity at $x = 0$ than the lower $Q^2$ Gargamelle sea is consistent with the QCD expectations illustrated in Fig. 11.

Returning to the three problems mentioned at the start of this section, we see that $Q^2$ dependent structure functions should indeed be used in the Drell-Yan annihilation formula. They should be the same ones measured in deep-inelastic reactions. Second, the various constituent scattering graphs drawn in Figs. 8 and 9 are not neglected. They generate the $Q^2$ dependence of structure functions, and they are automatically included in the annihilation term of Fig. 1 when $Q^2$ dependent structure functions are used. Finally, a unique prediction of QCD graphs, such as those drawn in Figs. 8 and 9, is that they generate relatively large transverse momenta. This leads to an answer to our third problem, as discussed in the next section, and relates the size of scaling violations to $\langle p_T \rangle$.

## V.  TRANSVERSE MOMENTUM DISTRIBUTIONS

In Section III.4 I discussed briefly the available data on trans-

verse momentum distributions and introduced possible interpretations.
Two contributions to the transverse momentum ($p_T$) of the lepton pairs
may be identified. I label one of these components the "confinement"
piece. Because the quark, antiquark, and gluon constituents are
confined in an initial hadron of finite size, they have some distri-
bution in their transverse momenta $k_T$, with $\langle k_T \rangle \approx 600$ MeV, as discussed in Sec. III.4. When these distributions in $k_T$ for the quark
and antiquark are introduced into Eq.(1), and convoluted as in Ref.14,
the lepton pairs are produced with non-zero $\langle p_T \rangle$, as shown in Figs.3

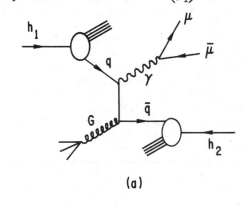

(a)

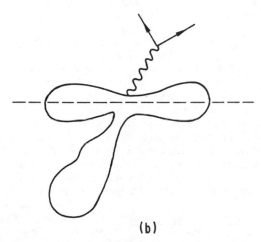

(b)

Fig.12   In part (a), the process is shown in which $h_1 h_2 \to \mu \bar{\mu} X$
via the constituent scattering process $q\bar{q} \to \gamma G$. The
transverse momentum of the $\gamma$ is balanced by a gluon jet.
In (b) the c.of.m. distribution of particles is shown
in the final state. The dashed line indicates the
longitudinal axis of the original collision. The forward and backward jets represent debris from the incident hadrons $h_1$ and $h_2$. A jet of hadrons from the
gluon dissociation is roughly back-to-back with the
virtual photon.

and 5. The second component of the $p_T$ distribution is the "hard scattering" part. If the constituents scatter "before" the virtual photon is emitted, as in Figs. 4, 8, and 9, the photon emerges with relatively large $p_T$. In the hard-scattering approach, the bulk of the transverse momentum of the photon is balanced by a jet of hadrons from the recoiling quarks or gluons, as sketched in Fig. 12. (If asymptotically free QCD can also be shown to lead to confinement, then the two components I have distinguished are really not distinguishable. Since proof of confinement does not exist, I will assume that the confinement and hard scattering contributions are physically distinct. They may also be distinguished experimentally).

1. <u>Confinement</u>. In Sec. III, Fermi motion arguments were used to estimate that confinement provides $\langle k_T \rangle_q \simeq 600$ MeV. Similar reasoning would suggest that quarks and antiquarks in the sea have greater values of $\langle k_T \rangle$ than the valence component. This idea could be tested most directly by a comparison of $\langle p_T \rangle$ in $\bar{p}N$ and $pN$ reactions. Because the $\bar{p}N$ process is dominated by valence-valence annihilation, we should find $\langle p_T \rangle_{\bar{p}p} < \langle p_T \rangle_{pp}$. Unfortunately, it is unlikely that data will be available soon from $\bar{p}N \rightarrow \mu\bar{\mu}X$ with values of $M > 5$ GeV. The reaction $\pi N \rightarrow \mu\bar{\mu}X$ is also dominated by valence-valence annihilation at large enough values of M. However, comparisons with $pN$ reactions are difficult because confinement estimates are likely to be different for pion and proton systems. The data[2] indicate that $\langle p_T \rangle_{\pi N} > \langle p_T \rangle_{pp}$ at the same lepton pair masses and beam energies (at 225 GeV/c, $\langle p_T \rangle_{\pi N} \simeq 1.2$ GeV; at 200 GeV/c, $\langle p_T \rangle_{pp} \simeq 1.0$ GeV, both for $M > 4$ GeV). Another question concerns the possible x dependence of $\langle k_T \rangle$. This uncertainty translates into uncertainty about the expected M and $x_F$ dependences of $\langle p_T \rangle$. The simple factorized form chosen in Eq. (6) is surely not correct, but its "prediction" that $\langle p_T \rangle$ is independent of both M and $x_F$ does agree with, for example, the $\pi p \rightarrow \mu\bar{\mu}X$ data[2] at 225 GeV, where $\langle p_T \rangle$ is independent of $x_F$ over the large range $0 < x_F < 0.6$. It would be valuable to have specific predictions from confinement (bag) model calculations to compare with the data.

One point on which the confinement and hard scattering approaches differ is in their expectations for the s dependence of the $p_T$ distribution. If both $x_F$ and $M/\sqrt{s}$ are fixed, then the quark and antiquark longitudinal fractions $x_1$ and $x_2$ are fixed in Eq. (1). Under these conditions, confinement models should predict that $\langle p_T \rangle$ is independent of s. This is not true in the hard scattering approach, as described below, and it appears not to be true in the data either[1]; c.f. Fig. 13.

2. <u>Hard Scattering Component</u>.[12,13] Because QCD is not a soft field theory, there is no cutoff in the model, and $\langle p_T \rangle$ increases in unbounded fashion with whatever momentum variable sets the dynamical scale. It is expected, therefore, that $\langle p_T^2 \rangle$ should have the form[13]

$$\langle p_T^2 \rangle = a + b\, M^2 / \log (M^2/\Lambda^2) \ . \tag{16}$$

Here a and b are functions of the dimensionless ratio $M/\sqrt{s}$. The function a is the confinement contribution, and the term $b\, M^2/\log (M^2/\Lambda^2)$ represents the QCD expectation. The scale parameter $\Lambda$ of the theory is in the range 0.5 to 1 GeV. The full M dependence of Eq.(16) requires knowledge of $b(M/\sqrt{s})$ which may be calculated explicitly from QCD diagrams, as shown below. Since $\langle p_T^2 \rangle$ is observed[1,2] to be nearly independent of M in the range $5 < M < 12$ GeV, b must be roughly of the form $(s/M^2)$ in this range of M.

Based on his QCD calculations, Politzer[13] proposed that quarks and antiquarks be assigned the mean transverse momentum

$$\langle k_T^2 \rangle_q \simeq 0.09 + \frac{M^2(1-x)}{16 \log (M^2/\Lambda^2)} , \tag{17}$$

with $\Lambda = 0.5$ GeV. At $x_F = 0$, $x = M/\sqrt{s}$, and, therefore, Politzer's approximation for b in Eq.(16) is

$$b = \frac{1}{8}(1 - M/\sqrt{s}) . \tag{18}$$

At 400 GeV/c and $x_F = 0$, this prediction provides a curve for $\langle p_T^2 \rangle$ versus M which rises almost linearly from 0.7 GeV$^2$ at $M = 5$ GeV to 1.9 GeV$^2$ at $M = 13$ GeV, in clear disagreement with the data shown in Fig.3(b). Politzer's approximation was based on an analytic approximation to the QCD graphs, valid only in the limit $x \to 1$. Since the data lie in the range $0.2 < x < 0.5$, the disagreement is not surprising. As described below, a complete numerical study of the same graphs leads to more satisfactory agreement with experiment.

Beginning with Eq.(16) and dropping the slowly varying $\log M^2$ factor, we may deduce that

$$\langle p_T^2 \rangle = a + \tilde{b} s , \tag{19}$$

where $\tilde{b} \equiv M^2 b/s$ is a new function of $M/\sqrt{s}$. We conclude that at fixed $M/\sqrt{s}$ the QCD portion of $\langle p_T^2 \rangle$ is predicted to grow linearly with s. This may be contrasted with the expectation of a constant $\langle p_T^2 \rangle$ from confinement.

Data available on the energy dependence of $\langle p_T \rangle$ in lepton pair production at $y \simeq 0$ are shown in Fig.13. The Fermilab data[1] from 200 to 400 GeV/c show the rise expected in the hard scattering approach. The ISR data[5] are taken at a different value of $M/\sqrt{s}$, and a direct comparison with the Fermilab results is therefore not possible. However, a rather large variation in the $M/\sqrt{s}$ dependence of $\tilde{b}$ in Eq.(19) would be required to accommodate both the FNAL and ISR results. It will be interesting to see whether the values of $\langle p_T \rangle$ observed at the ISR increase when greater statistics are accumulated.

3. <u>Explicit Calculations</u>. I turn now to an explicit calculation of the contributions of the hard-scattering graphs shown in Figs.4 and 8. Graphs similar to those in Fig.8 give rise to three-jet events in $e^+e^- \to$ hadrons and in deep inelastic processes such as $\mu p \to \mu' X$. Here I am interested in the contribution which the graphs make to the $p_T$ distribution of dileptons in $h_1 h_2 \to \ell^+ \ell^- X$.

After a sum over the spins of the final quark and of the two

leptons, an integral over the phase space of the lepton pair, and
an average over the initial quark and gluon spins, the absolute
square of the sum of the Compton scattering amplitudes in Figs.8(a)
and (b) is found to be[25]

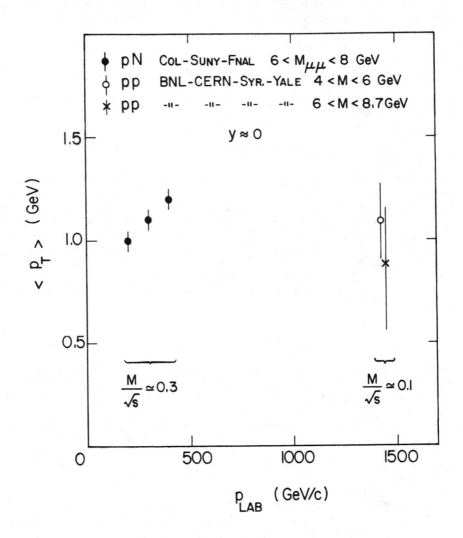

Fig.13.   ISR and Fermilab data on $\langle p_T \rangle$ are shown as a function
of lab momentum.  The FNAL data (from Ref.1) are from
the reaction $pN \to \mu\bar{\mu}X$ and are an average over the range
$0.2 < M/\sqrt{s} < 0.4$.   The ISR data (Ref.5) are from $pp \to$
$e^+e^-X$ and correspond to $M/\sqrt{s} \simeq 0.1$.

$$|A_C^i|^2 \equiv \frac{1}{4\pi} \int \overline{\sum_{\substack{spins, \\ color}}} |M_{COMPTON}|^2 \, d\Omega_{\ell^+\ell^-}$$

$$= \frac{4}{9} c_{qi}^2 (4\pi)^3 \alpha^2 \alpha_s \left[ \frac{-2M^4 + 2M^2(\hat{u} + \hat{s}) - (\hat{u}^2 + \hat{s}^2)}{\hat{s}\,\hat{u}\,M^2} \right]$$

(20)

Included in Eq. (20) is the appropriate factor (1/6) for the sum and average over color indices. The fractional charge $|c_{qi}|$ of the quark has the values $(\frac{2}{3},\frac{1}{3},\frac{1}{3})$ for the $i = (u,d,s)$ quarks. The variables $\hat{s}$ and $\hat{u}$ are indicated in Fig. 8(a): $\hat{s} = (p_q + p_G)^2$, and $\hat{u} = (p_\gamma - p_q)^2$. I assume that the quarks and the gluon are massless. The constant $\alpha = 1/137$, and in this report I fix $\alpha_s = 0.3$, independent of $Q^2$. My results are not changed in any significant way if I instead choose $\alpha_s \propto 1/\log(Q^2/\Lambda^2)$.

After a sum over final spins, an average over initial spins, and an integration over the phase space of the lepton pair, the absolute square of the sum of the annihilation amplitudes in Figs. 8(c) and 8(d) yields

$$|A_A^i|^2 = \frac{1}{4\pi} \int \overline{\sum_{\substack{spins, \\ color}}} |M_{Annih.}|^2 \, d\Omega_{\ell^+\ell^-}$$

$$= \frac{32}{27} c_{qi}^2 (4\pi)^3 \alpha^2 \alpha_s \left[ \frac{-2M^4 + 2M^2(\hat{t} + \hat{u}) - (\hat{u}^2 + \hat{t}^2)}{-\hat{t}\,\hat{u}\,M^2} \right].$$

(21)

Again, the appropriate factor (4/9) for the sum over color indices is included in Eq. (21). The variables $\hat{t}$ and $\hat{s}$ are indicated in Fig. 8(c); $\hat{s} + \hat{t} + \hat{u} = M^2$. Note that the definitions of $\hat{s}$, $\hat{t}$, and $\hat{u}$ differ in Figs. 8(a) and 8(c). The cross-section $d\sigma/dM^2 d\hat{u}$ for the process $q_i G \to (\ell^+\ell^-)q_i$ is obtained directly from Eq. (20) as

$$\frac{d^2\sigma_i^C}{dM^2 d\hat{u}} = \left(\frac{1}{2\pi}\right)^4 \frac{\pi}{16\hat{s}^2} |A_C^i|^2 .$$

(22)

Likewise, the cross-section for the process $q_i q_i \to (\ell^+\ell^-)G$ is obtained directly from Eq. (21) as

$$\frac{d^2\sigma_i^A}{dM^2 d\hat{t}} = \left(\frac{1}{2\pi}\right)^4 \frac{\pi}{16\hat{s}^2} |A_A^i|^2$$

(23)

Quark, lepton, and gluon masses have all be neglected in Eqs. (22)

and (23).

To find the contribution of the Compton graphs to $h_1 h_2 \rightarrow \ell^+ \ell^- X$, the expression in Eq.(22) must be multiplied by the probabilities that quark and gluon constituents in the initial hadrons carry longitudinal momentum fractions $x_1$ and $x_2$, respectively, and then an integral performed over inessential variables. I find

$$d\sigma(h_1 h_2 \rightarrow \ell^+ \ell^- X) = \int d^2 k_{T1} dx_1 d^2 k_{T2} dx_2 P_{qi/h1}(x_1, \vec{k}_{T1}) \tag{24}$$

$$P_{G/h2}(x_2, \vec{k}_{T2}) \left( \frac{d^2 \sigma_i^C}{dM^2 d\hat{u}} \right) dM^2 d\hat{u} + (1 \leftrightarrow 2) .$$

In discussing the hard scattering contribution in this report, I shall ignore the transverse momenta associated with confinement. I set $P_{q/h1} = q(x_1) \delta(\vec{k}_{T1})$ and $P_{G/h2} = G(x_2) \delta(\vec{k}_{T2})$. Smearing effects in the $p_{\vec{T}}$ spectra associated with the finite values of $\langle k_{T1} \rangle$ and $\langle k_{T2} \rangle$ are therefore ignored here. The gluon probability $G(x_2)$ is specified below.

Transforming variables in the integrand of Eq.(24), I find that

$$dx_1 dx_2 d\hat{u} = \frac{x_1 x_2 dx_F dp_T^2 dx_q}{\left[ x_F^2 + \dfrac{M^2 + p_T^2}{P^2} \right]^{\frac{1}{2}} \left[ x_q^2 + \dfrac{p_T^2}{P^2} \right]^{\frac{1}{2}}} . \tag{25}$$

Here $P$ is the c.of.m. momentum of the overall collision ($P \approx \sqrt{s}/2$), and $x_g$ is the fraction of longitudinal momentum carried by the <u>final</u> unobserved quark. The Eq.(25) may also be written as

$$dx_1 dx_2 d\hat{u} = \frac{x_1 x_2 dy dp_T^2 dx_q}{(x_q^2 + p_T^2/P^2)^{\frac{1}{2}}} \tag{26}$$

Combining Eqs.(20),(22), (24), and (26), I obtain the following contribution from the Compton graphs to $h_1 h_2 \rightarrow \ell^+ \ell^- X$

$$\frac{d\sigma^C}{dM^2 dp_T^2 dy} = \sum_{i=1}^{3} \int \frac{dx_q}{\left[ x_q^2 + p_T^2/P^2 \right]^{\frac{1}{2}}} x_1 \left[ q_i(x_1) + \bar{q}_i(x_1) \right] x_2 G(x_2) \frac{d^2 \sigma_i^C}{dM^2 d\hat{u}} \tag{27}$$

$$+ (1 \leftrightarrow 2)$$

In Eq.(27), M, $p_T$, and y are the mass, transverse momentum, and rapidity of the pair of leptons. Note that in Eq.(27) I include the contributions of both $\bar{q}G \rightarrow \gamma\bar{q}$ and $qG \rightarrow \gamma q$. The first term of the equation represents the process in which the quark or antiquark emanates from hadron 1 and the gluon from hadron 2. These roles are interchanged in the term $(1 \leftrightarrow 2)$. The variables $\hat{s} = x_1 x_2 s$ and $\hat{u}$ in the explicit expression for $d^2\sigma/dM^2 d\hat{u}$ in Eq. (27) may be reexpressed easily in terms of my chosen set $M^2$, $p_T^2$, $y$ and $x_q$. This transformation differs slightly for the second term $(1 \leftrightarrow 2)$. The integral over $x_q$ in Eq. (27) runs over both positive and negative values of $x_q$. With some care, this integration can be handled well numerically.

For the annihilation process, in which $h_1 h_2 \to \ell^+ \ell^- X$ via the process $q\bar{q} \to G\ell^+\ell^-$, I derive

$$\frac{d\sigma^A}{dM^2 dp_T^2 dy} = \sum_{i=1}^{3} \int \frac{dx_q}{\left[x_q^2 + p_T^2/P^2\right]^{\frac{1}{2}}} \; x_1 q_i(x_1) x_2 \bar{q}_i(x_2) \; \frac{d^2\sigma_i^A}{dM^2 d\hat{t}} \tag{28}$$

$$+ (1 \leftrightarrow 2)$$

The first term of Eq. (28) represents the process in which the quark emerges from hadron 1 and antiquark from hadron 2. These roles are reversed in the term $(1 \leftrightarrow 2)$.

Results in millibarn units are obtained after the factor 0.3893 is inserted on the right hand side of Eqs. (27) and (28).

4. Scaling Properties. It is instructive to examine the behavior of Eqs. (27) and (28) as functions of $s$ and $p_T$. For the Compton scattering process, after introducing the c.of.m. scattering angle $\theta$ between the initial quark and final virtual photon, I reexpress

$$\hat{u} = -\frac{1}{2} (\hat{s} - M^2)(1 - \cos\theta), \tag{29}$$

and

$$p_T = \frac{1}{2\sqrt{s}} (\hat{s} - M^2) \sin\theta . \tag{30}$$

The scaled transverse momentum $x_T$ is defined as

$$x_T = 2p_T/\sqrt{s} . \tag{31}$$

Note that $\sin\theta$ [or $\cos\theta$] is a function of the scaling variables $x_T$, $M/\sqrt{s}$, and of $x_1$ and $x_2$. Rewriting Eq. (29), I obtain

$$\hat{u} = -p_T^2 \; \frac{2\hat{s}}{(\hat{s} - M^2)(1 + \cos\theta)} . \tag{32}$$

The Compton scattering matrix element, Eq. (20) may therefore be expressed as

$$|A_C^i|^2 = \frac{1}{p_T^2} g_C(x_T, M/\sqrt{s}, x_1 x_2) , \tag{33}$$

where the function $g_C$ depends only on scaled quantities $x_T$, $M/\sqrt{s}$, $x_1$ and $x_2$. Likewise, the annihilation matrix element, Eq.(21), may be written as

$$|A_A^i|^2 = \frac{1}{p_T^2} g_A(x_T, M/\sqrt{s}, x_1 x_2) , \tag{34}$$

Analyzing the Eqs.(27) and (28) in similar fashion, I find that

$$\frac{d\sigma^C}{dM^2 dp_T^2 dy} = \frac{1}{s^2 p_T^2} f_C(x_T, M/\sqrt{s}, y) , \tag{35}$$

and

$$\frac{d\sigma^A}{dM^2 dp_T^2 dy} = \frac{1}{s^2 p_T^2} f_A(x_T, M/\sqrt{s}, y) . \tag{36}$$

The factor $s^{-2}$ in Eqs.(35) and (36) comes from the $\hat{s}^{-2}$ factor in Eqs.(22) and (23). The functions $f_C$ and $f_A$ depend only on $y$ and on the scaling variables $x_T$ and $M/\sqrt{s}$. Both $f_C$ and $f_A$ are regular as $p_T \to 0$.

The Eqs.(35) and (36) show that the Compton and annihilation cross-sections diverge as $p_T^{-2}$ as $p_T \to 0$. This (infra-red) behavior is evident in the explicit numerical results shown in Fig.14, and I will return to its ramifications below.

In the classical Drell-Yan model discussed in Secs.II and III, the scaling prediction is of the form:

$$\underline{\text{Classical Model Scaling}} \qquad s^2 \frac{d\sigma}{dM^2 dy} = f_0(M/\sqrt{s}, y) . \tag{37}$$

Presumably at fixed small $p_T$, in the region of transverse momentum where confinement effects control the $p_T$ spectrum, Eq.(37) is replaced by

$$s^2 \left. \frac{d\sigma}{dM^2 dy dp_T^2} \right|_{\text{small } p_T} = \tilde{f}_0(M/\sqrt{s}, y, p_T) . \tag{38}$$

By contrast, in the hard-scattering region, where the Compton and annihilation processes are dominant, the Eqs.(35) and (36) demonstrate a very different scaling form:

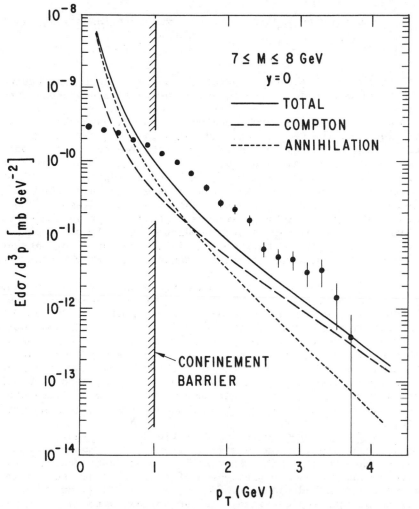

Fig.14. Data are shown from Ref.6 on the $p_T$ distribution of $pN \rightarrow \mu\bar{\mu}X$ at 400 GeV/c and y = 0 in the dimuon mass interval $7 \leq M \leq 8$ GeV. Shown also are calculations of the QCD expectations for this distribution. The theoretical curves are obtained from an evaluation of Eqs.(27) and (28) of the text. An integral was also performed over M to yield $Ed\sigma/d^3p$. The curves are normalized absolutely; $\alpha_s = 0.3$. The solid curve marked "total" is obtained from an incoherent addition of the Compton and annihilation contributions. Indicated by cross-hatching is the critical $p_T \simeq 1$ GeV below which the perturbation calculation is inapplicable, as discussed in Sec.V.5. The comparison of theory and experiment is similar for other values of M (not shown).

<u>Hard-Scattering Scaling</u>
$$s^3 \frac{d\sigma}{dM^2 dp_T^2 dy}\bigg|_{\text{large } p_T} = f_1(M/\sqrt{s}, y, x_T) \ . \quad (39)$$

The $s^2$ factor of Eq.(38) is replaced by $s^3$, and the $p_T$ dependence
on the right hand side of the equation enters as a dependence on
$x_T = 2p_T/\sqrt{s}$ .

The Eq.(39) is a general consequence of the hard-scattering
assumption, and its verification in the data for $p_T \gtrsim 1$ GeV is a
critical test of whether the hard-scattering mechanism is responsible for the "large" values of $\langle p_T \rangle$ seen in massive lepton pair production. In the "confinement" or "infra-red" region of $p_T \lesssim 1$ GeV,
the classical form Eq.(38) may hold. However, for $p_T > 1$ GeV the
hard-scattering expectation Eq.(39) should set in. Tests of the
hard-scattering prediction in massive lepton pair production should
be cleaner than in high-$p_T$ hadron production reactions because in
the lepton pair process the whole jet is always captured. Absent
are the complicated smearing effects associated with the quark decay
into hadrons.[26]

The different energy dependences represented by Eqs.(38) and
(39) are illustrated in Fig.15. I have plotted the explicit numerical results of my QCD calculations, which satisfy Eq.(39) perfectly,
as a function of $p_T$, for fixed $M/\sqrt{s} = 0.265$. If the QCD explanation
is correct, the cross-section $s^2 d\sigma/dM^2 dy dp_T^2$ in the reaction $pN \to \mu\bar{\mu}X$
at $M/\sqrt{s} = 0.265$ and $y \simeq 0$ should increase by a healthy factor of 3 at
$p_T = 2.5$ GeV when the lab momentum is increased from 200 to 400 GeV/c.
This dramatic prediction of the hard-scattering approach should be
verified soon. It is not subject to some of the ambiguities discussed below associated with QCD predictions for the moments $\langle p_T^2 \rangle$
and $\langle p_T \rangle$.

    5. <u>Infra-Red Divergence and Comparisons with Data</u>. I now
address two problems which beset all attempts to compare calculations of QCD processes with experiment. In QCD perturbation theory,
the quarks and gluons are treated as if they are free and can emerge
from their parent hadrons (c.f. Fig.4), whereas in Nature they
appear to be entirely confined. In comparisons with data, we must
deal somehow with the non-perturbative effects which provide or are
associated with confinement, or else seek tests of QCD which are
insensitive to the non-perturbative effects. As discussed earlier,
confinement effects are expected to be dominant at small values of
$p_T$, but they will also cause some smearing of the $p_T$ spectrum at
large $p_T$. Second, perturbative QCD is subject to infra-red divergences analogous to those which are present in QED. The infra-red
problem is manifested in Eqs.(35) and (36) by the $p_T^{-2}$ divergence of
the equations. When the momentum carried by the massless exchanged
quark in Figs.8(a) or 8(c) vanishes (i.e. when the quark goes on-
shell) the cross-section becomes infinite. These twin-problems of
confinement and infra-red divergence in QCD are not unique to the
lepton pair production process. They are also faced elsewhere, for
example, in calculations of jet effects in $e^+e^- \to$ hadrons.[27]

    One lesson of the solution of the infra-red problem in QED is that the divergences are mastered if we deal with cross-sections defined with suitable energy and angle cutoffs. This presumably will also be true in QCD.[28] When the virtual photon carries small $p_T$, events associated with the processes represented in Figs.4(a) and 4(b) cannot be distinguished experimentally from those due to Fig.1. At small $p_T$, the final quark in Fig.4(a), or the final gluon in Fig.4(b), is not outside the region of phase space populated by the

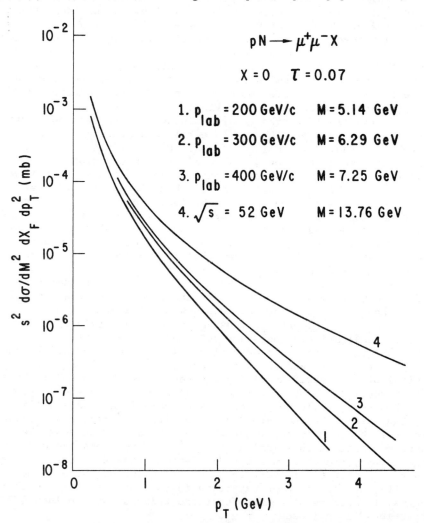

Fig.15.  Theoretical results on the energy dependence expected if QCD processes dominate the reaction $pN \to \mu\bar{\mu}X$. Curves are presented for three values of lab momentum in the Fermilab and CERN-SPS energy range and for one ISR momentum value. At all energies, $\tau = M^2/s = 0.07$, and $x = 0$.

constituents which have stayed behind and form the debris of the parent hadron $h_2$. The processes in Fig.1 and 4 are not incoherent at small $p_T$, and it would be improper to add cross-sections. Coherence effects are important, and one should deal with a sum of amplitudes — a problem of both infra-red and confinement complexity as yet beyond the reach of theorists.

The suitable "energy and angle cutoffs" in the problem of lepton pair production correspond to a selection of $p_T$ larger than some critical value. Below that value, cross-sections computed in QCD perturbation theory are inapplicable not only because of their infra-red divergences, but also because of the more serious (and related) neglect of coherence and confinement effects. When the virtual photon transverse momentum is above the critical $p_T$, the processes represented in Fig.4 should dominate over the rapidly decreasing small $p_T$ spectrum associated with Fig.1.

I add two final remarks before turning to a discussion of my numerical calculations. When an integral is done over $p_T^2$, the $p_T^{-2}$ divergence of Eqs.(35) and (36) at small $p_T$ gives rise to a logarithmically divergent $\alpha_s \log(Q^2/p^2)$ contribution to $d\sigma/dM^2 dy$. This is exactly the scaling violation contribution in first order in $\alpha_s$, discussed in Sec.IV. Second, in a somewhat different hard-scattering approach to massive-lepton pair production, Blankenbecler and collaborators[12] introduce a large quark mass (~1 GeV). They use diagrams similar to those in Fig.4, with the gluon replaced by a scalar meson. Owing to the large mass in the exchanged quark propagator, the $p_T^{-2}$ divergence of Eqs.(35) and (36) is avoided. While such a large quark mass is perhaps hard to motivate, the phenomenological result is to provide a $p_T$ distribution without a divergence near $p_T = 0$. The quark mass of 1 GeV plays the role in the Blankenbecler model of the ~1 GeV confinement cutoff which I use.

6. <u>Specific Parametrizations and Data</u>. To obtain specific numerical results it is necessary to choose expressions for the quark, antiquark, and gluon densities which appear in Eqs.(27) and (28). For the quark and antiquark densities I use the structure functions of the hybrid model I presented in Sec.III.6 and in Table 1. As discussed, these functions fit data in the range $M/\sqrt{s}$ from 0.2 to 0.5 GeV. While only functions of x, and thus of a scaling form, they are to be understood as effectively $Q^2$ dependent, valid in an average sense in the range $25 < Q^2 < 150$ GeV$^2$. Because the structure functions thus include some QCD "corrections" to order $\alpha_s$ and higher, the Compton and annihilation cross-sections I present are not purely of first order in $\alpha_s$ but also include some higher order effects. There is no way to avoid this situation since there is no way to measure structure functions which do not include QCD "corrections" to all orders in $\alpha_s$.

For the gluon density, G(x) in Eq.(27), I choose the parametrization

$$xG(x) = \frac{p+1}{2} (1-x)^p , \qquad (40)$$

where the power p is the "only free parameter" of my QCD calculations. The parametrization in Eq.(40) is normalized so that 50% of the nucleon's momentum is carried by gluons. To determine the power p, I require that the QCD model yield a variation of $\langle p_T^2 \rangle$ which is independent of lepton pair mass M in the range $5 < M < 10$ GeV, as is observed in the data.[6] This requirement fixes $p \simeq 5$ or 6 in Eq.(40). Higher (lower) powers of p result in a decreasing (rising) curve of $\langle p_T^2 \rangle$ vs. M. The calculation of $\langle p_T^2 \rangle$ is discussed below. For all results presented in this report I fix $p = 6$.

Results of my explicit evaluation of Eqs.(27) and (28) are compared with data in Fig.14. I show data only for the mass interval $7 \leq 8 \leq 9$ GeV, but the comparison is qualitatively similar in other regions of M. Evident in Fig.14 is the $p_T^{-2}$ divergence of the theoretical curves at small $p_T$. In the region $1 < p_T < 2$ GeV, the Compton and annihilation processes have comparable magnitudes; the Compton contribution is dominant at large $p_T$. The solid curve is obtained when the Compton and annihilation cross sections are added incoherently. As discussed in Sec.V.5, below a given critical $p_T$, which I expect to be about 1 GeV, the perturbation theory curves are inapplicable. In the region $p_T < 1$ GeV, confinement effects which are outside the scope of the theory should control the experimental distribution and also remove the infra-red divergence. In the region $p_T > 1$ GeV a comparison of theory and experiment should be meaningful. Two problems are obvious in Fig.14: the theory curves are a factor of two or more below the data in absolute normalization; the shape of the curves may be qualitatively incorrect, showing upward curvature instead of the downward trend of the data. In the calculation I assume that the initial constituents in the scattering carry no transverse momentum. It remains to be shown whether smearing effects in $p_T$ obtained by assigning non zero $\langle k_T \rangle$ to the initial quark, antiquark, and gluon constituents improve the agreement of theory and experiment significantly.

The energy dependence of the theoretical distribution is shown in Fig.15 at fixed $\tau = M^2/s = 0.07$. This value of $\tau$ is selected in order that the associated values of M be both accessible experimentally for laboratory momenta in the Fermilab and SPS energy ranges, and in the relevant continuum region of M between the $J/\psi$ and $\Upsilon$ families. As emphasized in Sec.V.4, the factor of three increase of $s^2 d\sigma/dx_F dM^2 dp_T^2$ at $p_T \simeq 2.5$ GeV predicted by the theory is a critical test of the hard-scattering assumption. In Fig.15 the solid curves result from addition of the Compton and annihilation cross-sections.

7. Moments. In the search for theoretical variables and distributions which are insensitive to the infra-red, confinement, and other non-perturbative problems of the small $p_T$ region, it is interesting to consider moments $\langle p_T^n \rangle$ of the $p_T$ distribution. Owing to the relatively large statistical errors in the data at large $p_T$, only the first few moments ($n = 1, 2$) are meaningful.

In a perturbation approach, it is expected that the cross-section $d\sigma/dM^2 dy$ has an expansion in $\alpha_s$ of the type

$$\sigma(M,y) \equiv \frac{d\sigma}{dM^2 dy} = \sigma_0(M,y) + \alpha_s \sigma_1(M,y) + \alpha_s^2 \sigma_2(M,y) + \ldots \qquad (41)$$

The first term in Eq. (41) represents the basic zero'th order classical Drell-Yan process of Fig. 1. We may also consider the integrals

$$\left\langle p_T^n \frac{d\sigma}{dM^2 dy dp_T^2} \right\rangle \equiv \int p_T^n dp_T^2 \left( \frac{d\sigma}{dM^2 dy dp_T^2} \right) . \tag{42}$$

As in Eq. (41), these have an expansion

$$\left\langle p_T^n \frac{d\sigma}{dM^2 dy dp_T^2} \right\rangle = \left\langle p_T^n \frac{d\sigma_0}{dM^2 dy dp_T^2} \right\rangle + \alpha_s \left\langle p_T^n \frac{d\sigma_1}{dM^2 dy dp_T^2} \right\rangle + O(\alpha_s^2) \tag{43}$$

In the approximation in which the hadron constituents carry no intrinsic transverse momentum, the first term on the right hand side of Eq. (43) is zero. The second term on the right hand side of Eqs. (41) and (43) is provided by the Compton and annihilation processes, Eqs. (27) and (28). In spite of the $p_T^{-2}$ divergence displayed in Eqs. (35) and (36), the second term on the right hand side of Eq. (43) is finite for all $n \geq 1$. Looking ahead to the possibility of adding later non-perturbative and confinement effects in quadrature, one is led to concentrate on the second moment, with $n = 2$. We may investigate the finite ratio $R_2$ defined by

$$R_2 = \frac{\alpha_s \left\langle p_T^2 \frac{d\sigma_1}{dM^2 dy dp_T^2} \right\rangle}{\sigma(M,y)} . \tag{44}$$

In this ratio, the denominator is the full cross-section given in Eq. (41); it is a measured quantity. Formally, we might consider replacing $\sigma(M,y)$ in the denominator of $R_2$ by the zero'th order term $\sigma_0(M,y)$ of Eq. (41). This would change the value of $R_2$ only to order $\alpha_s^2$. Thus, $R_2$ provides a fine definition of the first-order perturbative QCD contribution to the moment $\langle p_T^2 \rangle$.

Is $R_2$ a quantity which is insensitive to infra-red and non-perturbative problems? It is clear that there is no infra-red divergence in the calculation of $R_2$; the $p_T^{-2}$ behavior of the cross-section is exactly compensated by the $p_T^2$ insertion, and the integrand is well behaved for all $p_T$. However, it is not so easy to dispose of other non-perturbative problems. If the lower limit of integration in Eq. (42) extends to $p_T = 0$, as is usual in the definition of moments, then some contribution to the answer necessarily comes from the non-perturbative, confinement region of $p_T < p_{T,crit} \approx 1$ GeV where the perturbative calculation is inapplicable. One way to handle this problem would be to define incomplete moments in which the integral is done only for values of $p_T > p_{T,crit} \approx 1$ GeV. To make sense this would have to be done also for the cross-section

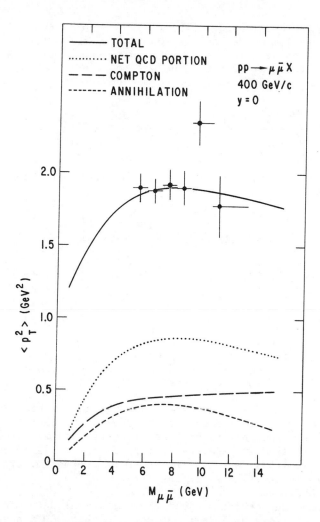

Fig. 16.   Shown are values of $\langle p_T^2 \rangle$ computed according to the
definition given in Eqs. (44) and (46) of the text.
The short dashed line illustrates the value obtained
from the annihilation process $q\bar{q} \to G\mu\bar{\mu}$, and the long
dashed line represents the contribution of the Compton
process $qG \to q\mu\bar{\mu}$.  The dotted line is the net QCD contribution to $\langle p_T^2 \rangle$ to first order in $\alpha_s$; it is obtained
by addition of the Compton and annihilation portions.
The solid line is obtained from the "net QCD" curve by
addition of the constant confinement contribution 1.04
$GeV^2$.  The data are from Ref. 6.  All curves are calculated for the process $pN \to \ell^+\ell^-X$ at $p_{lab} = 400$ GeV/c and
$y = 0$.  In this paper  N  is composed of 60% neutrons
and 40% protons.

appearing in the denominator of Eq.(44). In this report I set this confinement problem aside, and I adopt $R_2$ in Eq.(44) as the proper definition of $\langle p_T^2 \rangle$ in first order QCD, with the lower limit of integration in Eq.(42) extended all the way to $p_T = 0$.

Recalling Eqs.(35) and (36), we may derive

$$\alpha_s \left\langle p_T^2 \frac{d\sigma_1}{dM^2 dy dp_T^2} \right\rangle = \frac{1}{s^2} \int (f_C + f_A) s \frac{dp_T^2}{s} \equiv s^{-1} \tilde{f}_1(M/\sqrt{s}, y) . \qquad (45)$$

Inasmuch as the denominator of Eq.(44) satisfies the classical scaling property $s^2 \sigma(M,y) = f_0(M/\sqrt{s}, y)$, I conclude that

$$\langle p_T^2 \rangle_{QCD} \equiv R_2 = s \frac{\tilde{f}_1(M/\sqrt{s}, y)}{f_0(M/\sqrt{s}, y)} . \qquad (46)$$

This linear growth of $\langle p_T^2 \rangle_{QCD}$ with s at fixed $M/\sqrt{s}$ was discussed above on more general grounds. The present analysis provides specific (model-dependent) predictions for the coefficient b in the general formula $\langle p_T^2 \rangle = a + bs$.

The results I obtain for $\langle p_T^2 \rangle_{QCD}$ are shown as a function of mass in Fig.16 for the reaction $pN \to \mu\bar{\mu}X$ at 400 GeV/c and $y = 0$. For $M \approx 7$ GeV the Compton and annihilation processes each provide $\langle p_T^2 \rangle_{QCD} \approx 0.4$ GeV$^2$. As shown in the figure, the sum of the two QCD processes yields a curve of $\langle p_T^2 \rangle$ which rises rapidly with M, flattens off in the range $5 < M < 11$ GeV, and then begins to drop as M is increased further. As remarked earlier, the shape of the Compton contribution to $\langle p_T^2 \rangle_{QCD}$ vs. M is influenced by the choice of the power p in Eq.(40). If I select powers smaller (larger) than my value $p = 6$, the Compton contribution in Fig.16 will rise faster (fall faster) with M than the result I have shown. The shape of the annihilation contribution in Fig.16 is fixed since the structure functions $q(x)$ and $\bar{q}(x)$ are both fixed.

To my knowledge there are no other empirical determinations of the power p of the gluon distribution. If this QCD analysis of the M dependence of $\langle p_T^2 \rangle$ is accepted as a relevant constraint, the power $p = 5$ or 6 is determined for the first time. This power places the slope in x of the gluon distribution somewhere between those of the valence and sea quark distributions, which seems reasonable.

The net QCD contribution illustrated by the dotted line in Fig. 16 represents only about one-half the experimental value of $\langle p_T^2 \rangle$ at 400 GeV/c. To reproduce the data in the range $5 < M < 10$ GeV, it is necessary to add $\sim 1.04$ GeV$^2$ to the QCD contribution. This addition may be associated with the non-perturbative and "confinement" contributions which I set aside above. Thus, I write

$$\langle p_T^2 \rangle = 1.04 + \langle p_T^2 \rangle_{QCD} . \qquad (47)$$

The confinement portion may depend on the ratio $(M/\sqrt{s})$ but should be otherwise independent of s. In this report I take it to be a pure constant. The result of the addition of the QCD and confinement effects is shown as the solid line in Fig.16.

The test of these QCD calculations and speculations lies in the energy dependence of $\langle p_T^2 \rangle$. In Fig.17, I provide my expectations for $\langle p_T^2 \rangle_{QCD}$ in the Fermilab and SPS energy range. A rather flat behavior is obtained for the dependence of $\langle p_T^2 \rangle$ vs. M at all energies. The slope b which I calculate in the expression $\langle p_T^2 \rangle = a + bs$ is shown as a function of $M/\sqrt{s}$ in Fig.18. It has a maximum value of $b \simeq 1.15 \times 10^{-3}$ GeV$^{-2}$ for $M/\sqrt{s} \simeq 0.3$. The results of my explicit calculation of $b(M/\sqrt{s})$ do not resemble the analytic form $(1 - M/\sqrt{s})$ guessed by Politzer.[13] His form is valid perhaps in the neighborhood of $M/\sqrt{s} = 1$.

At a value of $M/\sqrt{s} \simeq 0.1$ typical of the ISR energy range, my slope is $b \simeq 0.74 \times 10^{-3}$ GeV$^{-2}$. Thus, at $\sqrt{s} = 52$ GeV and $M/\sqrt{s} \simeq 0.1$, I predict $\langle p_T^2 \rangle_{ISR} = 1.04 + 2.00 \cong 3$ GeV$^2$, and $\langle p_T \rangle_{ISR} \simeq 1.4$ to $1.5$ GeV. These predictions are about 50% higher than the ISR data shown in Fig.13. It is important to confirm the ISR measurements with data of higher statistics.

A semi-empirical method may be adopted to obtain the first moment $\langle p_T \rangle$. The experimenters[6,1] report that their data are well fitted by the expression

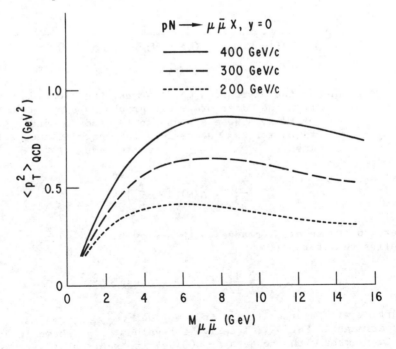

Fig.17. The first order (Compton plus annihilation) QCD contribution to $\langle p_T^2 \rangle$ is shown as a function of M for the process $pN \to \mu\bar{\mu}X$ at $y = 0$ for three values of lab momentum.

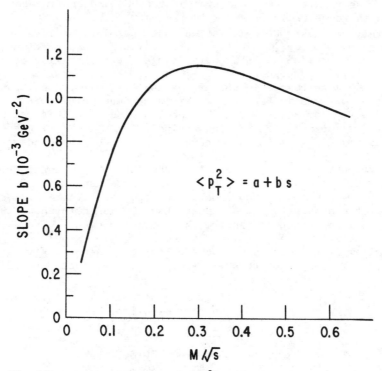

Fig.18.    In the expression $\langle p_T^2 \rangle = a + bs$, the slope b computed
from the first order QCD graphs is shown as a function
of $M/\sqrt{s}$.   These slopes can be used to obtain predictions for $\langle p_T^2 \rangle$ at various lab energies and lepton pair
masses.

$$E \frac{d\sigma}{d^3 p} = A(M)\left(1 + \frac{p_T^2}{\lambda^2}\right)^{-6} \tag{48}$$

where the value of $\lambda$ changes with energy but not with M.   This fit
implies that for all $\lambda$

$$\langle p_T \rangle = \frac{35\pi}{128} \langle p_T^2 \rangle^{\frac{1}{2}} = 0.859 \langle p_T^2 \rangle^{\frac{1}{2}} . \tag{49}$$

Starting with my results for $\langle p_T^2 \rangle = 1.04 + \langle p_T^2 \rangle_{QCD}$, and using the
"experimental" Eq.(49), I derive the values of $p_T$ shown in Fig.19.
The agreement with the data at 400 GeV is excellent (by construc-
tion).   The new feature of Fig.19 is the prediction of $\langle p_T \rangle_{th} \approx 1.03$
GeV at 200 GeV.   This is in fine agreement with the experimental
value[1] of $\langle p_T \rangle_{exp} \approx 1.00 \pm 0.05$ GeV.   This comparison suggests that
the QCD graphs reproduce the energy dependence of $\langle p_T \rangle$ very well,

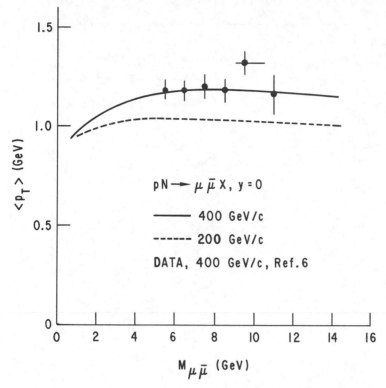

Fig. 19.  Shown are expectations for the mean transverse momen-
tum $\langle p_T \rangle$ of lepton pairs produced in $pN \to \mu\bar{\mu}X$ at $y = 0$
and $p_{lab} = 200$ and $400$ GeV/c. The curves are derived
from Eqs. (47) and (49) of the text.

and it encourages more precise tests of the energy dependence predicted by the model, as illustrated in Figs. 15 and 17.

It is relevant to ask whether the value 1.04 GeV$^2$ is a reasonable amount of $\langle p_T^2 \rangle$ to assign to "confinement" effects. Apportion-
ing it equally between the two initial constituents, I find that
each constituent carries $\langle k_T^2 \rangle \simeq 0.52$ GeV$^2$. This implies $\langle k_T \rangle \simeq 600$
MeV per constituent, similar to the value which I suggested earlier
should be provided by the fermi motion of quarks within a hadron.
The value 600 MeV is also approximately the mean transverse momentum
of $\rho$ mesons and of "clusters" produced in inclusive hadronic reactions.[29] Finally, it is approximately the amount of internal transverse momentum which is assigned to the constituents in attempts to
fit details of high $p_T$ hadronic data.[26] In view of these arguments,
the value of $\langle p_T^2 \rangle_{confinement} \simeq 1.04$ GeV$^2$ does not seem too large.

Although I have discussed moments at some length, I believe
that they do not provide a sensible test of QCD perturbation calculations. The answers in both the data and model calculations are
much too influenced by the non-perturbative region where $p_T < p_{T,crit}$
$\simeq 1$ GeV. If one is in need of a single parameter for confronting

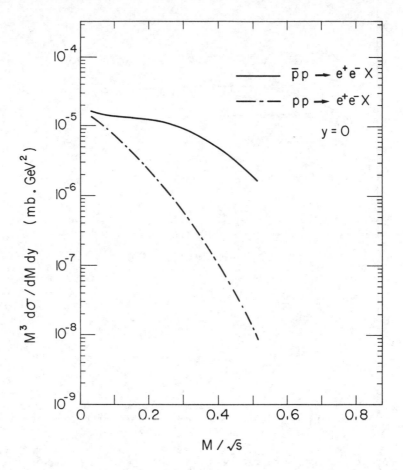

Fig.20.  Estimates of the lepton pair yield in antiproton-
proton collisions are presented as a function of $M/\sqrt{s}$
at $y = 0$ and are compared with yields in pp collisions.

theory with experiment, then perhaps it is best to fit $d\sigma/dM^2dydp_T^2$
to a simple form, say $\exp(-\lambda p_T)$, for $p_T > p_{T,crit}$, and to discuss
the dependence of the slope $\lambda$ on M, s, y, and $x_F$.  I would be glad
to supply interested readers with relevant predictions.

8.  <u>Antiproton Reactions.</u>  The reaction $\bar{p}p \to \ell^+\ell^-X$ serves in
many respects as the cleanest test of the classical Drell-Yan mechanism.  At large M, the cross-section is dominated by valence-valence annihilation, and it therefore measures the valence quark distribution almost directly.  Shown in Fig.20 are the scaling cross-
sections I compute with my structure functions for both $\bar{p}p \to \ell^+\ell^-X$
and $pp \to \ell^+\ell^-X$.  The computation of the transverse momentum distribution of the lepton pairs produced in $\bar{p}N$ reactions proceeds along
the same lines as for pN processes.  In Fig.21, I present the first
order QCD (Compton plus annihilation) contribution to $\langle p_T^2 \rangle$ for $\bar{p}N \to$

$\mu\bar{\mu}X$ at 400 GeV/c and $y = 0$. In the $\bar{p}N$ case, the annihilation process $q\bar{q} \to G\mu\bar{\mu}$ dominates, and the Compton process $qG \to q\mu\bar{\mu}$ contributes a negligible portion of $\langle p_T^2 \rangle_{QCD}$. This situation may be contrasted with the pN case in Fig.16 where Compton and annihilation contributions are comparable. The results in Fig.21 suggest that transverse momentum effects in $\bar{p}N$ reactions are almost totally insensitive to the gluonic distribution. This provides another argument for the study of lepton pair production in antiproton collisions. Both the integrated cross-section and the transverse momentum distributions are controlled by valence-valence annihilation diagrams.

The magnitude of $\langle p_T^2 \rangle_{QCD}$ for $\bar{p}N$ reactions is comparable to that of $\langle p_T^2 \rangle$ for pN collisions (c.f. Figs.16 and 21). I have not been able to trace the source of speculations that QCD predicts otherwise. In my results, $\langle p_T^2 \rangle_{QCD}^{\bar{p}N} > \langle p_T^2 \rangle_{QCD}^{pN}$ for M < 9 GeV. Above M > 9 GeV, $\langle p_T^2 \rangle^{pN}$ becomes larger, but this latter effect is sensitive to the choice of the gluon distribution. In attempting predictions for $\bar{p}N$ data, one

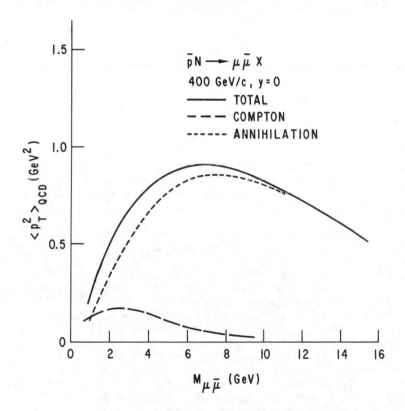

Fig.21. A prediction of the first order QCD expectation for $\langle p_T^2 \rangle$ in antiproton-nucleon collisions at 400 GeV/c and $y = 0$ for various masses of the lepton pair. Shown also is the breakdown of the answer into the contributions from the Compton and annihilation processes shown in Fig.4.

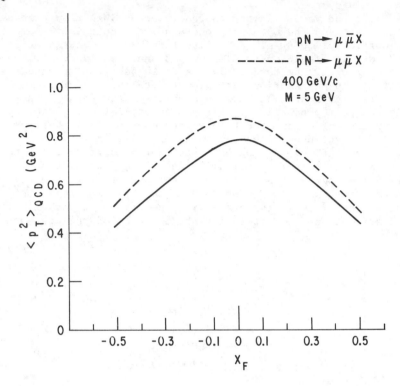

Fig. 22.  A prediction for the dependence of $\langle p_T^2 \rangle_{QCD}$ on the longitudinal momentum fraction $x_F$ of the lepton pair in
both pN and $\bar{p}$N collisions at 400 GeV/c and at the lepton pair mass M = 5 GeV.

must add a confinement contribution to the results of Fig. 21.  Since
the confinement portion of $\langle p_T^2 \rangle$ is not "understood", I have no reason
to believe that the same value of 1.04 GeV$^2$ deduced from the pN data
should apply also in the $\bar{p}$N case.

Just as in other branches of strong interaction phenomenology
the chance to compare results from pN and $\bar{p}$N processes in lepton
pair production would aid our theoretical understanding considerably.

9.  Dependence on $x_F$.  In Fig. 22 I present the calculated vari-
ation of $\langle p_T^2 \rangle_{QCD}$ expected when the longitudinal momentum $x_F$ of the
lepton pair is varied.  Results are shown for both $\bar{p}$N → $\mu\bar{\mu}$X and pN →
$\mu\bar{\mu}$X at 400 GeV/c and M = 5 GeV.  In the pN case the Compton and annihilation contributions are about equal for the values of $x_F$ shown;
this breakdown is not presented here.  To compare the predictions
in Fig. 22 with data, one must add a contribution for the confinement
effects.  This was found to be 1.04 GeV$^2$ in pN collisions at $x_F = 0$.
The observation[1] that $\langle p_T^2 \rangle$ seems to be independent of $x_F$ in the narrow range of available acceptance $0 < x_F < 0.2$ in pN collisions suggests that the confinement contribution may rise slightly with increasing $x_F$.  However, here again, I stress that $\langle p_T^2 \rangle$ is not a suit-

able variable for tests of QCD. This objection is especially relevant when the calculated values of $\langle p_T \rangle_{QCD} \lesssim 1$ GeV. The entire answer is dominated by values of $p_T < p_{T,crit} \simeq 1$ GeV where the perturbative calculation should not be used. It would be best to compare data directly with calculations of $d\sigma/dM^2 dx_F dp_T^2$ for values of $p_T > p_{T,crit}$.

10. $\pi N$ Collisions. A wealth of data will soon be available on lepton pair production in $\pi^{\pm}N$ reactions.[2] Once the structure functions $q(x)$ and $\bar{q}(x)$ appropriate for the pion are deduced from these data, predictions for the transverse momentum spectra will follow readily.

## VI.   IMPLICATIONS AND CONCLUSIONS

Observation of the intermediate vector bosons, $W^{\pm}$ and $Z^0$, the mediators of weak interactions is one goal of high energy experimentation which seems almost within reach. The classical Drell-Yan mechanism has been used to provide estimates for W and $Z^0$ yields.[30] New structure functions deduced from the most recent data[1] on $pN \rightarrow \mu\bar{\mu}X$ and scaling violations[7-9] modify these predictions somewhat. However, the rather large values predicted by QCD for $\langle p_T \rangle_W$ would seem to have the most substantial impact on the design of experiments. Values of the slope $b$ shown in Fig.18 can be used to provide the expected values of $\langle p_T^2 \rangle_W$ at energies at which ISABELLE or the FNAL collider may operate. For example, for pp collisions of 400 GeV/c on 400 GeV/c ($\sqrt{s} = 800$ GeV), and for $M_W = 60$ GeV ($M_W/\sqrt{s} = 0.075$), $b = 0.575 \times 10^{-3}$. Consequently, $\langle p_T^2 \rangle_W \simeq 368$ GeV$^2$, and $\langle p_T \rangle_W \simeq 17$ GeV. At energies of the proposed CERN $\bar{p}p$ collider,[31,32] $\sqrt{s} = 2 \times 270$ GeV, with $M_W/\sqrt{s} \simeq 0.11$, I calculate $b_{pp} = 0.79 \times 10^{-3}$ GeV$^{-2}$, and $b_{\bar{p}p} = 0.89 \times 10^{-3}$ GeV$^{-2}$. Therefore, I predict

$$\langle p_T^2 \rangle_{W,pp} \simeq 230 \text{ GeV}^2,$$

and

$$\langle p_T \rangle_{W,pp} \simeq 13 \text{ GeV} .$$

Likewise,

$$\langle p_T^2 \rangle_{W,\bar{p}p} \simeq 260 \text{ GeV} ,$$

and

$$\langle p_T \rangle_{W,\bar{p}p} \simeq 14 \text{ GeV} .$$

Obviously a considerable extrapolation has been made in energy to obtain these QCD predictions. The only basis for confidence in them is the fact the theory appears to reproduce data on $\langle p_T \rangle$ in the Fermilab energy range of 200 to 400 GeV/c. Restoration of the $\log(M^2/\Lambda^2)$ factor which I neglected after Eq.(16) would reduce my estimates slightly. On the other hand, experience with the $\Upsilon$ suggests that resonances are produced with larger $\langle p_T \rangle$ than the neighboring continuum.

222

A favored method for observing the W is to detect a sharp peak at $M_W/2$ in the single $\mu$ inclusive momentum spectrum. The peak signals the decay $W \to \mu\nu$. Once $\langle p_T \rangle_W$ exceeds ~10 GeV as I predict, this expected peak is substantially washed out, and other more difficult experimental methods to establish the W may have to be employed.[31,32]

There are many questions I have not addressed in this report which are nevertheless of considerable interest. Predictions have been made for various properties of the hadrons produced in association with a massive lepton pair.[33] The successful use of quark and antiquark distribution functions to explain the x dependence of the inclusive hadron yield $Ed\sigma/d^3p$ at small $p_T$ also deserves further study.[34] Polarization phenomena[35] in constituent scattering processes may yield new insight into the dynamics of quarks and gluons.

Massive lepton pair production has become an industry on its own. It provides tests of several important aspects of the parton model of interacting quarks and gluons. Now that the scaling predicted by the classical Drell-Yan model seems verified in the data, it is time to identify the scaling violations predicted by QCD. The process $pN \to \mu\bar{\mu}X$ serves to specify the average sea quark distribution in a region of $Q^2$ much higher than reached so far in inelastic neutrino reactions $\nu N \to \mu X$. To the extent that one accepts a QCD analysis of $\langle p_T^2 \rangle$ in $pN \to \mu\bar{\mu}X$, the average gluon distribution is also determined by this reaction. In Sec.V, I described in some detail the QCD approach for explaining the transverse momentum distribution of lepton pairs. The data from Fermilab on $\langle p_T^2 \rangle$ can be accommodated in a QCD calculation if we add an energy independent "confinement" contribution of 1.04 GeV$^2$ to the perturbative QCD prediction. It is desirable to understand this non-perturbative confinement portion in more detail. I mentioned a few important tests of the QCD calculation. In rough order of importance, I suggest:

1) verify the energy dependence predicted in Fig.15;
2) extend the data in Fig.14 to higher $p_T$ to see whether theory and data diverge or converge;
3) obtain high statistics data on $\langle p_T \rangle$ and $\langle p_T^2 \rangle$ at ISR energies;
4) verify the energy dependence displayed in Fig.17;
5) obtain data on $\bar{p}N \to \mu\bar{\mu}X$; and
6) understand the $x_F$ dependence of the $p_T$ distribution, both the confinement and QCD portions.

## ACKNOWLEDGMENTS

In preparing this report, I have benefited especially from discussions with S. Wolfram and M. Jacob. I acknowledge valuable conversations with J. Babcock, L. DiLella, J. Ellis, R. Petronzio, C. Sachrajda, J. C. Sens and T. M. Yan. I thank D. Sivers and G. Thomas for comments on the text. Some of the research I report here was done while I was a Visiting Scientist in the Theory Division at CERN. I am grateful to Jacques Prentki and others in the Division for their hospitality. This work was supported in part by the United States Department of Energy.

## REFERENCES

1. Columbia-Fermilab-Stony Brook Collaboration, R. Kephart, Proceedings of the 3rd International Conference at Vanderbilt University on New Results in High Energy Physics.
2. Chicago-Princeton Collaboration, K. J. Anderson, *ibid.*
3. S. D. Drell and T. M. Yan, Phys. Rev. Lett. <u>25</u>, 316 (1970); Ann. Phys. (N.Y.) <u>66</u>, 578 (1971). For a recent discussion of the subject, consult the review by T. M. Yan, presented at the VIII International Symposium on Multiparticle Dynamics, Kaysersberg, France, June, 1977.
4. D. Antreasyan et al Phys. Rev. Lett. <u>39</u>, 906 (1977).
5. J. H. Cobb et al, Phys. Lett. <u>72B</u>, 273 (1977).
6. D. M. Kaplan et al, Phys. Rev. Lett. <u>40</u>, 435 (1978).
7. I. Hinchliffe and C. H. Llewellyn Smith, Phys. Lett. <u>66B</u>, 281 (1977) and Nucl. Phys. <u>B128</u>, 93 (1977).
8. N. Cabibbo and R. Petronzio, CERN report Ref. TH. 2440 (1978).
9. J. Kogut and J. Shigemitsu, Nucl. Phys. <u>B129</u>, 461 (1977).
10. A goldmine of information on the past, present, and future of experiments on lepton pair production at Fermilab, CERN-SPS, and CERN-ISR energies is the CERN report ISR Workshop/2-8 (October, 1977), edited by M. Jacob. For a review of the experimental situation as of the Summer of 1977, consult the paper by M. J. Shochet, Univ. of Chicago report EFI 77-66, presented at the SLAC Summer Institute, July, 1977.
11. M. Duong-Van, SLAC report, SLAC-PUB-1819 (1976); J. Gunion, Phys. Rev. <u>D14</u>, 1400 (1976); F. Close, F. Halzen, and D. Scott, Phys. Lett. <u>68B</u>, 447 (1977); A. Davis and E. J. Squires, Phys. Lett. <u>69B</u>, 249 (1977); J. Bell and A.J.G. Hey, CERN report Ref. TH-2437 (1977); and R. C. Hwa, S. Matsuda, and R. G. Roberts CERN report Ref. TH-2456 (1978).
12. M. Duong-Van, K. V. Vasavada, and R. Blankenbecler, Phys. Rev. <u>D16</u>, 1389 (1977); M. Duong-Van and R. Blankenbecler, SLAC-Pub-2017 (Sept. 1977).
13. C. S. Lam and T. M. Yan, Phys. Lett. <u>71B</u>, 173 (1977); J. Kogut and J. Shigemitsu, Phys. Lett. <u>71B</u>, 165 (1977); H. D. Politzer, Nucl. Phys. <u>B129</u>, 301 (1977); H. Fritzsch and P. Minkowski Phys. Lett. <u>73B</u>, 80 (1978); G. Altarelli, G. Parisi, and R. Petronzio, CERN report Ref. TH. 2413 (October, 1977) and CERN report Ref. TH. 2450 (January, 1978); K. Kajantie and R. Raitio, Helsinki report HU-TFT-77-21 (October, 1977); F. Halzen and D. Scott, Univ. Wisconsin, Madison, report COO-881-21 (Feb., 1978); K. Kajantie, J. Lindfors, and R. Raitio, Helsinki report HU-TFT-78-5 (1978).
14. E. L. Berger, J. Donohue, and S. Wolfram, Phys. Rev. <u>D17</u>, 858 (1978).
15. K. V. Vasavada, Phys. Rev. <u>D16</u>, 146 (1977).
16. J. C. Collins and D. E. Soper Phys. Rev. <u>D16</u>, 2219 (1977).
17. Gargamelle Neutrino Collaboration, H. Deden et al Nucl. Phys. <u>B85</u>, 269 (1975).
18. R. D. Field and R. P. Feynman, Phys. Rev. <u>D15</u>, 2590 (1977).
19. I have not made a survey of all the different models for struc-

ture functions. A partial list of references includes P. V. Landshoff and J. C. Polkinghorne, Nucl. Phys. B33, 221 (1971); R. McElhaney and S. F. Tuan, Phys. Rev. D8, 2267 (1973); G. Farrar, Nucl. Phys. B77, 429 (1974); V. Barger and R.J.N. Phillips, Phys. Lett. 73B, 91 (1978); and F. T. Dao et al Phys. Rev. Lett. 39, 1388 (1977).

20. BEBC Neutrino collaborations, J. Mulvey, these proceedings.

21. H. L. Anderson et al, Phys. Rev. Lett. 38, 1450 (1977).

22. S. M. Berman, D. J. Levy, and T. L. Neff Phys. Rev. Lett. 23, 1363 (1969).

23. H. D. Politzer, Rev. 13; C. T. Sachrajda, Phys. Lett. B73, 185 (1978); J. B. Kogut, Phys. Lett 65B, 377 (1976); H. Georgi, Harvard report HUTP-78/A003 (1978 Orbis Scientiae, Coral Gables, Florida). K. H. Craig and C. H. Llewellyn Smith Oxford report TP 67/77 (Nov. 1977).

24. Consult, e.g., A. DeRujula, H. Georgi, and H. D. Politzer, Ann. Phys. (N.Y.) 103, 315 (1977); G. C. Fox, Nucl. Phys. B131, 107 (1977). I. Hinchliffe and C. H. Llewellyn Smith, Ref.7; A. J. Buras and K.J.F. Gaemers, CERN report Ref. TH-2322 (1977); G. Altarelli and G. Parisi, Nucl. Phys. B126, 298 (1977); C. H. Llewellyn Smith and S. Wolfram, Oxford report 6/78 (1978).

25. I am grateful to J. Babcock for help in the derivation of Eqs. (20) and (21). See also some of the papers listed in Ref.13.

26. R. P. Feynman, R. D. Field, and G. C. Fox, Nucl. Phys. B128, 1 (1977).

27. H. Georgi and M. Machacek, Phys. Rev. Lett. 39, 1237 (1977); E. Farhi, Phys. Rev. Lett. 39, 1587 (1977); and A. DeRujula, J. Ellis, E. G. Floratos, and M. K. Gaillard, CERN report Ref. TH 2455 (1978).

28. G. Sterman and S. Weinberg, Phys. Rev. Lett. 39, 1436 (1977).

29. D.R.O. Morrison, Review presented at the Symposium on Hadron Structure and Multiparticle Production, Kazimierz, Poland, May, 1977 CERN EP report (1977); and CERN-College de France-Heidelberg-Karlsruhe Collaboration, M. Della Negra et. al., CERN report CERN/EP/PHYS-77-30 (1977).

30. C. Quigg, Rev. Mod. Phys. 49, 297 (1977); R. F. Peierls, T. L. Trueman, and L. L. Wang, Phys. Rev. D16, 1397 (1977).

31. Aachen-Annecy-Birmingham-CERN-College de France-Queen Mary-Riverside-Rutherford-Saclay Collaboration Proposal, A. Astbury et. al, CERN/SPSC/78-06 (1978).

32. Saclay-CERN-Orsay Proposal, M. Banner et. al., CERN/SPSC/78-8 (1978).

33. T. A. DeGrand and H. I. Miettinen, Phys. Rev. Lett. 40, 612 (1978).

34. W. Ochs, Nucl. Phys. B118, 397 (1977); K. P. Das and R. C. Hwa, Phys. Lett. 68B, 459 (1977); F. C. Erné and J. C. Sens, CERN report, 1978, submitted to Phys. Rev. Letters.

35. F. Close and D. Sivers, Phys. Rev. Letters. 39, 1116 (1977); J. Babcock, E. Monsay, and D. Sivers, Phys. Rev. Lett. 40, 1161 (1978).

RECENT RESULTS FROM THE PLUTO AND DASP DETECTORS AT DORIS

presented by

U. Timm
Deutsches Elektronen-Synchrotron, DESY, Hamburg, F.R.G.

## ABSTRACT

Decays of the heavy lepton $\tau$ have now been observed in $e^+e^-$ collisions at the energy of the $\psi'$-resonance, which results in an improved mass determination $M_\tau = 1.807 \pm 0.020$ GeV. For the $\tau$-lifetime a new upper limit of $\tau_\tau \leq 3.5 \cdot 10^{-12}$ sec. is given. The evidence for the decay $\tau \to \nu\pi\rho$ has been strengthened, and the $\pi\rho$-system is found to be in an S-wave state with $J^P = 1^+$, proving the axial vector component in the leptonic coupling of the $\tau$ to hadrons. The branching ratio is given with $BR(\tau^- \to \nu\pi^-\rho^0) = 0.050 \pm 0.015$. The total hadronic cross section agrees within errors with the quark sum rule including QCD-corrections. The production and decay distributions of the decay $J/\psi \to f^0\gamma$ have been measured and found to agree well with a QCD calculation based on the exchange of two gluons.

## INTRODUCTION

The results reported here have been obtained by the two detectors PLUTO and DASP at the $e^+e^-$-storage ring DORIS in the period following the latest Lepton-Photon-Conference held in Hamburg, August 1977. The institutions and physicists involved in these experiments are listed in references [1,2]. My talk covers the following subjects :

- mass of the $\tau$
- lifetime of the $\tau$
- the decay $\tau \to \nu\pi\rho$
- the total $e^+e^-$ hadronic cross section
- angular distribution of the decay $J/\psi \to f^0\gamma$.

## MASS OF THE TAU

The first publication[3] concerning the observation of a new heavy lepton quoted a mass between 1.6 and 2.0 GeV. Since that time the mass value of the $\tau$ has narrowed down by measurements from PLUTO[4], $M_\tau = 1.79 - 1.92$ GeV, and, more recently by the LBL-SLAC lead glass wall detector[5] $M_\tau = 1.80 - 1.85$ GeV, where the $\tau$ has

ISSN:  0094-243X/78/225/$1.50  Copyright 1978 American Institute of Physics

226

been observed at the $\psi''$ resonance. The accuracy in the mass determination increases the more as we approach the production threshold. There is now an observation of τ-decays at the energy of the
$\psi'$, below the charm threshold.

    This experiment has been carried out by the DASP-collaborat-
ion [6] with the double arm spectrometer shown in fig. 1. The two
spectrometer arms shown cover 0.072 str. A particle emerging from
the $e^+e^-$ - interaction region passes three proportional wire chambers in front of the magnet, six spark chambers behind of it, time of
of flight- and shower-counters, steel absorbers and range counters.
A high momentum resolution is achieved, with $\Delta p/p = 0.007$ p (p measured in GeV/c), and an excellent separation of π, K, and p. The
inner part near the beam is free of field. It contains a detector
with four layers of sandwiches, consist-
ing of scintillators, lead and proportional tube chambers, and a lead scintillator shower counter. This part is shown
in fig. 2. It covers 0.70 x 4π str.. Two
Cerenkov counters have been inserted
to improve the elec-tron identification
in the outer arms in order to detect

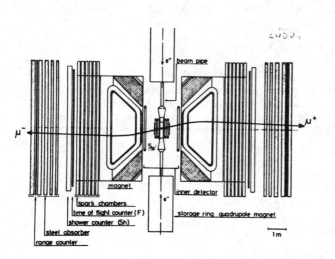

Fig. 1. Plan view of the double arm
        spectrometer DASP at DORIS.

electrons from τ- (and charm-) decays. The resulting suppression of
hadrons achieved is given by the probability of misidentification
for hadrons with $P_{h \to e} = 4 \cdot 10^{-4}$.

The detector is triggered on a single track in one of the outer arms, requiring a momentum $p \geq 0.4$ GeV/c at $\sqrt{s} = 3.684$ GeV, and $p \geq 0.1$ GeV/c at other energies, $3.1 \leq \sqrt{s} \leq 5.2$ GeV. The events are selected according to the following signatures :

$$e^+ e^- \rightarrow e^{\pm} + 1pr + n \cdot \gamma \qquad (1)$$
$$e^+ e^- \rightarrow \mu^{\pm} + 1pr + n \cdot \gamma$$
$$e^+ e^- \rightarrow e^{\pm} \mu^{\pm} + \text{nothing.}$$

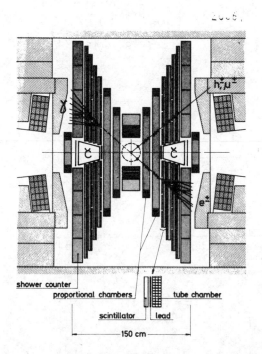

Fig. 2. View in beam direction of the inner DASP-detector.

shower counter
proportional chambers
scintillator    lead
tube chamber
150 cm

I restrict myself here to the data sample (1) at $\sqrt{s} = 3.864$ GeV, where $1400$ nb$^{-1}$ have been collected and 17 events been found. The momentum distribution of these events is shown in fig. 3. The cluster around 1.5 GeV/c is explained by the cascade decay $\psi' \rightarrow J/\psi + X \rightarrow e^+ e^- X$, where X are photons, and one electron is misidentified in the inner detector as hadron ($P_{e \rightarrow h} = 2 \cdot 10^{-2}$). The 9 events in the second cluster are compared by the hypothesis that these events come from the decay $\tau \rightarrow \nu e \nu$ with M $= 1.80$ GeV and a massless neutrino (solid line). Being on the $\psi'$ resonance, the production of pairs is enhanced over normal QED by a factor of 2.3 The effective luminosity was derived from the number of elastic muon pairs observed in the outer detector. The hypothesis clearly is compatible with the observed event distribution. The background to these events is estimated as follows :

| eeμμ | 0.6 evts. |
| beam-gas | < 0.1 evts. |
| h → e | 0.84 evts. |
| Dalitz pairs | 0.2 evts. |
| total | < 1.7 evts. |

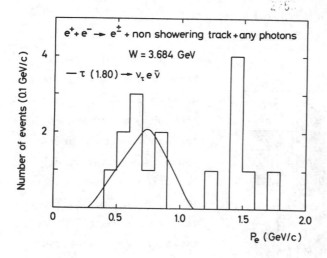

Fig. 3.
Electron momentum distribution observed at $\sqrt{s}$ = 3.864 GeV, data sample (1)

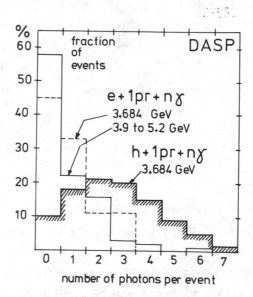

Fig. 4.

Photon multiplicity distributions for event class (1) compared to the equivalent hadron class at $\sqrt{s}$ = 3.864 GeV.

The conclusion is that a significant signal is observed in the momentum spectrum, which is compatible with the decay $\tau \to \nu e \nu$.

A further proof that these events come from τ-decays is demonstrated by the accompagnying photon multiplicities shown in
fig. 4. While the distribution for the electron class (1) observed
at the ψ' agrees well with the same class observed all the way
up to 5.2 GeV, it drasticly disagrees with the equivalent hadron
class at the ψ'. The conclusion is that the electron events seen
at the energy of the ψ' are in fact coming from the decay τ → νeν.

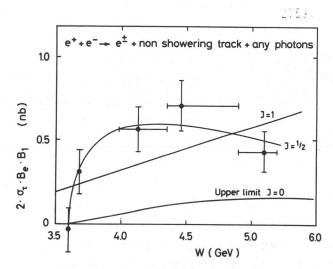

Fig. 5. Integrated inclusive cross section
for event class (1) as function
of √s.

This observation
of τ below the
charm threshold for
the first time sep-
arates unambiguous-
ly the phenomena of
charm and heavy
lepton.

The mass
determination then
includes the total
amount of data for
the event class (1)
taken up to √s =
5.2 GeV. The integ-
rated inclusive

cross section is plotted in fig. 5, where the solid line, marked
J = 1/2, is a fit of the data to the threshold function of $\tau\bar{\tau}$-prod-
uction $2\sigma_{\tau\tau} = \sigma_{\mu\mu}\beta(3 - \beta^2)$, with $\beta = (1 - (M_{\tau}/E_{beam})^2)^{1/2}$. The fit
yields a mass value of $M_{\tau} = 1.807 \pm 0.020$ GeV for the τ.


## LIFETIME OF THE TAU


From measurements with the PLUTO detector, a lifetime limit
was given of $\tau_{\tau} \leq 1 \cdot 10^{-11}$ sec at the time of the Hamburg-Con-
ference [7]. The lifetime expected on the basis of a sequential heavy

lepton is given by

$$\tau_o = BR(\mu) \cdot \tau_\mu \, (M_\mu/M_\tau)^5 = 2.7 \cdot 10^{-13} \text{ sec} \qquad (2)$$

with $BR(\mu) = 0.18$, $\tau_\mu$, $M_\mu$ the lifetime and mass of the muon, and $M_\tau = 1.807$ GeV. The analysis has been refined resulting in an improvement of the upper limit quoted earlier.

For the lifetime determination events of the two prong class (3) with one identified muon and no photons are studied :

$$e^+e^- \rightarrow \mu^\pm + 1pr + \text{missing mass} \qquad (3)$$

These events have been discussed in a previous publication[4] where it was shown to originate from $\tau$-decays. The data have been taken between $\sqrt{s} = 3.9$ and 5.0 GeV, with a total integrated luminosity of 5600 nb$^{-1}$. We found 131 events of type (3) with a momentum $p_\mu \geq 1$ GeV/c, 53 of which are used for the lifetime analysis. They have an average center of mass energy $\sqrt{s} = 4.5$ GeV.

The detector is shown in fig. 6. It is a $4\pi$-spectrometer of compact design, due to its superconductive coil, which produces a uniform magnetic field of 2 T. The space inside the coil, 1.4 m diameter and 1.05 m length, contains 14 cylindrical proportional chambers, interleaved with two lead converters of thickness 0.44 and

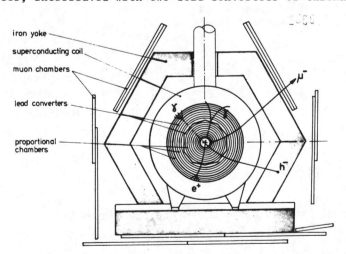

Fig. 6. The detector PLUTO, view in beam direction.

1.72 radiation lengths respectively. The converters allow the identification of photons and electrons through pair production,
Bremsstrahlung or showerdevelopment, which are detected in the subsequent chambers. The surrounding iron yoke acts as muon filter.
Hadrons are stopped by nuclear reactions and the penetrating muons,
$P_\mu \geq 1$ GeV/c, are detected in proportional tube chambers covering
the outer surface. Due to the iron thickness ($\sim 70$ cm average) and
the short decay length for $\pi \to \mu\nu$, $K \to \mu\nu$ the misidentification of
muons is low : $P_{h \to \mu} = 2.8 \quad 10^{-2}$.

To get an idea of the experimental approach to the problem,
the decay length expected from the lifetime (2) is given at
$\sqrt{s} = 2E_{beam} = 4.5$ GeV by

$$d = c\tau_o \beta\gamma = c\tau_o \left( (E_{beam}/M_\tau)^2 - 1 \right)^{1/2} = 0.067 \text{ mm}.$$

In view of the fact that the beam size at DORIS is three
times as large, $\sigma_x = \sigma_y = 0.2$ mm, and that the origin is reconstructed from tracks with a precision $\sigma_r = 0.5$ mm, it becomes evident that the experiment yields only an upper limit at the energies
available at present.

A quantity which is sensitive to the decay length d is the
distance $r_{min}$ of the extrapolated track from the origin, as indicated in fig. 7. The resulting Gaussian distribution of $r_{min}$ from

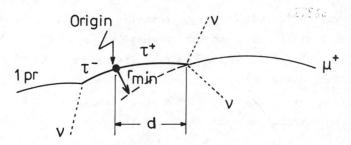

Fig. 7. Decay schematic for $e^+ e^- \to \tau^+ \tau^- \to \nu\mu\nu + \nu + 1pr$.

a set of events is characterized by $\sigma(r_{min}, d)$. As explained above,
the experiment yields this quantity only at zero decay length,

whereas the dependence on d has to be studied by a Monte Carlo program, which generates the final state (3) according to the production and decay of a τ-pair (see fig. 7). The program simulates the tracks and reanalyses them to obtain the distribution of $r_{min}$ by varying the decay length.

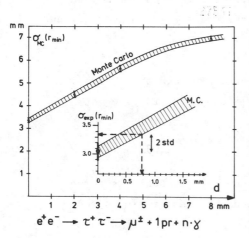

$$e^+e^- \longrightarrow \tau^+\tau^- \longrightarrow \mu^\pm + 1\,pr + n\cdot\gamma$$

Fig. 8. Monte-Carlo result $\sigma_{MC}(r_{min})$ as function of decay length d of the τ, compared to the experimental value $\sigma_{exp}(r_{min},0)$.

From extensive studies of the pure muon and pure hadron samples, and a comparison between them and the τ-sample as defined by equ. (3), we made sure that at zero decay length the value obtained from the Monte Carlo simulation, $\sigma_{MC}(r_{min},0)$, is well understood. This investigation becomes necessary since $\sigma_{exp}(r_{min},0)$ not only depends on the momenta, but also is somewhat sensitive to the difference in multiple scattering of muons and hadrons. The Monte Carlo results are given as function of the decay length in fig. 8. The width of the errors is indicated, and repeated in the insert, scaled to the statistically best experimental value at d = 0 of $\sigma_{exp}(r_{min},0)$ = 3.05 ± 0.15 mm deduced from the data sample (3). If its 2 standard deviations error is projected on to the lower error bound from Monte Carlo, a maximum decay length d ≤ 0.8 mm is obtained, corresponding to an upper limit

$$\tau_\tau \leq 3.5 \cdot 10^{-12} \text{ sec; } 95 \text{ \% CL.} \tag{4}$$

The new lifetime limit, which can also be expressed as a ratio, $\tau_\tau/\tau_0 \leq 13$, has the following bearing on the nature of the

$\tau$-neutrino :

(1)  If the $\nu_\tau$ is identical with the old muon neutrino $\nu_{\mu L}$, the $\tau$ can be produced in neutrino beams via the reaction

$\nu_{\mu L} + N \rightarrow \tau^- + \ldots \quad e^- + \ldots$, where the observation of soft electrons indicates the $\tau$-decay. Both FNAL and CERN bubble chamber groups [8,9], have looked for this process and quoted lower limits $\tau_\tau(\nu_\mu)/\tau_o > 40$ and 15 respectively. Thus the lefthanded muon neutrino is ruled out as possible $\tau$-neutrino.

(2)  In models [10], where the $\tau$-neutrino is of heavy mass, $M_{\nu_\tau} \gtrsim M_\tau$, the decay can only proceed via mixing to the old neutrinos,

$\nu_\tau + \epsilon_e \nu_e + \epsilon_\mu \nu_\mu$, where the limits on the mixing parameters, $\epsilon_e$, $\epsilon_\mu < 0.1$, are derived from nuclear reactions. The lower limit on the lifetime ratio for this model is then $\tau/\tau_o = (\epsilon_e^2 + \epsilon_\mu^2)^{-1} > 50$, which exceeds the PLUTO limit and excludes a heavy mass neutrino.

## THE DECAY $\tau \rightarrow \nu\pi\rho$

This decay was first reported on the Hamburg Conference [7] and has been published [11]. Since that time the statistics was increased by extending the $e\pi\pi$ sample and by including events of the type $\mu\pi\pi$.

The experiment has been carried out with the PLUTO detector by investigating the final state

$$e^+ e^- \rightarrow \ell\pi\pi\pi + \text{missing mass} \qquad (5)$$

where $\ell$ = e or $\mu$, the charge of the four particles is balanced and no photon is detected. It has to be shown, that the majority of the event sample (5) originates from the production and decay of $\tau$-pairs in the reaction

$$e^+ e^- \rightarrow \tau^+ \tau^- \rightarrow \ell\nu\nu + \pi\pi\pi\nu, \quad \ell = e,\mu \qquad (6)$$

234

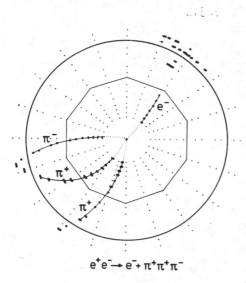

$$e^+e^- \to e^- + \pi^+\pi^+\pi^-$$

Fig. 9. Four prong event in PLUTO,
view along the axis. The
lead converters are indicated,
the electron produces a shower.

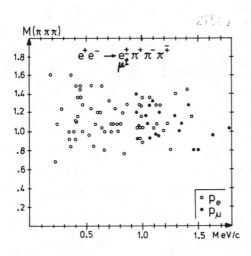

Fig. 10. Momentum distribution
of leptons versus the
invariant three pion mass, from
the event class (5).

An event of type (5) is shown in fig. 9. The electron and the three pion system emerge from the interaction region in opposite directions as would be expected if the event is interpreted according to reaction (6).

The analysis is based on the same total data set as quoted in the last chapter. Charge balanced 4 prong events without photons are selected out of the total data, and electrons identified by an analysis of the shower pattern in the proportional chambers behind the lead converter. The method yields a misidentification of hadrons $P_{h \to e}$ = 1.2 %, and a detection efficiency for electrons rising from 0.30 at 0.4 GeV/c to 0.65 for momenta $p_e \geq 1.0$ GeV/c. QED-events are removed by a missing mass cut $MM^2 \geq 0.9$ $GeV^2$. After a final cut in momenta, $p_e \geq 0.4$, $p_\mu \geq 1.0$ GeV/c, there remain 66 events. Fig. 10 shows the event sample before the momentum cut. The invariant $3\pi$-mass is plotted versus momentum. The plot demonstrates that the $3\pi$-mass is low, $\sim 1.1$ GeV, that the lepton spectrum is hard as

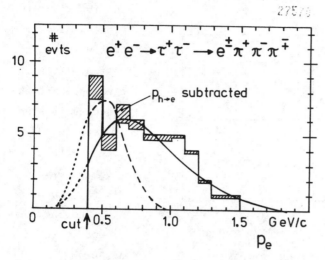

Fig. 11. Distribution of electron momenta.
Solid line : calculation for
$\tau \to \nu\nu e$, normalized. Dashed line : spectrum
from charm decays [12], not normalized.

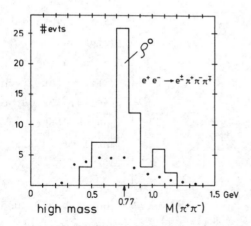

Fig. 12. Distribution of the invariant two
pion mass (high solution). The
black points are the experimental background,
calculated from the misidentification of
hadrons, $P_{h \to e}$.

expected from $\tau$ – decays, and that both quantities are uncorrelated. All this is consistent with the decay of a pair of heavy particles.

Consistency with reaction (6) is further verified by the momentum distribution of electrons shown in fig. 11. The spectrum agrees well with the calculated $\tau$-decay into electrons and neutrinos. An estimate of the contribution from charm decays yields at most 3 events and the spectrum is moreover soft. We therefore conclude that the three pions and the leptons are indeed coming from reaction (6).

In order to investigate the three pion system

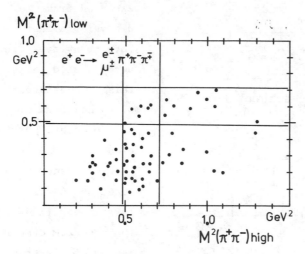

Fig. 13. Invariant mass squared,
$M(\pi^+\pi^-)^2$ in the Dalitz plot
representation. The lines define the
$\rho^0$-bands : $0.70 \le M(\pi^+\pi^-) \le 0.84$ GeV.

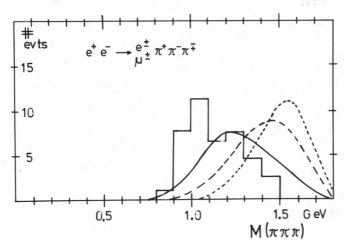

Fig. 14. Invariant 3π-mass distribution of
events within the rho-bands.
Calculated 3π-mass distributions are given
for $J^P = 1^+$, S-wave (solid line);
$J^P = 0^-$, P-wave (dashed); and $J^P = 1^+$,
D-wave (dotted).

the invariant two pion mass distribution is plotted in fig. 12. A clear peak at the rho-mass is seen. It contains 33 events within the rho-band $0.7 \le M(\pi^+\pi^-) \le 0.84$ GeV. The background, which is not peaked at the mass of the rho meson, contains 25 events, 6 of which lie within the rho-band. Both, high and low solutions are shown in the Dalitz plot given in fig. 13. Out of the 66 events, a total of 42 lie within the rho-bands. In view of the 25 background events we conclude that the final state must be a pure πρ-system. The branching ratio is then

calculated with the result $BR(\tau^- \to \nu\pi^-\rho^0) = 0.052 \pm 0.012$, consistent with the branching ratio expected on the basis of a
sequential heavy lepton of $1/2 \cdot BR(\tau^- \to \nu A_1^-) = 0.04$ [13]. Fig. 14
finally shows the three pion mass distribution compared with the
calculated distributions of a $\pi\rho$-system for the three lowest angular momentum states S, P, D, which have $J^P = 1^+$, $0^-$, $1^+$ respectively. Clearly, the state $J^P = 1^+$, $\ell = 0$ is favoured by the data,
with a resonance like peak at threshold. This result is supported
by a three dimensional Dalitz plot analysis. The $A_1$-meson is the
only known axial vector meson, $J^P = 1^+$, in this mass region. A
preliminary mass determination yields $M_{A1} = 1.0$ GeV, $\Gamma_{A1} = 0.47$ GeV.
The decay of the heavy lepton $\tau$ into the hadronic state $1^+$ establ-
ishes the presence of axial vector in the weak coupling.

## THE TOTAL $e^+e^-$ HADRONIC CROSS SECTION

The total cross section for hadron production, $\sigma_h(e^+e^- \to h)$,
measured with the PLUTO detector, is shown in fig. 15. The cross
section is given in units of the cross section for muon pair production, $R = \sigma_h/\sigma_{\mu\mu}$, and covers the energy range from 3.6 to 5.0 GeV,
with a gap between 3.7 and 4.0 GeV. Contributions to R come from
the cross section for one photon annihilation, $\sigma_{tot}$, and the cross
section for the $\tau$-pair production, which rises above the threshold
according to $\sigma_{\tau\tau} = \sigma_{\mu\mu} \cdot \beta \cdot (3 - \beta^2)/2$; $\beta$ denotes the velocity of
the $\tau$. $R_\tau$ which is shown by the dashed line in figure 15,
approaches one unit for $\beta \to 1$. It must be remembered that 85 %
of the decaying $\tau$-pairs have hadrons in the final state; the 15 %
of pure leptonic decays left have not been removed from the data
because of their acollinearity. All other QED events, however,
were separated by appropriate cuts. $\sigma_{tot}$ is predicted by the triplet quark sum rule. Including QCD-corrections [14] the following
theoretical expression for R is then expected :

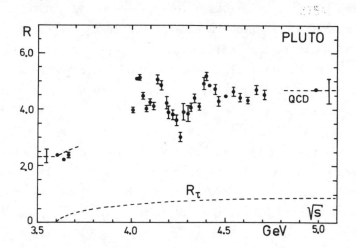

Fig. 15. Total hadronic cross section in units
of $\sigma_{\mu\mu}$ vs. center of mass energy.

$$R = \Sigma 3Q_i^2 \ (1 + \alpha(s)/\pi) + R_\tau \qquad\qquad (7)$$

$$Q_i = \text{quark charges (u, d, s, c)}$$

$$\alpha(s) = 12\pi/(25 \cdot \ln s/\Lambda^2)$$

The QCD-correction with the value $\Lambda$ = 0.73 GeV [15], amounts to
0.30 at $\sqrt{s}$ = 3.6 GeV, and 0.42 at $\sqrt{s}$ = 5.0 GeV. It can be seen
that the data agree well with the prediction, though a definite
conclusion is prohibited since the data have a systematic error of
10 % which is indicated in the figure.

The rise of the cross section seen above 4 GeV is now believed
to be well understood by the new channel which opens for charm production at $\sqrt{s}$ = 3.73 GeV. If the observed peaks are interpreted as
resonances (due to D, D$^*$, F, F$^*$ production), a fit to the data
yields the resonance parameters given in table I.

Table I   Resonance paramters

| M (GeV) | $\Gamma$ (keV) | $\Gamma_{ee}$ (keV) |
|---------|----------------|---------------------|
| 4.035   | 55 ± 5         | 0.7 ± 0.1           |
| 4.146   | 47 ± 11        | 0.4 ± 0.1           |
| 4.400   | 33 ± 9         | 0.3 ± 0.1           |

ANGULAR DISTRIBUTION OF THE DECAY $J/\psi \rightarrow f^{o}\gamma$

This process has been observed in the PLUTO detector [16] via the decay $f^{o} \rightarrow \pi^{+}\pi^{-}$ at the resonance energy $\sqrt{s}$ = 3.091 GeV. The signature of the selected events is thus :

$$e^{+}e^{-} \rightarrow J/\psi \rightarrow \pi^{+}\pi^{-}\gamma(\pi^{o}) \tag{8}$$

The momenta of the pions and the direction of the photon are measured. The detected photon can also be an unresolved photon pair from the decay of a neutral pion. The trigger efficiencies for $\pi^{+}\pi^{-}\gamma$ and $\pi^{+}\pi^{-}\pi^{o}$ are 73 % and 75 % respectively, as obtained from Monte-Carlo studies. Out of 84 000 $J/\psi$ hadronic events, 1650 survive a fit to the hypothesis (8) with $X^{2} \leq 20$. These are further reduced to 825 events by cuts removing QED background. The distribution of these events in the Dalitz plot is shown in fig. 16, where the three rho-bands are seen to be populated. To study the neutral $\pi^{+}\pi^{-}$ masses we impose cuts $0.6 \leq M(\pi^{+}x^{o}) \leq 1.0$ GeV and plot the invariant mass distribution $M(\pi^{+}\pi^{-})$ in fig. 17. In the figure the $\rho^{o}$ - and $f^{o}$-signals are clearly visible. The solid line is a resonance fit with a polynomial background, from which the resonance parameters $M_{f}$ = 1.23 ± 0.04 GeV, and $\Gamma_{f}$ = 0.13 ± 0.05 GeV are obtained. Several checks assure that the signal is neither $\varepsilon$ nor $\rho'(1250)$.

As the $J/\psi$ decays violate the OZI rule [17], it is of considerable interest to understand the nature of the decay mechanism.

240

Fig. 16.

Dalitz representation of reaction (8), $M(\pi^-X^o)^2$ vs. $M(\pi^+X^o)^2$, $X^o = \gamma$ or $\pi^o$

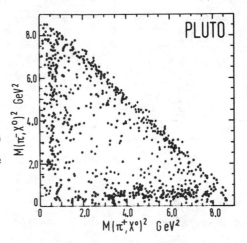

The branching ratio obtained for this decay is $BR(J/\psi \to f^o\gamma) = (2.0 \pm 0.7) \cdot 10^{-3}$. This rate can be compared with the decay into $f^o\omega$: $BR(J/\psi \to f^o\omega) = (4.0 \pm 1.4) \cdot 10^{-3}$, ref. [18] which has the same order of magnitude. This result would be incompatible with the decay diagram (a) of fig. 18, where we expect a ratio of order $10^{-3}$. However, the decay $J/\psi \to f^o\gamma$ may be dominated by a two gluon exchange, as given in fig. 18 (b), where photon couples directly to the $c(\bar{c})$ quark. Though a rate calculation does not exist for this diagram, there is a calculation of the angular distribution for the $f^o\gamma$ system in

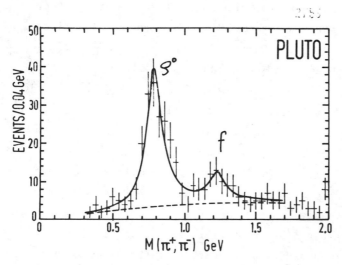

Fig. 17. Invariant mass distribution $M(\pi^+\pi^-)$

the framework of QCD [19], which offers the possibility for an experimental test [20].

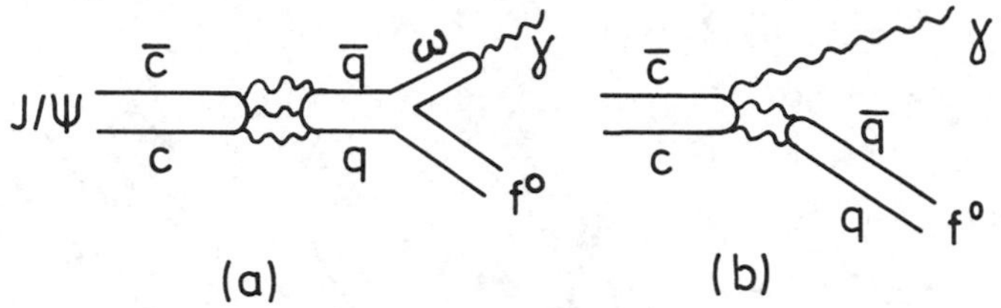

Fig. 18. Diagrams for the decay $J/\psi \rightarrow f^\circ \gamma$ with (a) three gluons, and (b) two gluons exchanged.

The angular distribution in terms of the $f^\circ$-helicity amplitudes $A_0$, $A_1$, $A_2$, with $x = A_1/A_0$, and $y = A_2/A_0$ is given by

$$W_f(\delta_P, \delta_M, \phi_M) \propto \tag{9}$$

$$3x^2 \sin^2\delta_P \sin^2 2\delta_M +$$

$$(1 + \cos^2\delta_P) \left[(3 \cos^2\delta_M - 1)^2 + \frac{3}{2} y^2 \sin^4\delta_M\right] -$$

$$\sqrt{3} x \sin 2\delta_P \sin 2\delta_M (3\cos^2\delta_M - 1 - \sqrt{\frac{3}{2}} y \sin^2\delta_M) \cos\phi_M +$$

$$\sqrt{6} y \sin^2\delta_P \sin^2\delta_M (3\cos^2\delta_M - 1) \cos 2\phi_M,$$

where $\delta_P$ is the production angle between the $f^\circ$-meson and the $e^+ -$ flight direction. The polar and azimuthal decay angles $\delta_M$ and $\phi_M$ are those of the $\pi^+$-meson measured in the helicity frame of the $\pi^+\pi^-$ center of mass system, where the z-axis is defined by the flight direction of the $\pi^+\pi^-$ - system.

The results of a fit to the data, the method of which is too elaborate to be discussed here, are given in fig. 19 by lines for constant $\chi^2$ in the x, y-plane. The $\chi^2$ values assigned to the contours are measured from the minimum $\chi^2 = 26$, marked by a cross, obtained for 25 degrees of freedom. Also shown is the QCD-prediction of $x = 0.76$, $y = 0.54$ [19]. The best experimental values of $x = 0.6 \pm 0.3$, $y = 0.3 {}^{+0.6}_{-1.6}$ are in very good agreement. The same is true for the nearby values of $x = 0.71$, $y = 0.41$ derived from a tensor dominance model [21].

To check the fitting results we repeated the fit by varying the background percentage in the $f^o$ mass band in the range of 40 - 60%, but no appreciable difference from the result given in fig. 19 were observed. A fit to only the background distributions, however, yields a rather different result, namely $x = -1.2$, $y = -0.4$ with $\chi^2 = 77$ for 25 degrees of freedom. A final check is made by applying the method to the angular distribution of the decay $J/\psi \to \rho^o \pi^o$. As fig. 17 shows

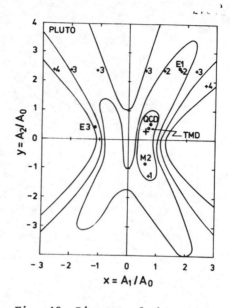

Fig. 19. Diagram of the constant $\chi^2$ in the x,y-plane from a fit of experimental data to distribution (9).

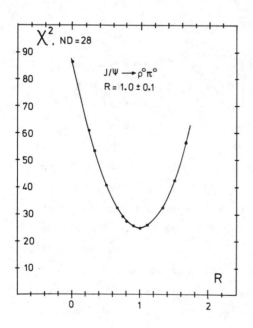

the $\rho^o$ is observed simultaneously with the $f^o$, but has to be associated with a $\pi^o$. The angular distribution of this process :

Fig. 20.
Fit to the angular distribution (10) of the decay $J/\psi \to \rho^o \pi^o$, $\chi^2$ vs. R.

$$W (\delta_P, \delta_M, \phi_M) \propto \qquad (10)$$

$$(1 + \cos^2 \delta_P) \sin^2 \delta_M + R \sin^2 \delta_P \sin^2 \delta_M \cos 2\phi_M$$

is described by only one independent helicity amplitude and is therefore uniquely fixed with R = +1. Our fitting procedure yields in fact R = 1.0 ± 0.1 with $\chi^2$ = 25 for 28 degrees of freedom, see fig. 20, a result which is in excellent agreement with R = 1 for the $J^P = 1^-$ assignment to the $J/\psi$ particle.

The decay $J/\psi \to f^o \gamma$ has also been observed in the DASP detector [22] by requiring one pion in the outer arm, and one track with a photon in the inner, non magnetic detector. The resulting Dalitz distribution, given in fig. 21 (a), is similar to the one given in fig. 16 for the PLUTO data. Fig. 21 (b) shows the resulting invariant mass distribution, where the $\rho^o$ and $f^o$ mass bands are populated. The branching ratio obtained in this experiment depends on the assumed angular distribution, which is unmeasurable. It is given in table II for the three radiative multipole transitions $E_1$, $M_2$, and $E_3$, which are also shown for comparison in fig. 19. It can be seen that only the $M_2$ transition lies within one standard deviation from the best value obtained by PLUTO. The branching ratio given for $M_2$ also agrees best with the respective value obtained in the PLUTO experiment.

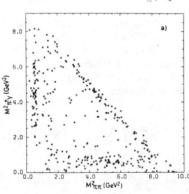

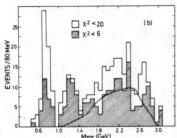

Fig. 21 (a) Dalitz plot representation of reaction (8),

(b) Invariant mass distribution $M(\pi^+\pi^-)$, DASP.

Table II  Branching Ratios for $J/\psi \rightarrow f^o\gamma$, DASP

| type of transition | branching ratio |
|---|---|
| $E_1$ | $(0.9 \pm 0.3) \cdot 10^{-3}$ |
| $M_2$ | $(1.5 \pm 0.4) \cdot 10^{-3}$ |
| $E_3$ | $(1.0 \pm 0.3) \cdot 10^{-3}$ |

Being aware that the storage ring is an integral part of $e^+e^-$ - physics, our thanks go to the collegues of the DORIS operation group for their continuous and effective efforts to optimize the running conditions for experiments.

## REFERENCES

1. PLUTO-Collaboration
   DESY  G. Alexander, L. Criegee, H.C. Dehne, K. Derikum, R. Devenish, G. Flügge, J.D. Fox, G. Franke, Ch. Gerke, E. Hackmack, P. Harms, G. Horlitz, Th. Kahl, G. Knies, H. Lehmann, R. Schmitz, R.L. Thompson, U. Timm, H. Wahl, P. Waloschek, G.G. Winter, S. Wolff, W. Zimmermann

   U. Aachen  W. Wagner

   U. Hamburg  V. Blobel, A. Garfinkel, B. Koppitz, E. Lohrmann, W. Lührsen

   U. Siegen  A. Bäcker, J. Bürger, C. Grupen, G. Zech

   U. Wuppertal  H. Meyer, M. Rössler, K. Wacker.

2. DASP-Collaboration
   DESY  D. Cords, R. Felst, R. Fries, E. Gadermann, H. Hultschig, P. Joos, W. Kock, U. Kötz, H. Krehbil, D. Kreinick, H.L. Lynch, W.A. McNeely, G. Mikenberg, K.C. Moffeit, D. Notz, R. Rüsch, M. Schliewa, B.H. Wiik, G. Wolf

   U. Aachen  R. Brandelik, W. Braunschweig, H.-U. Martyn, H.G. Sander, D. Schmitz, W. Sturm, W. Walraff

   U. Hamburg  G. Grindhammer, J. Ludwig, K.H. Mess, A. Petersen, G. Poelz, J. Ringel, O. Römer, K. Sauerberg, P. Schmüser

MPI München   W.de Boer, G. Buschhorn, W. Fues, Ch.v. Gagern,
              B. Gunderson, R. Kotthaus, H. Lierl, H. Oberlack

U. Tokyo      S. Orito, T. Suda, Y. Totsuka, S. Yamada.

3. M.L. Pearl et al., Phys.Rev.Lett. 35, 1489 (1975)

4. PLUTO-Collaboration, J. Burmester et al., Phys.Lett.68B, 297
   (1977), and DESY 77/24

5. A. Barbaro-Galtieri, Proc. 1977 Int.Symp. on Lepton and Photon
   Interactions at High Energ., Hamburg, Aug. 1977, page 21

6. DASP-Collaboration, R. Brandelik et al., Phys.Lett. 73B, 1o9,
   (1978), and DESY 77/81

7. G. Knies, Proc. 1977 Int. Symp. on Lepton and Photon Interactions at High En., Hamburg, Aug. 1977, page 93

8. M. Murtagh, Proc. 1977 Int. Symp. on Lepton and Photon Interactions at High En., Hamburg, Aug. 1977, page 405

9. K. Schultze, Proc. 1977 Int. Symp. on Lepton and Photon
   Interactions at High En., Hamburg, Aug. 1977, page 359

10. H. Fritzsch, Phys.Lett. 67B, 451 (1977)

11. PLUTO-Collaboration, G. Alexander et al., Phys.Lett. 73B, 99
    (1978), and DESY 77/78

12. DASP-Collaboration, R. Brandelik et al., Phys.Lett. 70B, 387,
    (1977), and DESY 77/41

13. Y.S. Tsai, Phys.Rev. D4, 2821 (1971)
    M.B. Thacker, J.J. Sakurai, Phys.Lett. 36B, 103 (1971)

14. A. DeRujula, H. Georgi, Phys.Rev. D13, 1297 (1976)
    E.C. Poggio, H.R. Quinn, S. Weinberg, Phys.Rev. D13, 1958(1976)
    R. Barbieri, R. Gatto, Phys.Lett. 66B, 181 (1977)

15. J.H. Mulvey, $\Lambda = 0.73 \pm 0.1$, Report on this Conference

16. PLUTO-Collaboration, G. Alexander et al., Phys.Lett. 72B,
    493 (1978), and DESY 77/72

17. S. Okubo, Phys.Lett. 5, 105 (1963)
    G. Zweig, CERN-Report TH 401,412 (1969)
    J. Iikuza, K. Okada, O. Shito, Proc.Theor.Phys.35, 1061 (1966)

18. PLUTO-Collaboration, J. Burmester et al., Phys.Lett. 72B, 135
    (1977), and DESY 77/50

19. M. Krammer, DESY 78/06, Phys.Lett. to be published

20. PLUTO-Collaboration, G. Alexander et al., Phys.Lett. to be
    published, and DESY 78/20

21. W. Gampp and H. Genz, Karlsruhe University Preprint,
    TKP/78-4, to be published

22. DASP-Collaboration, R. Brandelik et al., Phys.Lett.to be pub-
    lished, and DESY 78/01

# NEW RESULTS FROM DELCO

L. J. Nodulman[*]
University of California Los Angeles
Los Angeles, California 90024

## ABSTRACT

New results from the DELCO experiment at SPEAR are presented. New data and reanalysis give conclusive evidence for the τ heavy lepton with mass 1.777 + .005 - .009 GeV/$c^2$. Preliminary branching ratios for the τ are discussed, along with analysis of the electron momentum spectrum, favoring V-A. A preliminary look at D beta decay at ψ"(3770) in terms of K and K*(890) is also discussed.

## INTRODUCTION

DELCO[1] is an east pit experiment at SPEAR designed to study production of new particles tagged by direct electrons from weak decays. In practice we study charm[2] and the heavy lepton.[3] The physicists are a collaboration from Stanford, UCLA, Stony Brook, and U.C. Irvine.[4] I will describe the detector and briefly review old results presented last summer before describing our new results, which are of course, preliminary.

## THE DETECTOR

The detector is shown in Fig. 1. The interaction region is in a small volume of magnetic field (3.5kG). The beam passes through the poles of the magnet, and the return yoke is extended far up and down to avoid interfering with the detectors. Six cylinders of low mass wire proportional chambers with radii from 10-30 cm surround the vacuum pipe. Cathode strips aid in z reconstruction. Scintillation counters on the pole tips increase the solid angle for charge particle detection. The MWPC are surrounded by a segmented atmospheric ethane Cerenkov counter of one meter radiator. Two bounce optics focuses the Cerenkov light on five inch phototubes. Each of the sextents follows with two planes of XY magnetostrictive wire spark chambers and lead scintillator sandwich shower counters. The inner scintillator strips are timed at both ends. Over the summer the eight inch lead walls followed by spark chambers and scintillation counters were added to give muon identification over 20% of 4π. We do not have results to report yet on identified muons.

---

[*] Present mailing address: Stanford Linear Accelerator Center, Stanford, California 94305

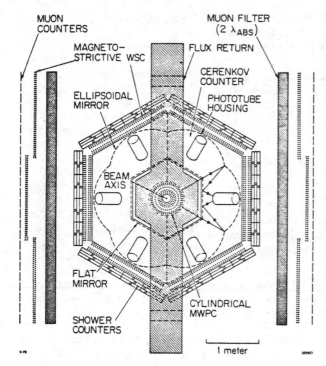

Our basic trigger requires two out of three layers in a shower counter in at least two sextants including at least one timed layer, as well as one or more tracks in the inner two layers of MWPC. There are neutral triggers which are currently being studied.

The detector gives some charged particle detection over 90% of 4π, and electron identification (hadron rejection <5x10$^{-4}$) and momentum measured (σ of δp/p is 8% x p (GeV)) over about 60% of 4π.

**Fig. 1.** The DELCO Apparatus

## OLD RESULTS

Quoted results from DELCO[1] include measurement of

$$R \equiv \left( \frac{\sigma\ e^+ e^- \to \text{hadrons}}{\sigma\ e^+ e^- \to \mu^+ \mu^-} \right)$$

in the charm threshold region with description of the ψ"(3770), the similar rate for multiprong events with an electron showing the charm resonance structure and, by comparison of the ψ", giving a D beta decay branching ratio of 11 ± 2%. A smooth excitation function for two prong events where one is identified as an electron and the other not (eX) was contrasted with the multiprong electron events as evidence for the τ.

## NEW RESULTS

New data points have been added to a reanalysis of eX events. Clean two prong events where both prongs could be identified electrons are selected in a physicist scan of reconstructed events. Any number of photons is allowed. The electron is required to have momentum greater than 200 MeV/c and the non electron greater than 300 MeV/c. These cuts are explained by the Cerenkov eff⁻¹⁻⁻ency determined by studying electron pairs, shown in Fig. 2.

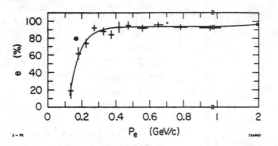

**Fig. 2.** Cerenkov efficiency. The data points are from analysis of electron pairs. The curve is a Monte Carlo calculation

The relative azimuthal angle must be less than 160°. Apparent eeγ topology events (15) are removed. A sample of 660 events results. To further eliminate electrons which have not been Cerenkov tagged, the X prong is required to have a shower counter pulse height less than 3.3 times minimum ionizing. This leaves 540 events with an estimated background of 15 events, which is

consistant with the rate of candidates seen at the ψ(3095) and $E_{CM}$ of 3.5 GeV. Two photon processes are estimated by comparing like sign events to opposite sign candidates, about 2%.

The first conclusion we reach is that any remaining doubts about the existence of the τ are dispelled. Not only is the excitation completely inconsistent with the charm associated structure, but also events are found at several energies below charm threshold: $E_{CM}$=3.72 and 3.625 GeV/c and the ψʹ. Candidates below charm threshold were reported previously.[6] The eX spectrum is shown in Fig. 3 for all candidates and events without photons. The distributions are fit to a

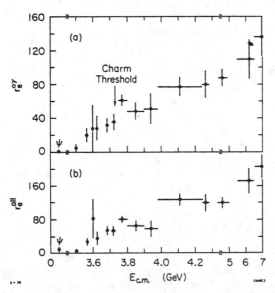

**Fig. 3.** Rate of eX production normalized to the muon pair rate for (a) events with no photons (b) all events. ($r=R_{eX}$x10$^{-3}$)

background plus heavy lepton. The point at the ψʺ is estimated to contain 30% background from charm and is eliminated from the fit. Charm background at other points is estimated to be less than our statistical error. The fits yield a heavy lepton mass and twice the branching ratio to electrons times the branching ratio to X. The results are summarized in Table I.[7] To interpret the branching ratio information we are required to use a model due to Gilman and Miller[8] yielding a branching ratio to electrons of 16±1% and a branching ratio to multiprongs of 32±3%, where the error is statistical only.

## TABLE I

| Sample | No. photons | All Events |
|---|---|---|
| Mass (GeV) | .1.777 + .005 − .009 | 1.780 + .003 − .006 |
| $2b_e b_x$ | .118 ± .008 | .170 ± .010 |
| $\chi^2/\text{NDF}$ | 9.9/11 | 19.2/11 |

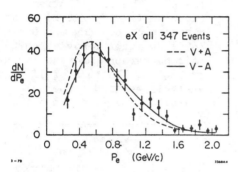

eX all 347 Events
--- V+A
— V−A

$\frac{dN}{dP_e}$    $P_e$ (GeV/c)

**Fig. 4.** Electron momentum spectrum from eX events, all energies. The curves are detailed Monte Carlo generated fits.

## TABLE II

| | $\rho$ | $\chi^2/\text{NDF}$ |
|---|---|---|
| 0 V+A | | 38/18 |
| .75 V−A | | 17.6/18 |
| .73 ± .15 | | 17.5/17 |

Next we study the electron momentum distribution. In complete analogy to muon decay, the momentum spectrum yields information about the helicity of the $\tau$ neutrino and thus the V−A or V+A characteristic of the decay. For maximum electron momentum the neutrinos are parallel and the combination is allowed for V−A and forbidden for V+A. We analyze the momentum distribution in terms of Michel parameter and also study the average of electron momentum divided by the beam energy, predicted to be .35 for V−A and .30 for V+A. The momentum distribution is shown in Fig. 4. The fits are summarized in Table II. Average scaled electron momentum is plotted in Fig. 5. We conclude that if V+A is not ruled out, it is at least very unlikely (<1% probability).

The effect of a finite neutrino mass is also to give lower electron momentum.[8] Fits to the momentum distribution give a 90% confidence level upper limit to the $\tau$ neutrino mass of 250 MeV/$c^2$.

Next we study the branching ratio of $\tau$ into multiprongs. Two methods are used. First we investigate multiprong electron events in the charm depleted regions at $E_{CM}$ of 3.72, 3.85 and 4.25 GeV. To eliminate residual charm background we require the electron momentum to be greater than one third the beam momentum. This gives 78 multiprongs compared to 29 eX events. After correction for relative detection efficiency the ratio is 2.1 ± .4, giving a multiprong branching ratio, assuming $b_e$, of 34 ± 6%. We can also use the fact that charm events do not give stiff electrons. As we cut progressively higher on the electron momentum the ratio of multiprongs to eX events

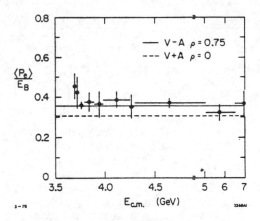

Fig. 5.   Average scaled electron momentum

should become constant. This distribution is shown in Fig.6. The ratio is 1.8 ± .3 gives a multiprong branching ratio of 35 ± 6%. This high multiprong branching ratio implies a 20% background from the τ in multiprong electrons at the ψ"!

Assuming we now reasonably understand the background from τ at the ψ" we can now investigate the electron spectrum at the ψ". The distribution is shown in Fig. 7, fit to a combination of D → Keν and D → K* (890) eν after removing soft pair background and the τ contribution. In this preliminary analysis we find comperable contributions from K and K*.

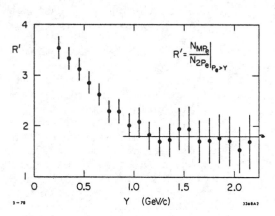

Fig. 6. Ratio of multiprong electrons to eX events versus minimum required electron momentum.

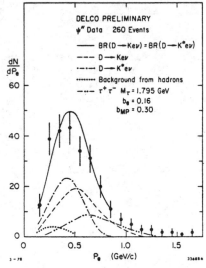

Fig. 7 Electron momentum distribution for multiprong electron events at the ψ".

## SPECULATION

Data taking continues and may include new points for the eX analysis. A reasonable point requires about a week of data. Further and more detailed analysis of the τ including muon information, and analysis of charm is continuing. Probably not all developments can be forseen.

## FOOTNOTES

1. The Direct Electron Counter experiment is supported in part by grants from the Department of Energy and the National Science Foundation.

2. S.L. Glashow, Illiopoulos and Maiani, Phys.Rev. D 2, 1285 (1970). G. Goldhaber et al., Phys. Rev. Letters 37, 1255 (1976).

3. M.L. Perl, et al., Phys. Rev. Letters 35, 1489 (1975).

4. DELCO Physicists include W. Bacino, R. Burns, P. Condon, P.Cowell, A. Diamant-Berger, T. Ferguson, A.Hall, J. Hauptman, G. Irwin, J. Kirkby, J. Kirz, J. Liu, F. Merritt, L. Nodulman, W. Slater, M. Schwartz, H. Ticho, and S. Wojcicki.

5. W. Bacino et al., Phys. Rev. Letters, 40, 671 (1978), J. Kirkby, SLAC-PUB-2040, 1977 International Symposium on Lepton and Photon Interactions at High Energies, Hamburg, Germany, August 25-31, 1977.

6. R. Brandelik et al., Phys. Letters, 37B, 109 (1978).

7. Heavy lepton mass results have been updated to include 9 events at $E_{CM}$=3.57, giving 1.75< $m_T$ <1.785 GeV/c$^2$.

8. F. Gilman and D.H. Miller, Phys. Rev. to be published, SLAC-PUB 2046 (1977).

9. The algorithm for including finite neutrino mass in our calculations was provided by Yung-Su Tsai.

HEAVY QUARK SYSTEMS

E. Eichten[*]
Lyman Laboratory of Physics
Harvard University
Cambridge, Massachusetts 02138

ABSTRACT

A critical summary of the status of nonrelativistic potential models for heavy quark systems is given.  Both the $\psi$ and $\Upsilon$ families of resonances are discussed.  The extension of potential models to include Zweig allowed hadronic decay is discussed in the context of interpreting the structure of R in the charm threshold region.

## I.   INTRODUCTION

In the nearly three and a half years since the discovery of the $\psi$/J resonance a vast amount of experimental information concerning the $\psi$ family of resonances has been obtained.[1]  It is remarkable that the interpretation of these resonances as charmed quark-antiquark $(c\bar{c})$ bound states interacting via a nonrelativistic potential has survived so successfully.

I take this occasion to summarize these successes and more importantly to emphasize what I consider the most significant inadequacies of these models.  The experimental information presently available on the $\Upsilon$ system[2,3] (presumably a family of even heavier quark-antiquark bound states) reflects on our understanding of the $(c\bar{c})$ system itself and, as I will try to show, complements that understanding.  Secondly, I hope to point out those aspects of the extension of potential models of $(c\bar{c})$ states to include Zweig allowed hadronic decays which bear significantly on interpreting the complex structure of the charm threshold region in $e^+e^-$ annihilation and lead to direct experimental tests.

I begin with a very brief summary[4] of the status of our understanding of the interactions between a very massive quark-antiquark pair within the context of quantum chromodynamics.  The basic criteria for quark confinement is the Wilson loop criteria[5] which I will restate as follows:[6]  Construct a color singlet gauge invariant state containing an infinitely massive quark and antiquark separated by a distance r at time -T/2; the probability amplitude (in Euclidean space) that at time +T/2 you find a state with the quark and antiquark separation again r behaves for $T \to \infty$ as $A_{if}\exp\{-E(r)T\}$.  $A_{if}$ depends on the particular initial and final state considered.  Confinement occurs if $E(r) \sim r$ as $r \to \infty$.  This procedure allows (at least in principle) the calculation of E(r).  For the phenomenology of heavy quark systems E(r) is to be understood as the nonrelativistic potential V(r).  The V(r) so defined does not have any contribution from internal quark pairs; we will return to this point later.

[*]Research supported in part by the National Science Foundation under Grant No. PHY77-22864.

For short distances $(r \to 0)$, or equivalently large momentum transfer $t = -Q^2$, $E(r)$ can be calculated in ordinary perturbation theory.[7] The result [8,9] through third order in the running coupling constant, $\alpha_s(Q^2)$, for the static potential between a quark and anti-quark in a total color singlet state is

$$V(Q^2) = - \frac{16\pi\,\alpha_s(Q^2)}{3Q^2}\left[1 + b\,\alpha_s^2(Q^2) + 0\left(\alpha_s^3(Q^2)\right)\right] \tag{1}$$

where[9] $b \approx +5$.

The potential is basically Coulombic with a small logarithmic $Q^2$ dependence. It is interesting to add that a potential exists only for total color singlet states.[9] The expected behavior, $E(r) \sim r$, at large distance has not yet been clearly established. However the discovery that there are nonperturbative Euclidean solutions to the field equations in QCD called instantons as discussed at this conference by C. Callan is certainly a hopeful development. Explicit calculations[10] of the contribution of instantons to $E(r)$ show that instanton effects are important at intermediate distance scales but do not provide for linear confinement at large distances. It has recently been suggested[10,11] that a nonperturbative configuration related to instantons called merons may actually provide the linear confinement at large distance. This conjectured situation is summarized in the plot of $E(r)$ in Fig. 1.

Within the near future we must be content with phenomenological models of $V(r)$. For most of this talk I will restrict my attention to the simple linear plus Coulomb potential

$$V(r) = - \frac{K}{r} + r/a^2 + V_0 \tag{2}$$

Two comments about the choice of this potential. First, although the form of $V(r)$ in Eq. (2) is consistent with our limited understanding of $E(r)$ it is in no way mandated. The important region of $r$ for $(c\bar{c})$ states is between 0.1 fm and 0.5 fm where the knowledge of $E(r)$ is most limited. Second, I will be interested in identifying those aspects of the charmonium system for which phenomenological potentials seem to fail seriously. These general features depend only on the gross structure of the potential.

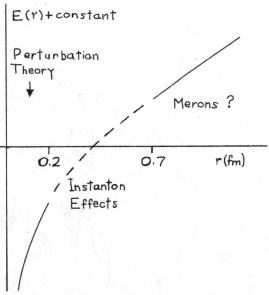

Figure 1. Behavior of $E(r)$. Possible major contributor on each distance scale indicated.

## II. COMPARISON OF POTENTIAL MODELS WITH
## EXPERIMENT: SPECTRUM AND TRANSITIONS

The potential in Eq. (2) has been studied in detail by K. Gott-fried, T. Kinoshita, K. D. Lane, T. M. Yan, and myself.[12] The param-eters $\kappa$, a, and the charm quark mass $m_c$ were fitted to the $\psi/J - \psi'$ splitting, the wavefunction at the origin of $\psi/J$. The remaining freedom is used to accommodate an acceptable fit to both the E1 tran-sitions and P states center of gravity. The following fit was ob-tained

$$V(r) = -\frac{0.132}{r} + \frac{r}{(2.07 \text{ GeV}^{-1})^2} - 0.872 \text{ GeV}. \tag{3}$$

The interpretation of the resonance structure from measurements of R alone is difficult. The four measurements of R in the charm threshold region are shown in Figs. 2a, 2b, 2c, and 2d.

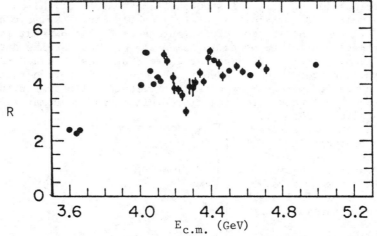

Figure 2a. R as measured by the PLUTO collaboration[13] at DESY.

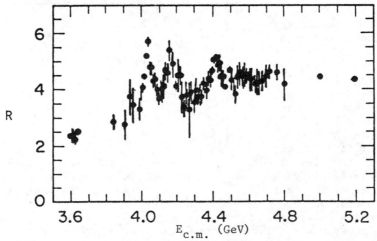

Figure 2b. R as measured by the DASP collaboration[14] at DESY.

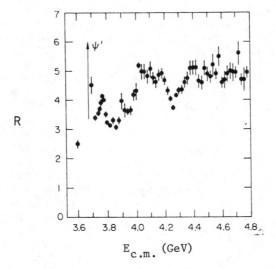

Figure 2c.  R as measured by the DELCO collaboration[15] at SLAC.

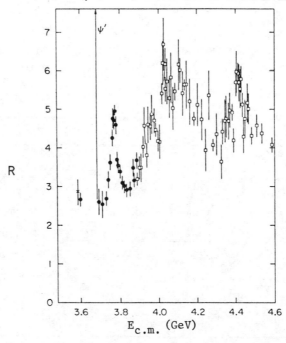

Figure 2d.  R as measured by the SLAC–LBL collaboration[16] at SLAC.

One can clearly identify resonances at 3.77 GeV, 4.03 GeV, and 4.41 GeV; there is also indication in all four measurements of an enhancement at 4.17 GeV.  There is also some indication of structure at 3.95 GeV; however this structure has a testable nonresonance

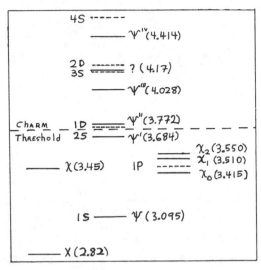

Figure 3. Spectrum of states. Observed states indicated with solid lines. The $(c\bar{c})$ states predicted by the potential in Eq. (3) is indicated with dashed lines. The $\psi/J$ and $\psi'$ masses are inputs to the model.

interpretation that I will discuss later.

These considerations suggest the spectrum of $(c\bar{c})$ states indicated in Fig. 3. States observed through the radiative transitions of $\psi/J$ and $\psi'$ are also indicated. The general agreement with the predictions of the potential in Eq. (3) is excellent. A comparison of the details[17,18] of the potential model above with the data for $^3P_J$ and $^3D_1$ states, and the El transition rates is given in Table I.

I am not concerned here with the specific agreement of this potential model with the data. However the dependence of the results on the choice of the Coulomb parameter $\kappa$ will be important. The results for the two values of $\kappa$ given in Table I indicate that the effect of increasing $\kappa$ is to improve agreement for

| | $\kappa = 0$ | $\kappa = 0.132$ | $\kappa = 0.132$ | exp. | Ref. |
|---|---|---|---|---|---|
| | | | Coupled Channels | | |
| Center of Gravity $1^3P_J$ Multiplet | 3439 | 3457 | 3446 | 3522 ± 5 MeV | 19 |
| Mass of $1^3D$ State | 3738 | 3755 | 3760 | 3772 ± 6 MeV | 16 |
| | | | El Transitions | | |
| $\Gamma\left(\psi' \rightarrow \gamma\chi(3.550)\right)$ | 23 | 27 | 19 | 16 ± 9 keV | 20 |
| $\Gamma\left(\psi' \rightarrow \gamma\chi(3.510)\right)$ | 32 | 38 | 27 | 16 ± 8 keV | 20 |
| $\Gamma\left(\psi' \rightarrow \gamma\chi(3.415)\right)$ | 37 | 44 | 37 | 16 ± 9 keV 17 ± 8 keV | 20 21 |

Table I. Comparison of the details of the mass spectrum and El transition rates. The results for a linear potential $\kappa = 0$, $a = 1.95$ GeV$^{-1}$, $m_c = 1.85$ GeV are displayed in the first column. The results for the potential of Eq. (3) are displayed in the second column. While in the third column the results of including coupling to decay channels for the second potential are given.

the mass spectrum at the expense of agreement for the E1 transition rates.

Now let us turn to the evident inadequacies of this simple potential model in an attempt to learn how to make improvements. Three general areas of disagreement with the experiment can be distinguished.

First, the observed states X(2.82)[14] and X(3.45)[19,14] have a natural interpretation within the (c$\bar{\text{c}}$) potential scheme as the pseudoscalar partners ($\eta_c$ and $\eta_c'$) of the $\psi$/J and $\psi'$ respectively. This interpretation, however, leads to disastrous predictions for M1 transition rates. Let me make the following assumptions:

(a)  The magnetic moment of the charmed quark is simply given by the Dirac moment

$$\frac{e_c}{2m_c} \text{ with } m_c = 1.65 \text{ (GeV)} .$$

(b)  The radial wavefunctions for any S state is spin-independent. In particular

$$\int d^3 r \, \psi^\dagger_{n\,^3S_1}(r) \psi_{m\,^1S_0}(r) = \delta_{mn}$$

for n,m = 1 or 2.

(c)  Asymptotic freedom arguments apply for calculating the behavior of the total rate of Zweig forbidden hadronic decays. This implies that for n = 1 or 2

$$\frac{\Gamma(n\,^1S_0 \to \text{hadrons})}{\Gamma(n\,^1S_0 \to \gamma\gamma)} = \frac{9}{8}\left(\frac{\alpha_s}{\alpha}\right)^2 .$$

These three assumptions may be used to calculate the M1 transition rates under the assumption that X(3.45) and X(2.82) are the pseudoscalar partners of $\psi'$ and $\psi$/J respectively. The comparison with available data is given in Table II.

| Branching Ratio | Experiment | Ref. | Theory | Assumptions Used |
|---|---|---|---|---|
| $\psi' \to \gamma X(2.82)$ | Not seen $\leqslant 1\%$ | 20 | 3% | a,b |
| $\psi \to \gamma X(2.82)$ | Seen in 3$\gamma$ mode<br>Not seen directly $\leqslant 1.7\%$ | 14,22<br>20 | 36% | a,b |
| $\psi' \to \gamma X(3.45)$ | Seen in $\psi' \to \gamma\gamma\psi$<br>Not seen directly $\leqslant 2.5\%$ | 20 | 5% | a,b |
| $\psi' \to \gamma X \to \gamma\gamma\psi$ | $0.6 \pm 0.4\%$ | 19 | $2 \times 10^{-3}\%$ | a,b,c |

Table II.  M1 Transition rates under the assumption that X(2.82) and X(3.45) are pseudoscalar (c$\bar{\text{c}}$) states.

The transition rate $\psi \to \gamma X(2.82)$ clearly illustrates the difficulty with assigning the X(2.82) as the pseudoscalar partner of the $\psi$. The transition would be an unhindered M1 and the rate calculated relies only on assumptions (a) and (b). These assumptions can be subject to independent experimental tests.

The rate of $D^{*+} \rightarrow \gamma D^+$ is quite sensitive to the magnetic moment of the charmed quark for values near the expected value. Parameterizing the charm quark moment as $e_c/2m_c^*$ ($m_c^* = m_c$ recovers the Dirac moment) and noting that the momentum of the photon in the decays $D^{*+} \rightarrow \gamma D^+$ and $D^{*0} \rightarrow \gamma D^0$ is nearly equal, we have

$$\frac{\Gamma(D^{*+} \rightarrow \gamma D^+)}{\Gamma(D^{*0} \rightarrow \gamma D^0)} = \frac{(k_\gamma^+)^3}{(k_\gamma^0)^3} \frac{\left[\frac{e_c}{2m_c^*} + \frac{e_d}{2m_d}\right]^2}{\left[\frac{e_c}{2m_c^*} + \frac{e_u}{2m_u}\right]^2} \simeq \left(\frac{m_d - m_c^*/2}{m_u + m_c^*}\right)^2 \left(\frac{m_u}{m_d}\right)^2 \tag{4}$$

where $e_u$ and $m_u$ and $e_d$ and $m_d$ are the charge and constituent mass of the up and down quarks respectively. Also

$$\frac{\Gamma(D^{*+} \rightarrow \gamma D^+)}{\Gamma(D^{*0} \rightarrow \gamma D^0)} = \frac{\Gamma(D^{*+} \rightarrow \gamma D^+)}{\Gamma(D^{*+} \rightarrow \pi^0 D^+)} \left(\frac{\Gamma(D^{*0} \rightarrow \gamma D^0)}{\Gamma(D^{*0} \rightarrow \pi^0 D^0)}\right)^{-1} \frac{\Gamma(D^{*+} \rightarrow \pi^0 D^+)}{\Gamma(D^{*0} \rightarrow \pi^0 D^0)} \ . \tag{5}$$

From M1 transitions among the ordinary mesons we know $m_u \approx m_d \approx 0.34$ GeV. Simple phase space[23] and isospin arguments determine

$$\frac{\Gamma(D^{*+} \rightarrow \pi^0 D^+)}{\Gamma(D^{*0} \rightarrow \pi^0 D^0)} = \frac{(k_{D^{*+} \rightarrow \pi^0 D^+})^3}{(k_{D^{*0} \rightarrow \pi^0 D^0})^3} = \left(\frac{36.6 \pm 2.5 \text{ MeV}}{45.3 \pm 1.9 \text{ MeV}}\right)^3 \ . \tag{6}$$

The ratio $\Gamma(D^{*0} \rightarrow \gamma D^0)/\Gamma(D^{*0} \rightarrow \pi^0 D^0)$ has been experimentally determined[23] to be $(55 \pm 15)/(45 \pm 15)$. The same ratio for $D^{*+}$ decays, $\Gamma(D^{*+} \rightarrow \gamma D^+)/\Gamma(D^{*+} \rightarrow \pi^0 D^+) \equiv 1/x$ has not yet been measured. We can use Eqs. (4), (5), and (6) to obtain $m_c^*$ in terms of the unknown x

$$m_c^* = \frac{\sqrt{x/a} + 1}{\sqrt{x/4a} - 1} (0.34) \text{ GeV} \quad \text{where} \quad a = 0.4 \pm 0.1 \ . \tag{7}$$

The sensitivity of $m_c^*$ on x is clear from Eq. (7); for $m_c^* = 1.65$ GeV we require $x \simeq 12$ while the value of $m_c^*$ needed to give agreement with the M1 rates ($m_c^* \simeq 5$ GeV) requires $x \simeq 3$.

The assumption (b) is tested by the failure of DASP to observe $\psi' \rightarrow \gamma X \rightarrow \gamma\gamma\gamma$.[19] This sets a limit on the rate

$$\frac{\Gamma(\psi' \rightarrow \gamma X)}{\Gamma(\psi \rightarrow \gamma X)} \leqslant 10 \qquad 90\% \text{ C.L.}$$

When we take into account the large difference in phase space we obtain a limit on the wavefunction overlaps

$$|\langle\psi'|X\rangle|^2 / |\langle\psi|X\rangle|^2 \leq 0.23 \ . \tag{8}$$

Within any potential model the radial wavefunction of the $1^1S_0$ state may be expressed by completeness as a sum of $n^3S_1$ states radial wavefunctions ($n = 1, 2, \ldots$). To obtain agreement for the M1 transition rate we would require

$$\int d^3 r \psi_{1S_0}(r) \psi_{3S_1}(r) \approx 0.25 . \qquad (9)$$

This would require a value of 0.7 for the sum of the squares of the overlap of the nodeless wavefunction of the X state with the $n^3S_1$ states ($n \geq 3$). This unreasonable situation certainly suggests that any failure of assumption (b) is not simply due to a large spin-spin force but that additional degrees of freedom are needed to describe the $\psi$ system.

The success of the assumptions similar to (a) and (b) for M1 transitions involving the ordinary mesons[24] makes the disagreement here puzzling indeed. Hopefully the status and quantum numbers ($J^P$) of the X and X(3.45) states will be firmly established in the near future as this problem of the pseudoscalar states is the most serious failure of the simple potential picture.

The second basic inadequacy of the model of Eq. (3) is that no spin-dependent splittings are included. These effects are a relativistic correction to the nonrelativistic limit of an interaction. Since we are without an understanding of the relativistic form of the interaction in QCD extensions to include spin-dependent forces rely on the analogy of the perturbative sector of QCD with QED and use a Breit-Fermi form for the spin-dependent terms. The general form assumed may be parameterized as

$$\begin{aligned}
V_{spin}(r) = \frac{1}{2m_c}&\left\{\frac{4\kappa}{r^3} + 4(1+\lambda)\frac{1}{r}\frac{dV_1}{dr} - \frac{1}{r}\frac{dV_0}{dr}\right\}\vec{L}\cdot\vec{S} \\
+ \frac{2}{3m_c^2}&\left\{4\pi\kappa\,\delta(\vec{r}) + (1+\lambda)\nabla^2 V_1\right\}\vec{S}_1\cdot\vec{S}_2 \\
+ \frac{1}{3m_c^2}&\left\{\frac{3\kappa}{r^3} + \frac{1}{r}\frac{dV_1}{dr} - \frac{d^2V_1}{dr^2}\right\}[3\vec{S}_1\cdot\hat{r}\vec{S}_2\cdot\hat{r} - \vec{S}_1\cdot\vec{S}_2] .
\end{aligned} \qquad (10)$$

This form arises from the nonrelativistic limit of the Bethe-Salpeter equation with the following kernel:

$$\begin{aligned}
V_{B-S}(k^2) = \;& \gamma_c^{\mu_1} \otimes \gamma_{\bar{c}}^{\mu_2} V_{coulomb}(k^2) \\
& + \left[\gamma^{\mu_1} - \frac{i\lambda}{2m_c}\sigma^{\mu_1\nu_1}k_{\nu_1}\right]_c \otimes \left[\gamma^{\mu_2} - \frac{i\lambda}{2m_c}\sigma^{\mu_2\nu_2}k_{\nu_2}\right]_{\bar{c}} V_1(k^2) \\
& + \mathbb{1}_c \otimes \mathbb{1}_{\bar{c}} V_0(k^2) .
\end{aligned}$$

The B-S kernel is arranged to obtain the linear plus Coulomb form for the spin-independent part of the nonrelativistic interaction,

$$V_{coulomb}(r) = \frac{-\kappa}{r} \qquad\qquad V_1 = \eta(r/a^2) \qquad\qquad V_0 = (1-\eta)(r/a^2) .$$

This parameterization has been used by T. Appelquist, R. M. Barnett, and K. D. Lane[18] to analyze the various calculations of spin-dependent forces. Their results are listed in Table III.

Recently Callan, Dashen, Gross, Wilczek, and Zee[25] have calculated the spin-dependent effects of instantons in heavy quark systems. The spin-spin interaction is sufficiently strong to account for the X-$\psi$

| Author; Parameters | $M_{\chi_2} - M_{\chi_1}$ | $M_{\chi_1} - M_{\chi_0}$ | $M_\psi - M_{\eta_c}$ | $M_{\psi'} - M_{\eta_c'}$ |
|---|---|---|---|---|
| Experiment | $44 \pm 6$ | $95 \pm 5$ | $265 \pm 14$ | $230 \pm 7$ |
| Schnitzer (Ref. 87) [0.2, 0.19, 1.6, 1.0, 0.0] | 87 | 63 | 70 | 58 |
| Pumplin et al. (Ref. 88) [0.0, 0.30, 1.5, 1.0, 0.0] | 152 | 117 | 119 | 92 |
| Henriques et al. (Ref. 89) [0.8, 0.18, 1.6, 0.0, 0.0] | 40 (input) | 80 | 95 | -- |
| Schnitzer (Ref. 87) [0.2, 0.19, 1.6, 1.0, 1.1] | 182 | 170 | 268 (input) | 225 |
| Chan (Ref. 89) [0.2, 0.15, 1.6, 0.12, 5.0] | 40 (input) | 90 | 262 (input) | 225 |
| Carlson and Gross (Ref. 89) [0.27, 0.20, 1.37, 0.08, 4.4] | 41 (input) | 98 | 265 (input) | 181 |
| Celmaster et al. (Ref. 86) [-, -, 1.98, 1.0, 0.0] | 92 | 100 | 150 | 80 |

Table III. Spin-dependent splittings in charmonium taken from the analysis of Ref. 23. The numbers in brackets are values of $\kappa$, $1/a^2$ ($\text{GeV}^2$), $m_c$(GeV), $\eta$ and $\lambda$ used by the various authors. The references to the specific papers involved may be found in Ref. 23.

splitting. It is interesting to note that they also obtain a spin-dependent interaction of the form of Eq. (10) although the potentials $V_1$ and $V_0$ are different.

The failure of any of these models to give satisfactory agreement with the experiment is evident in Table III. However, these detailed questions of spin-dependent forces may comfortably wait until the nonrelativistic potential itself is better understood. Possibly at least the general structure of the spin-dependent interaction given in Eq. (10) could be justified within QCD.

The third area of possible disagreement is the extension to even heavier quark systems. From our understanding of the potential in QCD, we expect $V(r)$ to be independent of the quark mass and thus to be equally valid for any quark system heavier than the $\psi$ family.

Assuming that the $\mu^+\mu^-$ enhancement seen in the Lederman experiment[2,3] at Fermilab is due to three narrow-width resonances of a new heavier quark-antiquark system[26] the data gives the following masses.

$$m(\text{T}) = 9.37 \text{ GeV}$$
$$m(\text{T}') = 9.94 \text{ GeV}$$
$$m(\text{T}'') = 10.32 \text{ GeV}$$

The spacings $m(\Upsilon') - m(\Upsilon) \simeq 0.57$ GeV and $m(\Upsilon'') - m(\Upsilon') = 0.38$ GeV are very close to the observed splittings in the $\psi$ system. This led Quigg and Rosner[27] to consider a purely logarithmic potential since a logarithmic potential has the property that its excitation spectrum is mass independent. The values for the splittings with the linear plus Coulomb potential in Eq. (3) are $m(\Upsilon') - m(\Upsilon) = 0.440$ GeV and $m(\Upsilon'') - m(\Upsilon') = 0.345$ GeV for a quark mass of 4.5 GeV.[28] Thus the potential used for the $(c\bar{c})$ system seems to produce systematically smaller splittings than observed. Here I would like to pursue a different line than modifying the form of the potential. The splittings of both the $\psi$ and $\Upsilon$ family of resonances can be used to fit the parameters of a potential of the form of Eq. (2). The resulting parameters[29] are

$$\kappa = 0.41 \qquad\qquad a = 2.32 \ (GeV)^{-1} \qquad\qquad (11)$$

As can be seen in Fig. 4 the potential is very similar in shape to the logarithmic potential of Quigg and Rosner for $0.05 \leqslant r \leqslant 0.4$ fm. Many of the properties of low lying states would be indistinguishably different for a potential defined by Eq. (11) and a logarithmic potential. The spacing of higher excitations is however much larger for a linear plus Coulomb potential $(E_n \sim n^{2/3})$ than for a logarithmic potential $(E_n \sim \log n)$ and this will distinguish these alternatives experimentally.

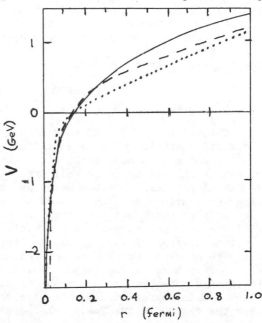

Figure 4. Various potentials as a function of r. Solid line is logarithmic potential of Quigg and Rosner. Dotted line is the potential given in Eq. (3). The dashed line gives the linear plus Coulomb potential with the parameters of Eq. (11). The overall constants of the last two potentials are adjusted so all three potentials cross the r axis at the same point.

Returning to the $(c\bar{c})$ system for the values of $\kappa$ and a in Eq. (11), we actually have some improvement in the $(c\bar{c})$ mass spectrum. The center of gravity of the P states moves to 3.50 GeV in better agreement with the experimental value (see Table I) and the 3S and 4S states are at 4.12 GeV and 4.56 GeV respectively. But as noted before the increase of Coulomb parameter $\kappa$ improves the mass spectrum

at the expense of the El transitions.  Actually in fitting the spectrum using the $\Upsilon$ system the mass of the charmed quark was simply
chosen to be $M_\psi/2$ and similarly the mass of the new quark to be $M_\Upsilon/2$.
This gives a value for $\Gamma(\psi \to e^+ e^-) = 8.0$ keV as compared with an observed value of $4.8 \pm 0.6$ keV.  If we attempt to include the leptonic
width of the $\psi/J$ into the fit as was done originally for the potential of Eq. (3) we are driven to a very light charm quark mass $m_c \approx$
1.2 GeV and this is a disaster for the validity of a nonrelativistic approximation, $<v^2/c^2>_\psi \sim 0.8$, and El transitions rates which worsen by a
factor of roughly two.  One finds similar results for the logarithmic
potential although the problem is not quite as severe there since the
short distance behavior is softer.  The problems we have found in
fitting the $\Upsilon$ system spectrum are just reflections of problems internal to the $c\bar{c}$ system of simultaneously fitting the spectrum and
internal properties of the ($c\bar{c}$) states.  A possible resolution to
this dilemma has been suggested by K. Johnson[30] where in the context
of the MIT bag model he has calculated the effective potential between
a heavy quark-antiquark pair obtaining a form well approximated by Eq.
(2) with parameters similar to those in Eq. (11).  There this problem
is resolved by realizing that for $r \leqslant 1/m_c$, the Compton wavelength of
the charmed quark, a nonrelativistic picture must necessarily fail but
it fails in a calculable way since QCD is calculable at short distances.  This leads to a modification of the relationship between the
leptonic width and the wavefunction at the origin of the nonrelativistic potential.  So fitting the wavefunction of the origin through
the leptonic width of $\psi$ may be a very unreliable procedure especially
for a large Coulomb parameter.

### III.  INCLUDING QUARK PAIR PRODUCTION
### - COUPLING TO DECAY CHANNELS

The potential $V(r)$ discussed in the introduction corresponds to
interaction within QCD between infinitely massive quarks.  In a more
realistic limit we must include the possibility of light quark pairs.
Of course the existence of pair production invalidates the notion of a
confining potential between heavy quarks if treated exactly since at
large separation  the heavy quark system has sufficient energy to produce a light quark pair and then the heavy quarks can separate to in-
finity each dragging a light quark along with itself.  In terms of
physical states this process is simply two mesons being separated to
infinite distance which does not require infinite energy.  However
the duality between the field theoretic picture and the description
in terms of the physical spectrum of strongly interacting particles
allows an interpretation of the effects of light quarks on states
near the threshold for Zweig allowed decays.

Specializing to the $\psi$ family of resonances, the effect on the
($c\bar{c}$) states below charmed threshold of virtual light quark pairs is
to modify the effective potential between the charmed quarks.  From a
field theoretic point of view these physical $\psi$-like states are mix-
tures of the basic ($c\bar{c}$) potential state and states containing in addition to the ($c\bar{c}$) pair any number of light quark pairs.  Near charm
threshold we would expect to consider only the contribution from a

single additional light quark pair. In terms of physical states this assumption about how to include light quark effects is equivalent to considering the contribution to the $\psi$-like states due to virtual decays into pairs of charmed mesons. Furthermore we would expect that near threshold only the lowest mass charmed meson pairs would give a significant contribution. Above charm threshold these same corrections to the effective potential exist and more importantly real pair creation can occur allowing for the hadronic decay of the physical state into charmed mesons.

The model of K. Gottfried, T. Kinoshita, K. D. Lane, T.-M. Yan, and myself[12] includes these light quark effects perturbatively by computing the off energy shell decay amplitudes for any $(c,\bar{c})$ state coupling to a pair of charmed mesons $(c,\ell)$ and $(\bar{c},\ell)$ ($\ell = u$, d, or s quark) and then by formal methods summing the effects of the real and virtual charmed meson intermediate states. The decay amplitude is given to lowest order in pair creation by a transition matrix element whose initial and final states are a charmonium bound state and a two charmed meson state respectively, and whose interaction creates or destroys a light quark pair and is simply related to the basic non-relativistic potential of Eq. (3). For more details I refer you to their article. The most useful feature of this model is that without introducing many new free parameters it can be used to calculate the complex effects of decay channels on the properties of the states below charm threshold and the production rate of each exclusive charm meson channel in the threshold region of $e^+e^-$ annihilation.

The contribution to R due to charm production, denoted $\Delta R_c$, can be extracted from the data by subtracting a background value of R due to light quarks which produce uncharmed hadrons and the contribution of the new lepton, $\tau$.[31] It is also simple to estimate where the various charmed meson thresholds will open. The naive excitation spectrum can be obtained from the potential of Eq. (3) for a system of reduced mass $\mu = (m_c m_u)/(m_c+m_u)$. The spin splittings may also be estimated from comparison with the K meson system treating both the charmed mesons and the strange mesons as systems with the light quark (up or down) moving in the influence of a fixed heavy quark (strange or charmed).

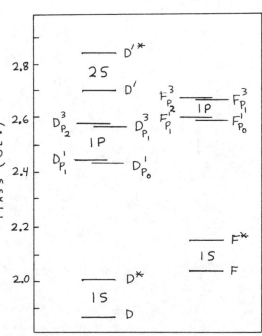

Figure 5. Expected mass spectrum for charmed mesons. The notation $D_{P_J}^{2j_\ell}$ denotes a P state charmed meson with total angular momentum J and angular momentum for the light quark $j_\ell$.

Making use of the experimental values of the D, D*, F, and F* masses[23] we obtain the spectrum shown in Fig. 5. Figure 6 shows the extracted $\Delta R_c$ with the various two body charmed meson thresholds indicated.

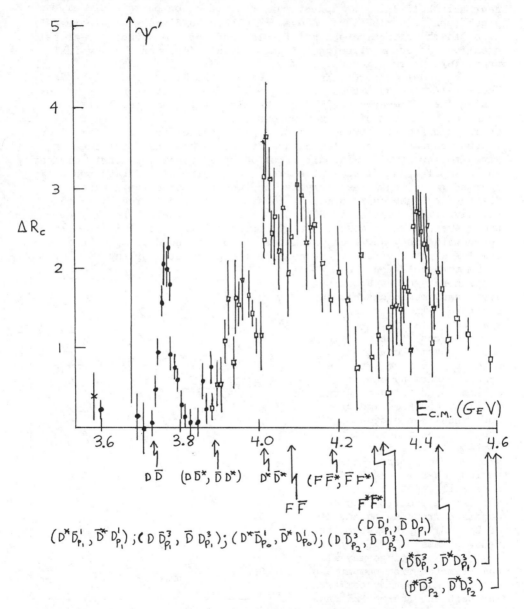

Figure 6. $\Delta R$ due to charm production. A background value of R = 2.4 due to the production of ordinary hadrons (old physics) and the contribution due to the new heavy lepton τ with mass 1.8 GeV have been subtracted from the SLAC-LBL data for R.

In the explicit calculations of the Cornell Group[12] for the exclusive channels only the $D^0\overline{D}^0$, $D^0\overline{D}^{*0} + D^{*0}\overline{D}^{*0}$, $D^+D^-$, $D^+D^{*-} + D^{*+}D^-$, $D^{*+}D^{*-}$, $F^+F^-$, $F^{*+}F^- + F^+F^{*-}$, and $F^{*-}F^{*+}$ thresholds were considered. The parameters $m_u$, $m_d$, $m_s$, $m(D^0)$, $m(D^+)$, $m(D^{*0})$, $M(D^{*+})$, $m(F^+)$, $m(F^{*+})$, $\kappa = 0.132 = 4\alpha_s/3$ were fixed throughout; $m_c$ and a were adjusted so that poles of $\psi'$ and $\psi$ and $\Gamma_{e^+e^-}(\psi)$ agreed with experiment yielding $m_c = 1.68$ GeV$^{-1}$ and a $= 1.797$ GeV$^{-1}$. The resulting curve of R charm in the 3.726 – 3.860 GeV energy range is shown in Fig. 7. We can compare the model with the

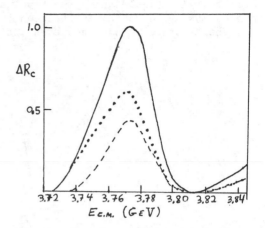

parameters of 3.772 GeV resonance. The model predicts the mass of 1D state to be 3.760 GeV. When the mass is shifted to agree with the observed value of 3.772 GeV, the total width at half maximum is 34 MeV as compared to $28 \pm 5$ MeV measured by the SLAC-LBL[16] group and $\Gamma_{e^+e^-}$ is .2 keV compared to $0.37 \pm 0.09$ keV measured by the SLAC-LBL group[16] and $0.18 \pm 0.06$ keV as measured by DELCO.[15] Three points should be noted about the model: First, the D state had no wavefunction at the origin and so does not couple directly to the photon within the model. The coupling is due to an induced tensor force caused by using nondegenerate D and $D^*$ masses. Second, since charm production occurs only through the coupling of $(c\overline{c})$ $1^{--}$ states to the photon there is no continuum contribution; thus where there are no sufficiently nearby resonances $\Delta R_c$ drops to zero. We

Figure 7. The predicted behavior of $\Delta R_c$ in the 3.7 GeV to 3.85 GeV energy region. The resonance is the $1^1D_3(c\overline{c})$ state. The dashed line is the $D^+D^-$ exclusive channel and the dotted line is the $D^0\overline{D}^0$ exclusive channel. The total $\Delta R_c$ is represented by the solid line.

see this as a feature both in the date in Fig. 6 and in the model in the 3.80 – 3.83 GeV region. Third, within the model each of the exclusive channels has a complicated form factor behavior which will be discussed below. This form factor effect is so strong that the ratio of $D^0\overline{D}^0$ to $D^+D^-$ production at 3.772 GeV differs by approximately 20% from the true threshold behavior (the ratio of momenta cubed).

The success in the 3.7 – 3.9 GeV region gives some confidence in applying the model to the complicated 3.9 – 4.3 GeV region. Here a number of testable features of the behavior of exclusive charm meson production are obtained which aid in the interpretation of the underlying resonance structure in this region. The masses of the 3S and 4S states computed in the model are higher than the observed peaks in R at 4.028 GeV and 4.414 GeV by roughly 130 MeV and 150 MeV respectively. So to compare the structure in this region the 3S and 2D states were shifted down by 130 MeV and the 4S was shifted down by 150 MeV.

It is easy to understand the discrepancy in these masses within the model. We have kept only the contribution to the (c̄c) states of the lowest mass charmed meson intermediate states. Even though only nearby thresholds would be expected to contribute significantly to the properties of the (c̄c) states and thus the lowest states would be sufficient for the $\psi$ and $\psi'$ as we move higher in energy the effects of the intermediate states we have ignored become more important. For the 3S state there are no new open channels but the higher states (in particular charmed P states) will have significant virtual effects. They would add a negative mass shift to the 3S and 2D states making the agreement between theory and experiment better. For the 4S state we have not even included all the open channels.

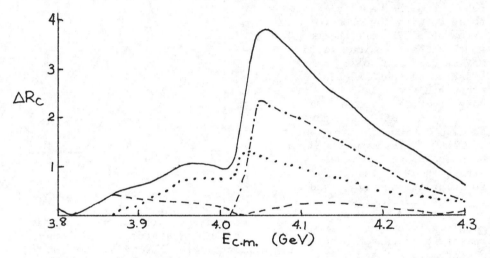

Figure 8. The behavior of the exclusive channels in 3.8 GeV to 4.3 GeV energy region. The dashed line is the sum of the $D\bar{D}$ exclusive channels ($D^0\bar{D}^0 + D^+D^-$). The dotted line is the sum of the $D\bar{D}^* + D^*\bar{D}$ exclusive channels ($D^0\bar{D}^{*0} + \bar{D}^0D^{*0} + D^+D^{*-} + D^{*+}D^{*-}$). The dot-dash line is the sum of the $D^*\bar{D}^*$ exclusive channels ($D^{*0}\bar{D}^{*0} + D^{*+}D^{*-}$). The total $\Delta R_c$ is represented by a solid line.

A representative curve of R with the masses of the 3S and 2D state shifted as indicated above is shown in Fig. 8. The exclusive channels indicate the complex structure of the model. At 4.028 GeV the $D\bar{D}$ which has the greatest available phase space is in fact very small while the $D^*\bar{D}^*$ channel gives a large contribution even though it has just opened. To understand this seeming unnatural behavior of the model we must look in detail at the structure of the various decay amplitudes. The basic decay amplitude for the $1^3S_1$ and $3^3S_1$ states into a pair of ground state charmed mesons is plotted in Fig. 9. The $1^3S_1$ decay amplitude has a completely conventional shape. The threshold behavior is linear in momentum as expected for a P wave decay and there is a smooth form factor dependence suppressing large momentum decays. The $3^3S_1$ state decay amplitude has a new feature

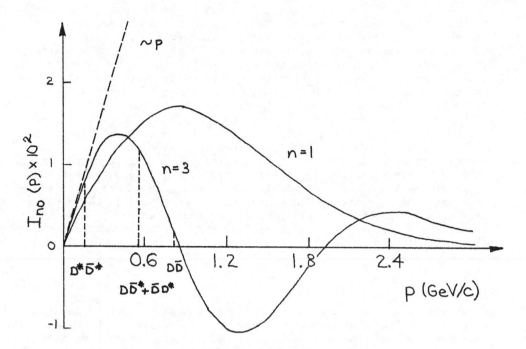

Figure 9. The decay amplitude, $I_n(p)$, for a $n^3S_1$ ($c\bar{c}$)
state decaying into a (D or D*) as a function of the outgoing momentum, p, of one of the charmed mesons. Both
the decay amplitude for the n = 1 ($\psi$) state and the n = 3
state is shown. The dashed line shows the expected linear
dependence of $I_n(p)$ on p near threshold for a P wave decay.
The dotted lines indicate the momenta for the decays of a
$3^3S_1$ state at 4.028 GeV into the $D\bar{D}$, $D\bar{D}^* + D^*\bar{D}$, and $D^*\bar{D}^*$
final states.

which is crucial to understanding the behavior of exclusive channels in the region of energy near the 3S resonance; that feature is
the existence of zeros in the decay amplitude. These zeros arise in
the model as a consequence of the nodes in the radial wavefunction
of S states with n > 1. In fact the number of nodes in the decay amplitude for the $n^3S_1$ state is (n–1). To make clear how important
this feature is to understanding the ratios of exclusive channels in
this region one needs only to compare the ratios $D\bar{D}$ : ($D\bar{D}^* + D^*\bar{D}$) : $D^*\bar{D}^*$
for a state at 4.028 GeV under three possible assumptions for the
behavior of the decay amplitudes:
(1)   The amplitude has its threshold behavior for all momenta in this
      range. This yields ratios 1 : 2.4 : 0.1.
(2)   The amplitude has a normal form factor dependence. Here we can
      use the $1^3S_1$ decay amplitude to estimate ratios obtaining
      1 : 2.7 : 0.05.
(3)   The amplitude is as shown in Fig. 9 for n = 3. Then we have
      ratios 1 : 12.5 : 9.5.

Returning to Fig. 8 there are two other important features of the model which can now be understood. There is some structure in the $3.9 - 4.0$ GeV region not arising from a new resonance but the opening of the $D\overline{D}^* + D^*\overline{D}$ channel and nearby decrease in the $D\overline{D}$ channel due to the zero in 3S decay amplitude. This structure is much less pronounced than the observed structure in $\Delta R_c$ shown in Fig. 6 but strongly suggests that the enhancement at 3.95 GeV should not be considered a new resonance. This interpretation can easily be tested by measuring the ratios of $D\overline{D}$ to $D\overline{D}^* + D^*\overline{D}$ exclusive channels in the $3.9 - 4.1$ GeV region where a dip through a near zero should occur.

The other prominent feature of Fig. 8 is the absence of any enhancement in $\Delta R_c$ associated with the $2^1D_3$ resonance. The induced S-D mixing which caused the $1^1D_3$ to appear in $\Delta R_c$ is too small at these energies to make the $2^1D_3$ resonance visible. There is an enhancement at 4.17 GeV in all measurements of R which would be a natural candidate for the $2^1D_3$ resonance within the model. The nature of this enhancement can be tested. The decay amplitudes depend on the orbital angular momentum, L, of initial state and thus distinguishes an S state from a D state. In particular at the same momentum the ratio of decay rates for $D\overline{D}$ to $D\overline{D}^* + D^*\overline{D}$ is given by

$$B_D \equiv \frac{\text{rate for } D\overline{D}}{\text{rate for } D\overline{D}^* + \overline{D}D^*} = \begin{matrix} \tfrac{1}{4} \text{ for } L = 0 \text{ (S state)} \\ 1 \text{ for } L = 2 \text{ (D state)} \end{matrix} \quad . \tag{12}$$

This test is complicated by the difference in D and $D^*$ masses which leads to different momentum for the two decays at the same total energy. The 4.028 state however fits well as the 3S state when the momentum difference is taken into account. Then the expected value of $B_D \sim 1/10$ for an S state and $B_D \sim 1/3$ for a D state, while the measured ratio[23,33] at 4.028 GeV is $B_D \lesssim 1/10$.

The 4.17 GeV structure has three *a priori* reasonable interpretations:
(1)   as the 2D state expected in the potential model of Eq. (3),
(2)   as the 4S state favored with logarithmic potential, or
(3)   as a threshold structure similar to the structure at 3.95 GeV.
These possibilities can be distinguished by the considerations given in Table IV.

| Interpretation of 4.17 GeV Structure | 2D State | 3S State | Threshold Structure |
|---|---|---|---|
| Behavior of $B_D$ in the 4.1 to 4.2 GeV region | increases sharply by factor $\sim$3-4 | slowly varying | slowly varying |
| Minimum $\Delta R_c$ between 4.03 and 4.17 GeV | cannot vanish | may vanish | may vanish |
| Is the opening of a new threshold observed | not necessary | not necessary | must be a major fraction of charmed states at 4.17 GeV |

Table IV. Distinguishing various interpretations of the 4.17 GeV structure in R.

Finally turning briefly to the structure in the 4.4 GeV region, the dip in $\Delta R_c$ at 4.28 GeV seen in Fig. 6 suggests that this structure is a new S state resonance not strongly coupled to the resonance structure in the 3.9 – 4.2 GeV region. As we also see in Fig. 6, there are many new thresholds opening in this energy region; however there is one channel which due to a large statistical enhancement and available phase space might be expected to be prominent. That is the final states $D_{P_1}^1 \bar{D}$ and $\bar{D}_{P_1}^1 D$ where $D_{P_{J=1}}^{2j_\ell=1}$ denotes an L = 1 charmed meson with total angular momentum J = 1 and total momentum for the light quark $j_\ell = \frac{1}{2}$. The

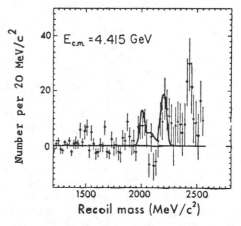

Figure 10. Recoil mass spectrum of $D^0$ at a center-of-mass energy of 4.415 GeV.

$D_{P_1}^1$ decays strongly into $D^*\pi$ and thus would be expected to have a normal hadronic width. There is such an enhancement observed[33] in the recoil spectrum of the $D^0$ at 4.415 GeV (see Fig. 10) at a mass of 2.45 GeV (roughly the expected mass of the $D_{P_1}^1$ state shown in Fig. 5).

## SUMMARY AND CONCLUSIONS

A simple linear plus Coulomb potential gives good agreement with the observed spectrum of $(c\bar{c})$ states and acceptable E1 transition rates. The three general difficulties with this simple potential model are:
(1) The failure to account even qualitatively for the M1 transitions involving the X(2.82) and $\chi$(3.45) states under the assumption that these states are the pseudoscalar partners of the $\psi$ and $\psi'$ respectively,
(2) The failure of any completely adequate model of spin-dependent forces, and
(3) The small spacing $m(T') - m(T) = 0.440$ GeV and $m(T'') - m(T') = 0.345$ GeV which results when this potential is extended to the T family of resonances.
Of these difficulties, the first is the most serious. In particular the assumption that the X(2.82) state is the pseudoscalar partner of $\psi$ leads to an unhindered M1 rate approximately 20 times larger than the experimental upper limit. The assumptions which went into this calculation of rate work reasonably well for the ordinary mesons where the spin splitting are equally large and nonrelativistic approximation is far worse. Furthermore these assumptions are subject to independent experimental tests.

The second difficulty is troublesome but since we are still so far from an understanding of nonrelativistic limit within QCD it may be premature to be too concerned with the precise nature of relativistic corrections. We should not overlook the experimental agreement for the signs and rough order of magnitudes of these forces.

The final problem seems to be a difficulty of a single model potential [given in Eq. (3)]. However as I have tried to show this is actually not true. If one is only concerned with fitting the observed mass spectrum of the $\psi$ and $\Upsilon$ families of resonances it is straight-forward to find parameters $\kappa$ and $a$ which give good agreement for both mass spectra [see Eq. (11)]. The resulting potential is similar to the logarithmic potential of Quigg and Rosner for distances between 0.05 fm and 0.4 fm. Further, the agreement for the center of gravity of $1P_J$ states, the 3S state, and the 4S state of the $\psi$ family is actually better than the potential of Eq. (3). If, however, the leptonic width of the $\psi$ is used as an additional constraint on the fit it leads to a value for the charmed quark mass $m_c \sim 1.2$ GeV which because of its small value gives rise to states which are much more relativistic and furthermore leads to considerably larger E1 transition rates. This inherent difficulty of these simple potentials to fully reconcile the mass spectrum with the internal properties of the states in the ($c\bar{c}$) system is what leads to the choice of the compromise parameters of Eq. (3), and this compromise in turn leads to some disagreement for the spectrum of masses in the $\Upsilon$ family.

One possible resolution of this problem in the ($c\bar{c}$) system is to abandon use of the leptonic decay rate of $\psi$ in fitting the parameters of the potential. Use instead the mass splittings $m(\psi') - m(\psi)$ and $m(C.O.G.1^3P_J) - m(\psi)$ and possibly the E1 transition rates for $\psi'$ to the $\chi_0(3.41)$ state since it is not affected greatly by coupling to decay channels. Parameters similar to those in Eq. (11) will result; the leptonic widths would however all be larger than experiment by roughly 50-60%. Since such a fit has a large Coulomb parameter it is not un-reasonable to expect that the nonrelativistic relation between the leptonic width and the wavefunction at the origin might have corrections of this magnitude.

Turning to the structure above threshold for $c\bar{c}$ systems, one finds that including the coupling to decay channels into the model, the structure of $\Delta R$ charm and the various exclusive channels in the region 3.7 – 4.3 GeV can be consistently interpreted. A number of simple experimental tests allow the determination of the nature of the structure at 4.17 GeV. The possibilities that this enhancement is the 2D state, the 4S state, or a threshold structure can be distinguished experimentally.

## ACKNOWLEDGEMENTS

I would like to thank my colleagues K. Gottfried, T. Kinoshita, K. D. Lane, and T.-M. Yan whose work I have used extensively in the preparation of this talk.

## REFERENCES

1. For summaries of the experimental data see: G. J. Feldman, M. L. Perl, Phys. Rpt. 19C, 234 (1975); R. F. Schwitters, K. Strauch, Ann. Rev. Nucl. Sci. 26, 89 (1976); B. H. Wiik, G. Wolf, DESY Report 77/01 (1977); and G. J. Feldman, M. L. Perl, Phys. Rpt. 33C, 285 (1977).
2. S. W. Herb *et al.*, Phys. Rev. Lett. 39, 252 (1977); W. R. Innes

*et al.*, Phys. Rev. Lett. <u>39</u>, 1440 (1977).

3. Also see the talk at this conference of R. Kephart, "Status Report on Upsilon and High Mass Dimuon Continuum".

4. See also L. Susskind, "Coarse Grained Quantum Chromodynamics", lectures given at the Les Houches School on High Energy Physics (1976); Ref. 10; and Ref. 6.

5. K. Wilson, Phys. Rev. D <u>10</u>, 2445 (1974).

6. This formulation of Wilson's criteria may be found in S. Coleman, "The Uses of Instantons", Harvard Preprint HUTP-78/A004 (1978), lectures delivered at the 1977 International School of Subnuclear Physics, Ettore, Majorana; also see Ref. 10.

7. The results through second order in $\alpha_s$ were first obtained by A. Duncan, Phys. Rev. D <u>13</u>, 2866 (1976).

8. T. Appelquist, M. Dine, and I. Muzinich, Phys. Lett. <u>69B</u>, 231 (1977); *ibid*. Phys. Rev. D to be published; W. Fischler, Nucl. Phys. <u>B129</u>, 157 (1977).

9. F. L. Feinberg, Phys. Rev. Lett. <u>39</u>, 316 (1977); *ibid*. M.I.T. Preprint C.T.P. 687 (1978). The explicit value of the numerical constant is given by S. Davis and F. L. Feinberg, M.I.T. Preprint C.T.P. 710 (1978).

10. C. Callan, R. Dashen, and D. Gross, Institute for Advanced Study Preprint COO-2220-115 (1977).

11. J. Glimm and A. Jaffe, Harvard Preprint HUTMP-77/B54.

12. E. Eichten, K. Gottfried, T. Kinoshita, K. D. Lane, and T.-M. Yan, Cornell Preprint CLNS-375 (1978).

13. J. Burmester *et al.*, Phys. Lett. <u>66B</u>, 395 (1977); G. Knies, Proc. 1977 International Symposium on Lepton and Photon Interactions at High Energies, Hamburg, Germany, August 25-31, p. 93.

14. S. Yamada, Proc. 1977 International Symposium on Lepton and Photon Interactions at High Energies, Hamburg, Germany, August 25-31, p.69.

15. J. Kirby, *ibid*. p. 3.

16. P. Rapidis *et al.*, Phys. Rev. Lett. <u>39</u>, 526 (1977).

17. K. Gottfried, Proc. 1977 International Symposium on Lepton and Photon Interactions at High Energies, Hamburg, Germany, August 25-31, p. 667.

18. T. Appelquist, R. M. Barnett, and K. D. Lane, SLAC Preprint SLAC-PUB-2100 (1978), to appear in Annual Review of Nuclear and Particle Science, Vol. 28.

19. See the review of G. J. Feldman and M. L. Perl, Phys. Rpt. <u>33C</u>, 285 (1977).

20. C. J. Biddick *et al.*, Phys. Rev. Lett. <u>38</u>, 1324 (1977).

21. W. Tanenbaum *et al.*, Phys. Rev. D <u>17</u>, 1731 (1978).

22. W. Braunschweig *et al.*, Phys. Lett. <u>67B</u>, 243 (1977).

23. For details of charmed meson masses and decays see: G. J. Feldman, SLAC-PUB-2068 (1977), lectures at the Banff Summer Institute on Particles and Fields, Banff, Alberta, Canada, August 26 - September 3, 1977.

24. N. Isgur, Phys. Rev. Lett. <u>36</u>, 1262 (1976).

25. C. Callan, R. Dashen, D. Gross, F. Wilczek, and A. Zee, Institute for Advanced Study Preprint COO-2220-132 (1978).

26. This interpretation has gained strong support recently by the observation at DESY by both the PLUTO and DASP groups of a narrow

resonance in $e^+e^-$ annihilation at 9.46 GeV. Ch. Berger *et al.*, DESY Preprint DESY 78/21 (1978).

27. C. Quigg, J. Rosner, Fermilab Report PUB 77/82-THY (1977).

28. The values appearing here differ slightly from those in E. Eichten, K. Gottfried, Phys. Lett. <u>66B</u>, 286 (1977) as they used a somewhat different potential than Eq. (3).

29. These parameters were obtained by assuming $m_c = M_\psi/2$ and $m_b = M_T/2$ and then fitting the values of $\kappa$ and a by the spacings $m(\psi') - m(\psi)$ and $m(T') - m(T)$. A similar fit was obtained in Ref. 27.

30. K. Johnson, M.I.T. Preprint CTP 725 (1978).

31. The mass of the new lepton has been measured by the PLUTO Group at DESY, see U. Timm, these proceedings; and by the DELCO Group at SLAC, see W. Bacino *et al.*, Phys. Rev. Lett. <u>41</u>, 13 (1978).

32. The details of these estimates will appear in the second part of the work on charmonium by K. Gottfried, T. Kinoshita, K. D. Lane, and myself (in preparation). The first part of this work appears in Ref. 12.

33. G. Goldhaber *et al.*, Phys. Lett. <u>69B</u>, 502 (1977).

## NEW RESULTS ON DIMUON PRODUCTION BY HIGH ENERGY NEUTRINOS AND ANTINEUTRINOS*

T. Y. Ling

Department of Physics, The Ohio State University
Columbus, Ohio  43210

### INTRODUCTION

The subject of multimuon production by high energy neutrinos and antineutrinos began more than three years ago with the observation of dimuons by the HPWF collaboration at Fermilab[1].  The opposite sign dimuon events were interpreted as evidence for a new hadronic quantum number-charm.[2,3]  Much has been learned since then. Recent experiments[4] not only confirmed the earlier observation but also supported the 'charm' interpretation.  The origin of the like-sign dimuons ($\mu^-\mu^-$) remain, however, unknown.  Because of the smaller observed rate for these events, decays of pions and kaons from ordinary deep inelastic neutrino interactions might account for a large fraction, if not all, of the observed events.  It is important to determine whether 'prompt' like-sign dimuons exist, for the rate and nature of the prompt $\mu^-\mu^-$ events would provide important clues to the overall understanding of another facet of multimuon phenomena, in particular, the recently discovered trimuon events.[5]

In this talk I would like to report on recent dimuon data from the FHOPRW collaboration (E-310) at Fermilab.  The issues I will concentrate on are:  Is there a prompt like-sign dimuon signal? What are the nature of these like-sign events as compared to the opposite sign events?  How consistent are the rates and properties of the opposite sign events with charm particle production and their semi-leptonic decay?  The trimuon events observed during the same runs from which the present dimuon samples were obtained have already been published and reported at various other conferences and therefore will not be discussed here.

### BEAMS

The data samples reported here were acquired at Fermilab in three runs, using a quadrupole triplet (QT) and sign-selected bare target (SSBT) beams.  In the QT beam the secondary hadrons produced in the proton-target collision were focussed by a quadrupole triplet and left to decay without charge selection.[6]  The resultant neutrino flux contains a mixture of $\nu_\mu$ and $\bar\nu_\mu$.  The SSBT beams employed no focussing elements but did charge selection of the secondary hadrons by means of a "dog-leg" arrangement of bending magnets.[7]  Hence the resultant beams contain only $\nu_\mu$ or $\bar\nu_\mu$ depending on the selected sign of the parent hadrons.  These will be referred to as the SSBT$\nu$ and SSBT$\bar\nu$ beams.  The calculated spectra of $\nu_\mu$ and $\bar\nu_\mu$ from these beams are shown in Fig. 1.  The QT and SSBT$\nu$ runs yielded 199 $\mu^-\mu^+$

ISSN:  0094-243X/78/273./$.50  Copyright 1978 American Institute of Physics

events and 46 $\mu^-\mu^-$ events.  The SSBT$\bar{\nu}$ run yielded 49 $\mu^+\mu^-$ events and
2 $\mu^+\mu^+$ events.

## DETECTOR

The detector of E-310 is shown in Fig. 2.  It is an enlarged
version of the earlier detector of E-1A with important modifications.  i)  A target-detector of three parts, an iron target (FeT),
a liquid scintillator calorimeter (LiqC), and an iron-plate (4"
thick plates) calorimeter (FeC), each part of different density.
This makes possible an empirical determination of the pion and kaon
decay background in the multimuon data.  Table I shows the hadronic
absorption length and fiducial masses of the three target-detectors.
ii)  A large solid angle muon spectrometer consists of three large
area toroidal magnets (24' diameter) in addition to the existing 12'
diameter toroids.  Muons of angles up to 500 mrad relative to the $\nu_\mu$
beam direction can be detected, compared with the limited angle of
225 mrad of EIA.

## LIKE-SIGN DIMUONS

1.  Existence of a Prompt Signal?
    The issue of foremost importance is to determine whether the
like-sign dimuons are indeed all due to pion and kaon decays.  This
can in principle be inferred from the observed $\mu^-\mu^-$ rate produced
in each target.  The absolute rate of $\mu^-\mu^-$ events are difficult to
determine because of the differences in acceptance, trigger requirements, etc between the targets.  The ratio $N(\mu^-\mu^-)/N(\mu^-\mu^+)$ is, however, insensitive to these target dependent systematic effects, for
to a good approximation these are the same for both the $\mu^-\mu^+$ and
$\mu^-\mu^-$ events.  The numbers of observed $\mu^-\mu^+$ and $\mu^-\mu^-$ events are shown
in columns 3 and 5 of Table I for muon momentum cuts $p \geq 5$ and 10
Gev/c respectively.  The observed ratios $N^{OBS}(\mu^-\mu^+)/N^{OBS}(\mu^-\mu^+)$ are
plotted against absorption length in Figs. 3a and 3b.  To simplify
the interpretation, we subtract from $N^{OBS}(\mu^-\mu^+)$ the calculated numbers of the $\mu\bar{\mu}^+$ events resulting from $\pi$ and K decays.[8]  The resulting numbers of prompt $\mu^-\mu^+$ events are presented in column 4 of
Table I.  In Figs. 3c and 3d, we show the ratios $N^{OBS}(\mu^-\mu^-)/$
$N^{prompt}(\mu^-\mu^+)$, again for the muon momentum cuts $p \geq 5$ and 10 Gev/c.
We observe that:  i) pion and kaon decays account for a significant
fraction of the $\mu^-\mu^-$ events, particularly in the lower density
targets, as exhibited by the steep slope shown in Fig. 3c; ii) the
extrapolation of the linear fit in Fig. 3c gives an intercept of
0.10 ± .07, a little more than one standard deviation from zero;
iii) the slope decreases as the momentum requirement for muons is
raised from 5 to 10 Gev/c, resulting from the reduction of pion
and kaon decay contributions, but the intercept remains unchanged

within error.  We conclude that although statistically still incon-
clusive, these data at least suggest the existence of a prompt $\mu^-\mu^-$
signal.  There exist in recent literature other data on like-sign
dimuon production by neutrinos.  In a run using the dichromatic
beam at CERN, the CDHS collaboration reported the observation of
257 $\mu^-\mu^+$ events and 47 $\mu^-\mu^-$ events which satisfy the muon momentum
cutoff of 4.5 Gev/c.[9]  The ratio $N^{OBS}(\mu^-\mu^-)/N^{PROMPT}(\mu^-\mu^+)$ for $p_\mu > 5$
Gev, obtained using the reported[9] $\pi$,K decay contributions and the
slow muon momentum spectra, is 0.18 ± 0.03.  The target detector of
the CDHS experiment is primarily iron with an average hadronic absorption length of 30cm.  The CDHS result is plotted on Fig. 3c for
comparison with this experiment.  The agreement between the two experiment is very good.  The intercept obtained from the fit
including the CDHS data is 0.09 ± 0.05.

It should be noted that although the analysis described above
depend to some extent on the correction to subtract $\pi$, K decay contribution from $N^{OBS}(\mu^-\mu^+)$, the value of the intercept from the extrapolation is relatively insensitive to this correction.  The
reason for this is that $\pi$ and K decay contributions are relatively
much smaller fraction of the $\mu^-\mu^+$ events.  This is true especially
for the FeT point which carries a lot of weight in determining the
value of the intercept.  Nonetheless, we have checked the reliability of the $\pi$, K decay calculation by comparing the density dependence of $N^{OBS}(\mu^-\mu^+)/N(\mu^-)$.  This comparison is shown in Fig. 4.  The
calculated slope is found to be very consistent with the data as
shown.  This gave us further confidence in this analysis.

The two $\mu^+\mu^+$ events observed in the SSBT($\bar{\nu}$) run have very low
energy muons (<5 Gev/c).  Pion and kaon decays are estimated to
yield 4 ± 2 events.  Hence the two observed events are consistent
with being backgrounds.  If a prompt $\mu^+\mu^+$ signal were to exist at
the same rate relative to the $\mu^+\mu^-$ event as the $\mu^-\mu^-$ signal, we
would expect to observe 5 events in addition.  So from this sample
there is as yet no clear evidence of a prompt $\mu^+\mu^+$ signal.

2.  Properties of the $\mu^-\mu^-$ events

The important properties of the $\mu^-\mu^-$ events are shown in
Figures 5 – 9.  Figure 5 shows the scatter plot of the momentum of
the fast $\mu^-$ against the slow $\mu^-$.  A rather large momentum asymmetry
between the two muons is observed, similar to that observed in the
$\mu^-\mu^+$ data as shown in Fig. 6.  The distribution in the azimuthal
angle between the two muons is shown in Fig. 7a and 7b for the $\mu^-\mu^-$
and $\mu^-\mu^+$ events.  Both distributions tend to peak at $\Delta\phi = 180^\circ$,
suggestive of a hadronic origin for the second muon.  Figure 8a
shows the distribution in $p_{\perp W}^s$, the transverse momentum of the slow
$\mu^-$ with respect to the direction of the W-boson which is defined by
the directions of the incident neutrino and the fast $\mu^-$.  Again for
comparison we show in Fig. 8b the distribution in $p_{\perp W}^+$ of the $\mu^+$ for
the $\mu^-\mu^+$ events.  No distinctive differences can be seen.  The dis-

tribution in the $E_{vis}$ distribution are also similar between the $\mu^-\mu^-$ and $\mu^-\mu^+$ events as shown in Fig. 9a and 9b.

It could be argued that since a significant fraction of the $\mu^-\mu^-$ events are in fact from $\pi$ or K decays, the properties of the prompt events could be largely masked. For this reason, the $\mu^-\mu^-$ events which satisfy the 10 Gev momentum cut are shaded for comparison. Contamination of pion and kaon decays are much smaller in this case. We note that although the statistics is very limited, the distribution exhibit the same general features. Properties of energetic $\mu^-\mu^-$ events where both muons have $p_\mu > 15$ Gev are shown in Table 2.

## OPPOSITE-SIGN DIMUONS

We now turn to the measurements of the rates of opposite-sign dimuons. The observed $E_{vis}$ ($\equiv E_h + E_{\mu^-} + E_{\mu^+}$) distribution for the $\mu^-\mu^+$ events from the QT run is shown in Fig. 10a. The energy spectrum of the single muon events from the same run is shown in Fig. 10b. Figure 11 shows the relative rate $R(\mu^-\mu^+)/R(\mu^-)$ as a function of $E_\nu$. Two things were further taken into account to obtain the data points shown in Fig. 11. First, the dimuon events were individually weighted by a factor which corrected for the geometric acceptance and triggering biases. This weighting factor was calculated for each of the observed events by azimuthally rotating the event in the detector. Secondly, the contribution from $\pi$ and K decays had been subtracted from the data. Figs. 12 and 13 show the corresponding data for $\bar{\nu}$ induced $\mu^+\mu^-$ events from the SSBT($\bar{\nu}$) run. We note that the apparent rise in the dimuon rate relative to single muon rate with energy for both the $\nu$ and $\bar{\nu}$ samples are predominantly the result of the 5 Gev/c momentum cut. The relative rates averaged over the corresponding beam energy spectra are

$$\left[ R(\mu^-\mu^+)/R(\mu^-) \right] \quad \nu(30\text{-}200 \text{ Gev}) = (0.4 \pm 0.08) \times 10^{-2}$$

and

$$\left[ R(\mu^-\mu^+)/R(\mu^+) \right] \quad \bar{\nu}(30\text{-}200 \text{ Gev}) = (0.27 \pm 0.09) \times 10^{-2}$$

The average $\bar{\nu}$ dimuon rate is lower because the SSBT($\bar{\nu}$) spectrum is substantially softer than the QT neutrino spectrum. For a given energy bin, we observe that the $\mu^-\mu^+$ rate relative to single $\mu$ rate is approximately the same for the $\nu$ and $\bar{\nu}$ data. At high energy, say $E_{vis} > 80$ Gev, where acceptance due to kinematic cutoff is less limited we have

$$\left[ R(_{\mu^-}{}^+)/R(_\mu{}^-) \right] \quad \nu(>80 \text{ Gev}) = (0.65 \pm 0.13) \times 10^{-2}$$

and

$$\left[ R(_\mu{}^+{}_\mu{}^-)/R(_\mu{}^+) \right] \quad \bar{\nu}(>80 \text{ Gev}) = (0.70 \pm 0.25) \times 10^{-2}$$

If we assume $\sigma^{\bar{\nu}}/\sigma^{\nu} \approx 0.5$, it then follows from the data that

$$R^{\nu}(\mu^-\mu^+)/R^{\bar{\nu}}(\mu^+\mu^-) \approx 2.$$

To compare this with the prediction of the GIM model, we note that in that model, charm quarks are produced by neutrino through their interaction either with the D(valence) or with the strange (sea) quarks, namely,

$$\nu + d \rightarrow \mu^- + c \quad , \quad (\sigma \alpha \sin^2\theta_c)$$

and $\qquad\qquad \nu + s \rightarrow \mu^- + c \quad , \quad (\sigma \alpha f_s \cos^2\theta_c),$

where $\theta_c$ is the Cabbibo angle and $f_s$ is the fraction of momentum carried by the s(or $\bar{s}$) quarks relative to that carried by the d quarks in the nucleon. In antineutrino interactions, however, the charm quark can only be produced by the process

$$\bar{\nu} + \bar{s} \rightarrow \mu^+ + \bar{c} \quad . \quad (\sigma \alpha f_s \cos^2\theta_c).$$

Therefore $\sigma^{\nu}(\mu^-\mu^+)/\sigma^{\bar{\nu}}(\mu^+\mu^-) \approx 2$ provided that $f_s \approx \tan^2\theta_c = 0.05.$

The $X_{vis}$ and $Y_{vis}$ distributions of the $\nu$ and $\bar{\nu}$ induced $\mu^-\mu^+$ events are shown in Figs. 14 and 15. The $Y_{vis}$ distributions are similar for $\nu$ and $\bar{\nu}$ and are consistent with kinematic and acceptance cut-off in the low and high-y regions. The $\bar{\nu}$ induced dimuons, however, have a sharper $X_{vis}$ distribution when compared to that of the $\nu$ data. This is also consistent with the GIM model since only the quark ($\bar{s}$) play the role for charm production by antineutrinos.

To give an overall perspective of the multimuon physics, we present in Table 3 the rates for neutrino induced $\mu^-\mu^+$, $\mu^-\mu^-$ and $\mu^-\mu^-\mu^+$, all relative to the charge-current interaction rate.

## SUMMARY AND CONCLUDING REMARKS

In summary, we have shown that the data suggest that a $\nu$-induced prompt $\mu^-\mu^-$ signal may in fact exist. The rate of the prompt $\mu^-\mu^-$ events is measured to be $0.10 \pm 0.07$ relative to the prompt $\mu^-\mu^+$ rate. The properties of the $\mu^-\mu^-$ events are rather similar to that of the $\mu^-\mu^+$ events. No clear evidence has as yet been established for prompt $\mu^+\mu^+$ events from $\bar{\nu}$ interactions.

What are the origins of the prompt $\mu^-\mu^-$ events? We remark that only $\mu^-\mu^+$ events are expected if charm particles are singly produced by neutrinos. Mechanisms to explain the $\mu^-\mu^-$ events which invokes new physics beyond charm must be measured against the following alternatives: (a) radiative or direct muon pair production in deep inelastic charged-current interactions, (b) associated production of charmed particles. Only trimuons can in principle be produced by mechanism (a). However, $\mu^-\mu^-$ events could result from this source if the $\mu^+$ escapes experimental detection. Then one would expect $R(\mu^-\mu^-)/R(\mu^-\mu^-\mu^+) \ll 1$, contrary to the experimental observation. Therefore mechanism (a) is unlikely to be the dominant source for the $\mu^-\mu^-$ events. In associated charm production, both $\mu^-\mu^-$ and $\mu^-\mu^-\mu^+$ are expected. The ratio $R(\mu^-\mu^-)/R(\mu^-\mu^-\mu^+)$ should just be $[BR(c \to \mu + x)]^{-1} \sim 10$. The properties of the $\mu^-\mu^-$ events shown earlier are qualitatively compatible with the mechanism. The problem may lie in the absolute $\mu^-\mu^-$ rate.[10] If the measured $\mu^-\mu^-$ rate are confirmed with more data, then a large fraction, if not all, of the trimuons have to be attributed to associated charm production. More trimuon data is required to check consistency.

We have shown that both the rates and properties of the opposite-sign dimuon events are consistent with the GIM model. Based on this model, we may use the $\mu^-\mu^+$ data to determine $f_s$, the amount of strange quark relative to valence d-quark in the nucleon. The data gives approximately $f_s \sim \tan^2\theta_c = 0.05$.

The data shown in this talk is the result of a collaboration of physicists from Fermilab, Harvard, Ohio State, Pennsylvania, Rutgers and Wisconsin, Individual members of the collaboration are: A. Benvenuti, F. Bobisut, D. Cline, P. Cooper, M.G.D. Gilchriese, M. Heagy, R. Imlay, M. Johnson, T.Y. Ling, R. Lundy, A.K. Mann, P. McIntyre, S. Mori, D.D. Reeder, J. Rich, R. Stefanski and D. Winn.

### REFERENCES AND FOOTNOTES

*Work supported in part by the Department of Energy.

1. A. Benvenuti et al., Phys. Rev. Lett. 34, 419 (1975).
2. B. J. Bjorken and S. L. Glashaw, Phys. Lett. 11, 255 (1964); S. L. Glashaw, J. Iliopoulous and L. Maiani, Phys. Rev. D2, 1285 (1970).
3. A. Benvenuti et al., Phys. Rev. Lett. 35, 1199 (1975).
4. B. C. Barish et al., Phys. Rev. Lett. 36, 939 (1976); M. Holder et al., Phys. Lett. 69B, 377 (1977).
5. B. C. Barish et al., Phys. Rev. Lett. 38, 577 (1977); A. Benvenuti et al., Phys. Rev. Lett. 38, 1110 (1977) and 40, 488 (1978).
6. A. Skuja, R. Stefanski and A. Windelbon, FNAL Technical Note TM469 (1974).
7. R. Stefanski and H. B. White, FNAL Technical Note TM626A (1976).
8. R. Imlay, Calculations of background from pion and kaon decays in neutrino interactions. (unpublished)
9. M. Holder et al., Phys. Lett. 70B, 396 (1977).
10. H. Goldberg, Phys. Rev. Lett. 39, 1598 (1977).

| TARGET | FIDUCIAL MASS (Tons) | ABS. LENGTH (cm) | OBS N ($\mu^-\mu^+$) | | PROMPT N ($\mu^-\mu^+$) | | OBS N ($\mu^-\mu^-$) | |
|---|---|---|---|---|---|---|---|---|
| | | | P>5 GeV | P>10 GeV | P>5 GeV | P>10 GeV | P>5 GeV | P>10 GeV |
| IRON (Fet) | 198 | 31 | 75 | 50 | 66.7 | 46.5 | 12 | 8 |
| IRON CAL. (FeC) | 42 | 63 | 42 | 23 | 33.3 | 20.4 | 10 | 4 |
| LIQ. CAL. (LiqC) | 36 | 116 | 56 | 32 | 36.4 | 24.3 | 16 | 6 |

TABLE 1.   Fiducial masses, absorption lengths and
numbers of like-sign and opposite-sign
dimuon events.

| RUN FRAME | TGT MOD | ENERGY (GeV) | | TRACK 1 | | | TRACK 2 | | |
|---|---|---|---|---|---|---|---|---|---|
| | | $E_{vis}$ | $E_h$ | QP | Px | Py | QP | Px | Py |
| 117 15365 | Liq 13 | 127 | 24 | −86 ±7.3 | 1.87 ±.13 | .21 ±.04 | −17 ±.7 | −1.42 ±.06 | .07 ±.05 |
| 120 18783 | FeC 21 | 114 | 19 | −57 ±5.7 | .27 ±.13 | −1.22 ±.23 | −38 ±4.4 | .52 ±.17 | 1.21 ±.11 |
| 137 31940 | FeT 2 | >110 | >46 | −48 ±3.2 | .50 ±.04 | 1.05 ±.11 | −16 ±1.2 | −.23 ±.08 | .27 ±.09 |
| 141 35757 | FeT 1 | > 48 | − | −33 ±12.8 | .87 ±.23 | −2.02 ±.65 | −15 ±1.2 | −.35 ±.14 | .71 .05 |
| 146 39767 | FeT 2 | > 67 | − | −49 ±3.8 | 1.94 ±.11 | −.09 ±.06 | −18 ±1.5 | −.82 ±.06 | .15 ±.06 |
| 279 145416 | FeC 20 | 12 | 83 | −26 ±2.2 | −.69 ±.12 | 1.34 ±.13 | −16 ±1.3 | .52 ±.05 | −.59 ±.05 |
| 282 149196 | FeT 3 | >58 | >11 | −28 ±1.4 | .96 ±.05 | −1.33 ±.04 | −19 ±1.1 | −.83 ±.04 | .15 ±.04 |

TABLE 2. Properties of the energetic $\mu^-\mu^-$ events ($p_\mu$ > 15 GeV). Units of all momenta are in GeV.

| | No E$_\nu$ cut[††] | E$_\nu$ > 100 GeV |
|---|---|---|
| $R(\mu^-\mu^+)/R(\mu^-)$ | $(4.0 \pm 0.8) \times 10^{-3}$ | $(6.5 \pm 1.3) \times 10^{-3}$ |
| $R(\mu^-\mu^-)/R(\mu^-)$[†] | $(4 \pm 2) \times 10^{-4}$ | $(6.5 \pm 3.5 \times 10^{-4}$ |
| $R(\mu^-\mu^-\mu^+)/R(\mu^-)$ | $(9 \pm 5) \times 10^{-5}$ | $(2.6 \pm 1.5) \times 10^{-4}$ |

TABLE 3.  Multimuon rates relative to the rate of deep

inelastic single muon events.

† Obtained using $R(\mu^-\mu^-)/R(\mu^-\mu^+)$ = 0.10 $\pm$ 0.05

††Averaged over the Quadrupole Triplet Spectrum

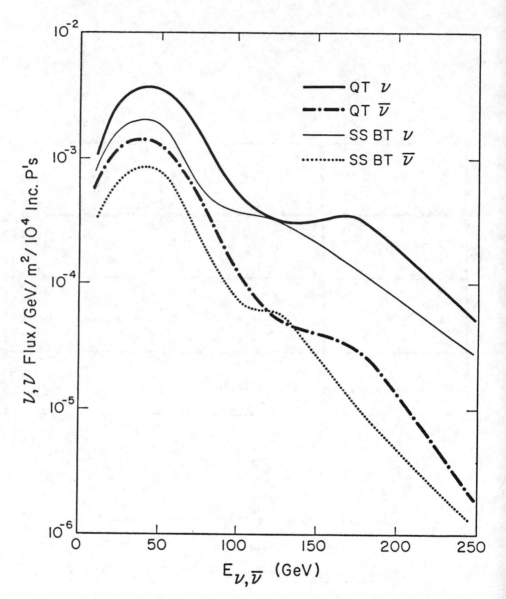

Fig. 1    Calculated neutrino and antineutrino spectra for the Quadrupole

Triplet and Bare Target Selected Beams.

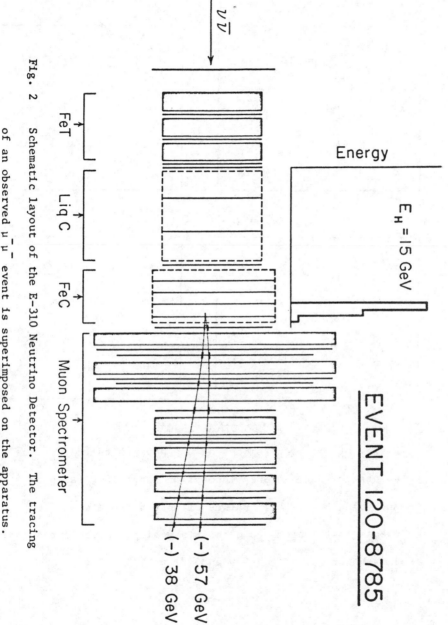

Fig. 2    Schematic layout of the E-310 Neutrino Detector. The tracing
of an observed μ⁻μ⁻ event is superimposed on the apparatus.

284

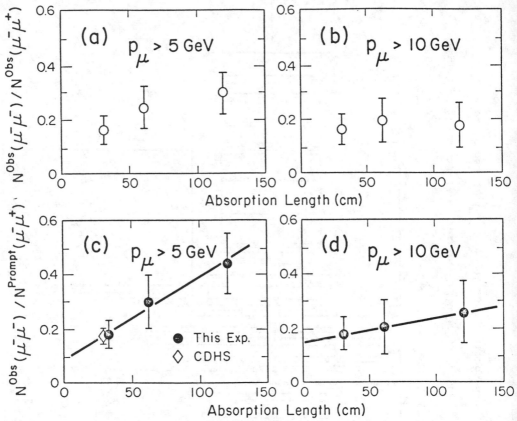

Fig. 3     Ratios of observed numbers of $\mu^-\mu^-$ events to that of the $\mu^-\mu^+$ events against hadronic absorption length for (a) $p_\mu$ > 5 Gev and (b) $p_\mu$ > 10 Gev. Ratios of observed numbers of $\mu^-\mu^-$ events to the <u>prompt</u> $\mu^-\mu^+$ events against hadronic absorption length for (c) $p_\mu$ > 5 Gev and (d) $p_\mu$ > 10 Gev.

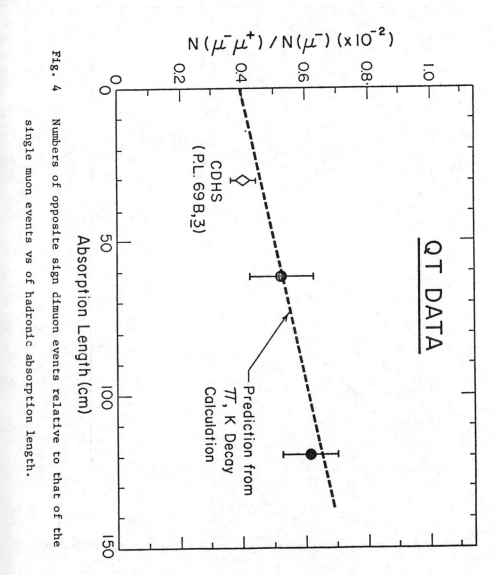

Fig. 4    Numbers of opposite sign dimuon events relative to that of the single muon events vs of hadronic absorption length.

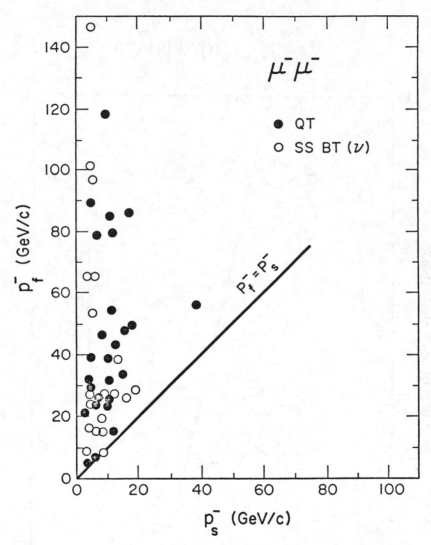

Fig. 5     Scatter plot of the muon momenta for the like-sign dimuon events.

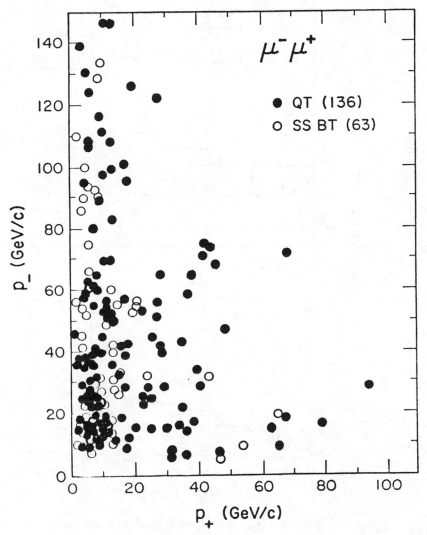

Fig. 6    Scatter plot of the muon momenta for the opposite sign dimuon

events.

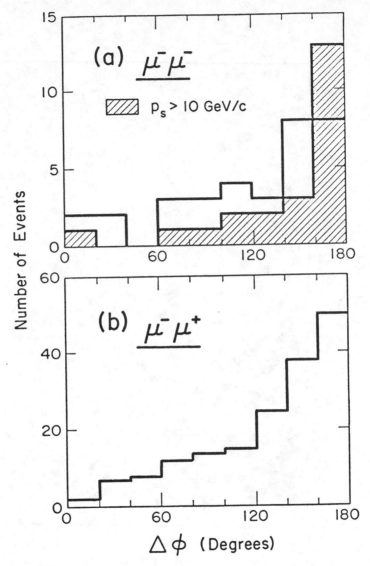

Fig. 7    Distributions in the relative azimuthal angle, ΔΦ, between the

two muons for (a) the like-sign dimuons and (b) the opposite-

sign dimuons.

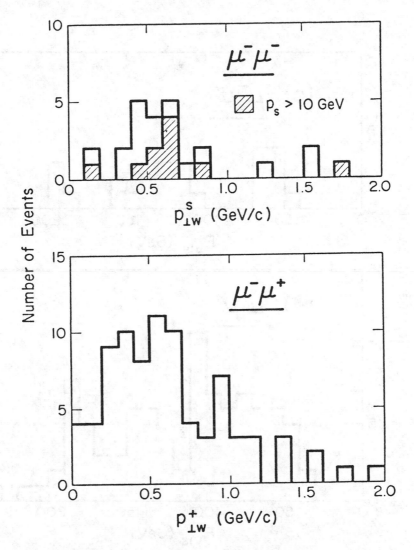

Fig. 8    Distributions in the transverse momentum relative to the
W-direction for the (a) like-sign dimuons and (b) opposite-sign
dimuons.

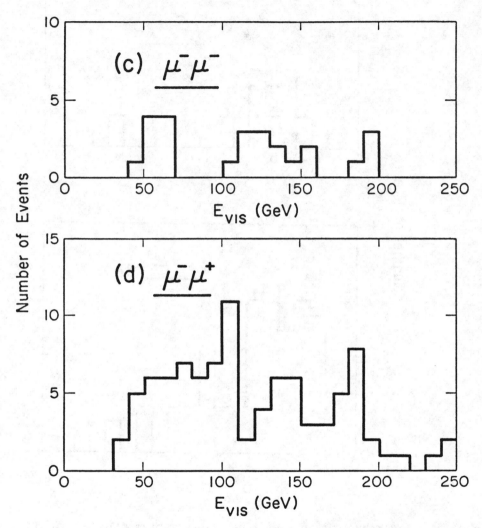

Fig. 9    Distributions of total visible energy for (a) like-sign dimuons and (b) opposite-sign dimuons.

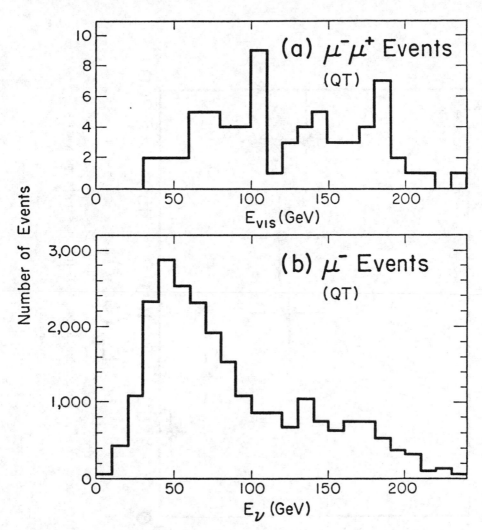

Fig. 10

(a)   Distributions of total visible energy for the  ν-induced

opposite-sign dimuon events.

(b)   Distributions of neutrino energy for the ν-induced single muon

events.

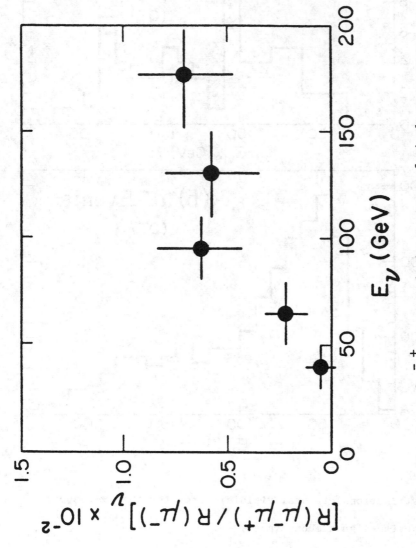

Fig. 11    Rate of $\mu^-\mu^+$ events relative to rate of single muon as a function of energy for neutrinos.

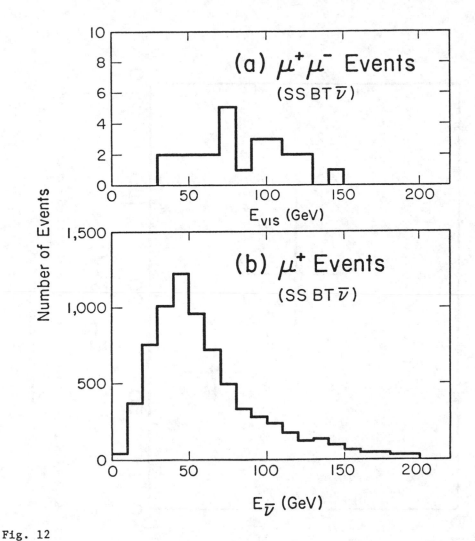

Fig. 12

(a)    Distributions of total visible energy for the $\bar{\nu}$-induced opposite sign dimuon events, and

(b)    distributions of antinuetrino energy for the $\bar{\nu}$-induced single muon events.

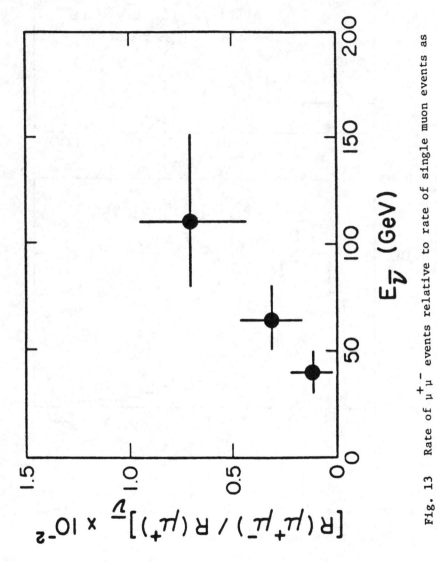

Fig. 13　Rate of $\mu^+\mu^-$ events relative to rate of single muon events as a function of energy for antineutrinos.

295

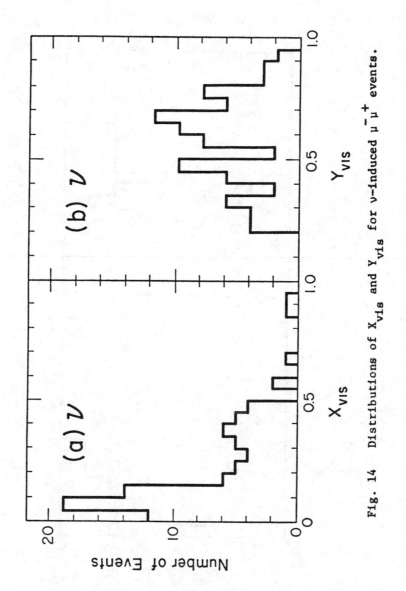

Fig. 14  Distributions of $X_{vis}$ and $Y_{vis}$ for $\nu$-induced $\mu^-\mu^+$ events.

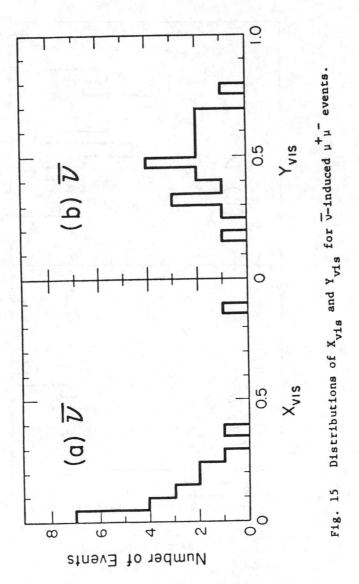

Fig. 15    Distributions of $X_{vis}$ and $Y_{vis}$ for $\bar{\nu}$-induced $\mu^+\mu^-$ events.

# ORIGIN OF TRIMUON EVENTS IN HIGH-ENERGY NEUTRINO INTERACTIONS

T. Hansl, M. Holder, J. Knobloch, J. May, H. P. Paar, P. Palazzi,
A. Para, F. Ranjard, D. Schlatter, J. Steinberger, H. Suter,
W. von Rüden, H. Wahl, S. Whitaker and E. G. H. Williams
CERN, Geneva, Switzerland

F. Eisele, K. Kleinknecht, H. Lierly, G. Spahn
and H. J. Willutzki
Institut für Physik[*)] der Universität, Dortmund, Germany

W. Dorth, F. Dydak, C. Geweniger, V. Hepp, K. Tittel and J. Wotsckack
Institut für Hochenergiephysik[*)] der Universität, Heidelberg, Germany

P. Bloch, B. Devaux, S. Loucatos, J. Maillard,
B. Peyaud, J. Rander, A. Savoy-Navarro and R. Turlay
D.Ph.P.E., CEN-Saclay, France

F. L. Navarria
Istituto di Fisica dell'Università, Bologna, Italy

## ABSTRACT

The properties of 76 neutrino-initiated $\mu^- \mu^- \mu^+$ events observed
in the CDHS detector in the 350 GeV and 400 GeV wide-band beams at
the CERN SPS are discussed. For neutrino energies > 30 GeV and muon
momenta $\gtrsim$ 4.5 GeV, the average trimuon rate is $(3.0 \pm 0.4) \times 10^{-5}$ of
the single-muon event rate. The data are in agreement with normal
charged-current interactions with the additional production of a
muon pair by both hadronic and radiative processes. No evidence is
found for either heavy-lepton or heavy-quark cascades. Upper limits
for these two possibilities are established.

[*)] Supported by the Bundesministerium für Forschung and Technologie.

In this paper we discuss the origin of trimuon events observed in high-energy neutrino interactions. A large variety of models for trimuon production mechanisms has been published*). Among those which involve new phenomena there are in principle two different models:

  i)  the production and cascade decay of new heavy leptons at the lepton vertex:

$$\nu N \rightarrow L^- + X$$
$$\phantom{\nu N \rightarrow}\hookrightarrow \mu^- \nu L^0$$
$$\phantom{\nu N \rightarrow \mu^- \nu}\hookrightarrow \mu^- \mu^+ \nu \ ;$$

 ii)  the production and subsequent decay of quarks with new flavour in the shower, as, for example:

$$\nu N \rightarrow \mu + \bar{b} + X$$
$$\phantom{\nu N \rightarrow}\hookrightarrow \bar{c} \mu^+ \nu$$
$$\phantom{\nu N \rightarrow \bar{c}\mu^+}\hookrightarrow \bar{s} \mu^- \nu,$$

   where $\bar{b}$, $\bar{c}$, and $\bar{s}$ indicate mesons containing bottom, charm, and strange antiquarks, respectively.

More conventional sources for trimuons are normal charged-current events with an additinal muon pair. Again there are two main possibilities:

 iii)  muon pairs from internal bremsstrahlung from either the leading muon or from the quarks;

  iv)  hadronic production of the muon-pair in the shower.

Since the total number of trimuons observed in neutrino interactions[2-4] had so far been rather limited, the question of their origin was still unclear.

In two runs in the CERN wide-band neutrino beam at 350 GeV and 400 GeV proton energy, we have measured the distributions of 76 $\mu^- \mu^- \mu^+$ events and 4 $\mu^- \mu^+ \mu^+$ events in a sample of 2.3 x $10^6$ single-muon events. The detector is described in detail elsewhere[5] and a full discussion of the characteristics of the trimuon events is published in a longer paper[6].

The relative rate of $\mu^- \mu^- \mu^+$ events to single-muon events above 30 GeV visible energy is $(3.0 \pm 0.4) \times 10^{-5}$, after having subtracted 8% for a $(\pi/K) \rightarrow \mu\nu$ decay background. The four $\mu^- \mu^+ \mu^+$ events are consistent with coming from this background, in rate as well as in the muon energy distribution. The energy dependence of the rate of the $--+$ events is shown in Fig. 1. It increases by a factor of 10 from 30 GeV to 130 GeV visible energy. This is mainly due to the muon detection threshold of $\sim$ 4.5 GeV, as is illustrated in the same figure by the efficiency curves for the different trimuon production models (i) to (iv).

---

*) A summary of trimuon production models is given in Albright et al., 1.

The analysis of the trimuon events is complicated by the presence of two negative muons. Two better demonstrate the properties
of trimuon events we have chosen the following definition of a
leading muon: it is that negative muon ($\mu_1$) for which the sum of
the absolute values of the transverse momenta of the other negative
muon ($\mu_2$) and the positive muon ($\mu_3$) with respect to the direction
of the W boson, $\vec{W} = \vec{\nu} - \vec{\mu}_1$, is minimal.

Themost instructive distribution of the trimuons is the scatter
plot of the two $\mu^-\mu^+$ mass combinations against the trimuon mass, as
shown in Fig. 2a. There are two distinct bands. One, with small
dimuon masses (< 1.6 GeV) for all trimuon masses, contains the mass
combination with the non-leading negative muon $\mu_2$. The projection
of $m_{23}$ is shown in Fig. 2b. The second band contains the mass $m_{13}$,
which increases as $m_{123}$ increases. The scatter plot, Fig. 2a, shows
that the bulk of the events contain a low-mass muon pair linked to
the hadron shower.

This is also reflected in the distribution of the azimuthal
angle $\phi_{1,23}$ between the leading muon and the sum of $\mu_2$ and $\mu_3$ in the
plane perpendicular to the neutrino direction. Figure 3a shows a
clear peak at $180^o$ due to muon pairs close to the shower direction.
But there is an additional, smaller peak at $0^o$. The latter indicates
that there is an additional source of muon pairs originating at the
lepton vertex.

In Fig. 3b we present the distribution in $p^s_{T23}$, the transverse
momentum of the pair with respect to the W direction. The larger
values for $p^s_{T23}$ are correlated with events having $\phi_{1,23} < 60^o$
(hatched histogram), which are probably coming from the lepton vertex. Omitting these events the average transverse momentum of the
hadronic events is 0.7 GeV/c.

In the case of theheavy lepton cascade[*] interpretation one
has to set the masses of thenew leptons at $M(L^-) \simeq 9$ GeV and $M(L^0)$
$\simeq 1.5$ GeV in order to minimize the disagreement with the observed
mass spectra. The general agreement with the data is not good. In
particular much larger transverse momenta $p^s_{T23}$ are expected in this
model (full line in Fig. 3b). We have no positive evidence for the
lepton cascade process, and only a small part of the observed events
could have such an origin. An upper limit of $\sim 10^{-5}$ for the production cross-section times the branching ratio for this process relative to the charged current process can be set.

The predictions from a quark cascade with new flavours[**] lead
to much larger pair masses $m_{23}$ than measured, even for a decay of

---

[*] In our Monte Carlo simulation of the lepton cascade, we used a constant decay matrix and assumed a flat production cross-section
above threshold and the same distributions in the Bjorken scaling
variables x and y as found in charged current interactions.

[**] We have simulated the production and decay of a 4.5 GeV meson
with the assumption that the x- and y-distributions are those from
the sea quarks $\bar{c}$ and $\bar{u}$.

the lightest possible new-flavoured meson of 4.5 GeV. The dotted line in Fig. 2b shows the result from a simulation of this model. The upper limit for this process times the two branching ratios is $Br^2 \times \sigma_{prod}(3\mu) / \sigma(1\mu) = 0.7 \times 10^{-5}$.

Calculations [7-9] of the radiative production of muon pairs in association with ordinary charged current interactions exhibit a very particular $\phi$-angle distribution, which is peaked at $0^o$ and at $180^o$. Attributing the observed events for $\phi_{1,23} < 60^o$ to the internal bremsstrahlung process from the leading muon, one gets a rate of $R(3\mu/1\mu) = (0.8 \pm 0.4) \times 10^{-5}$ for the total radiative pair contribution. This is not in disagreement with the predictions of $0.7 \times 10^{-5}$ to $2.0 \quad 10^{-}$ [7-9]†.

It is well known that muon pairs are also produced non-radiatively in pion-nucleon collisions. We have related this process to W-nucleon interactions, assuming that the total moun-pair rate depends only on the c.m.s. energy in the hadron system††). To obtain agreement with the data we changed the Feynman x-distribution, using and invariant cross-section proportional to $x_F(1 - x_F)$ which is flatter than observed in pion-nucleon interactions. The agreement with the data is in general very good; in particular the low moun pair masses and the peaking of the $\phi_{1,23}$ distribution at $180^o$ are well reporduced (see dashed line in Figs. 2 and 3). We expect a rate of $R(3\mu/1\mu) = 2.0 \times 10^{-5}$ for $E_\nu > 30$ GeV, in agreement with the observed number $R(3\mu/1\mu) = (2.2 \pm 0.4) \times 10^{-5}$ after subtraction of the contribution from the internal bremsstrahlung. The data can be attributed to these two conventional muon-pair sources in a relative rate of $\sim 1:3$ (see dashed-dotted lines in Figs. 3a and 3b).

In conclusion, trimuoun events of the charge configuration $\mu^- \mu^- \mu^+$ are produced at a rate of $(3.0 \pm 0.4) \times 10^{-5}$ of the single-muon events for $E_\nu > 30$ GeV. These events seem to be charged current interactions with additional low-mass muon pairs from either internal bremsstrahlung at a rate of $(0.8 \pm 0.4) \times 10^{-5}$, or from hadronic production in the shower at a rate of $(2.2 \pm 0.4) \times 10^{-5}$ of single muons above $E_\nu = 30$ GeV. No positive evidence for heavy lepton or quark cascades has been observed, and such processes cannot account for more than $\sim 15\%$ of the observed trimuon production.

We wish to acknowledge helpful discussions with J. Ellis and J. Smith, and thank our many technical collaborators for their capable assistance.

---

†) We actually used the simulation of electromagnetic muon pairs of Smith[10]. The expected rate from this calculation is $(0.7 \pm 0.07) \times 10^{-5}$.

††) As an input for the simulation we used the data of Anderson et al.[11] and an s-dependence as suggested by Bourquin and Gaillard[12].

REFERENCES

1.  C. H. Albright et al., Fermilab preprint 78/14 THY.
2.  B. C. Barish et al., Phys. Rev. Letters 38 (1977) 577.
3.  A. Benvenuti et al., Phys. Rev. Letters 38 (1977) 1110;
    A. Benvenuti et al., Phys. Rev. Letters 38 (1977) 1183.
4.  M. Holder et al., Phys. Letters 70B (1977) 393.
    P. C. Bosetti et al., Phys. Letters 73B (1978) 380.
5.  M. Holder et al., Nuclear Instrum. Methods 148 (1978) 235.
6.  T. Hansl et al., Characteristics of trimuon events observed in
    high energy neutrino interactions, submitted to Nuclear Phys.
7.  J. Smith et al., Stony Brook preprint ITP-SB 77-66.
8.  R. M. Barnett et al., Origin of neutrino events with three muons,
    submitted to Phys. Rev. D.
9.  V. Barger et al., Electromagnetic muon pair contributions in
    neutrino trimuon production, submitted to Phys. Rev. D.
10. J. Smith, private communication.
11. V. F. Anderson et al., Phys. Rev. Letters 37 (1976) 799.
12. M. Bourquin and J. M. Gaillard, Nuclear Phys. 114B (1976) 334.

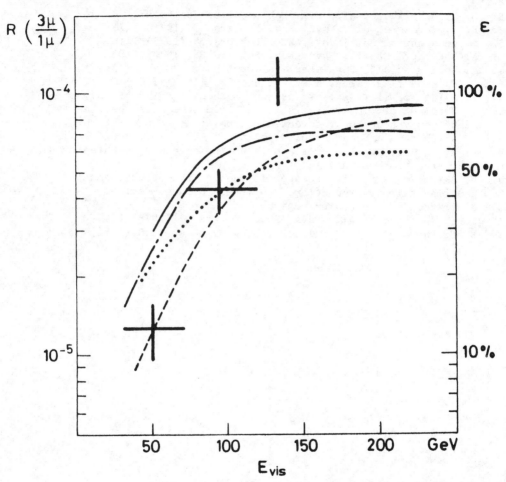

$R\left(\dfrac{3\mu}{1\mu}\right)$        $\varepsilon$

Fig. 1 : Relative rate of trimuon to single-muon production as a
function of visible energy.  The (π/K)-decay background
is subtracted.  The curves show the acceptance ε for the
following models: ———— heavy lepton cascade;
...... heavy quark cascade; ———— hadronic muon pairs;
– . – . internal bremsstrahlung.

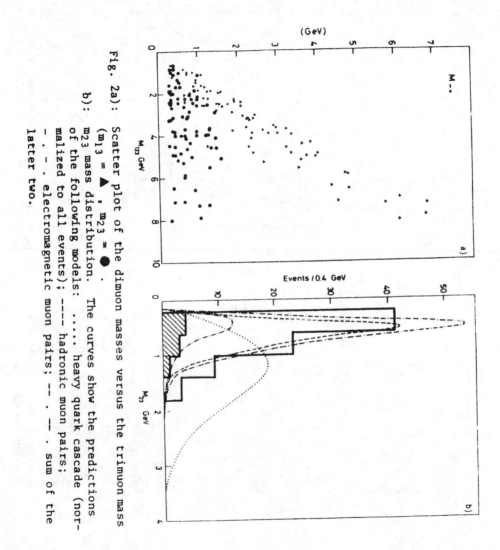

Fig. 2a): Scatter plot of the dimuon masses versus the trimuon mass
($m_{13}$ = ▲ , $m_{23}$ = ● .

b): $m_{23}$ mass distribution. The curves show the predictions
of the following models: ..... heavy quark cascade (normalized to all events); ---- hadronic muon pairs;
— . — . electromagnetic muon pairs; — . — . sum of the
latter two.

304

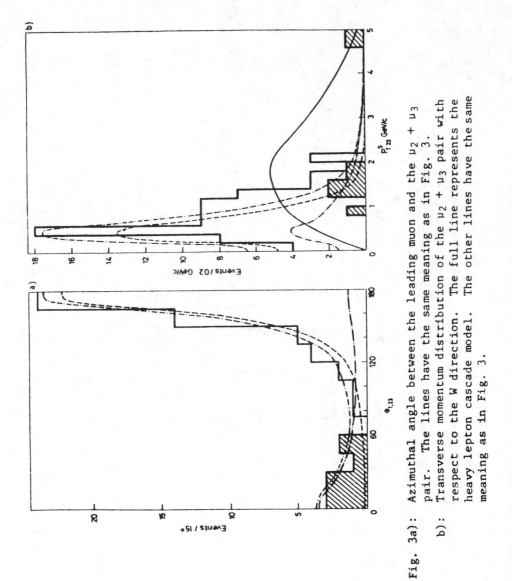

Fig. 3a): Azimuthal angle between the leading muon and the $\mu_2 + \mu_3$ pair. The lines have the same meaning as in Fig. 3.

b): Transverse momentum distribution of the $\mu_2 + \mu_3$ pair with respect to the W direction. The full line represents the heavy lepton cascade model. The other lines have the same meaning as in Fig. 3.

# CHARGED CURRENT NEUTRINO AND ANTINEUTRINO CROSS SECTION

# RESULTS FROM THE CITFR EXPERIMENT[*]

B. C. Barish, J. F. Bartlett,[a] A. Bodek,[b] K. W. Brown,
D. Buchholz,[c] Y. K. Chu, F. Sciulli, E. Siskind,
and L. Stutte[a]
California Institute of Technology
Pasadena, California 91125

and

H. E. Fisk, G. Krafczyk, and D. Nease
Fermi National Accelerator Laboratory
Batavia, Illinois 60510

and

O. Fackler
Rockefeller University
New York, New York 10021

[*]Work supported by the U.S. Energy Research and Development Administration.

## INTRODUCTION

In the Summer and Fall of 1975 the CITFR group had a major data taking run to measure total charged current cross sections using a dichromatic beam. Much of the data from that run are published in the September 9 and December 19, 1977 Physical Review Letters. The objectives of that run included tests of Bjorken scaling in deep inelastic neutrino and antineutrino scattering. These tests are conventionally discussed in terms of the scaling variables x, y.[1] If the usual assumptions about scaling, V-A weak charged currents, isoscalar target, $M_W$ large, etc., are made then:

$$\frac{d^2\sigma}{dxdy}^{\nu,\bar{\nu}} = \frac{G^2 ME_\nu}{\pi} \{ (1-y) F_2^{\nu,\bar{\nu}}(x) + y^2/2 \; 2xF_1^{\nu,\bar{\nu}}(x)$$

$$\pm y(1-y/2) F_3^{\nu,\bar{\nu}}(x) \}, \tag{1}$$

$$Q^2 = 4 E_\nu E_\mu \sin^2 \theta_\mu/2,$$

$$x = Q^2/2ME_H,$$

$$y = E_H/E_\nu.$$

We report here only on the observed y dependence of the cross sections. Consequently the following integrals over the structure functions are defined:

$$f_1 \equiv \int xF_1(x)dx; \quad f_2 \equiv \int F_2(x)dx; \quad f_3 \equiv \int xF_3(x)dx. \tag{2}$$

The cross section expression then becomes:

$$\frac{d\sigma}{dy}^{\nu,\bar{\nu}} = \frac{G^2 ME_\nu}{\pi} \{ (1-y)f_2 + (y^2/2) 2f_1 \pm y(1-y/2) f_3 \} \tag{3}$$

The thrust of our experimental effort was:

(1) To measure $\sigma_o \equiv \frac{d\sigma}{dy}\Big|_{y=0}$ and test the charge symmetry invariance of $f_2$ which implies $f_2^\nu = f_2^{\bar{\nu}}$.

(2)  To check the linear rise of $\sigma^{\nu,\bar{\nu}}$ with energy
     and the ratio $R_c = \sigma_{\bar{\nu}}/\sigma_{\nu}$. $R_c$ was reported to
     increase from 0.4 at less than 12 GeV to 0.6 or
     0.7 at 60 GeV by the HPWF group.[2]
(3)  To measure < y > for the neutrino and antineutrino
     data. Some authors[3] have indicated a statistically
     significant increase in < y > for antineutrino
     interactions at 60 GeV which might indicate scal-
     ing violations or new particle production while
     other recently published data[4] shows no energy
     dependence in < y >.
(4)  To check the energy independence of the integrated
     structure functions and the Callen-Gross relation
     $(2f_1 = f_2)$.
(5)  To look for possible evidence which might suggest
     the existence of new massive quarks which couple
     right-handedly to neutrinos.

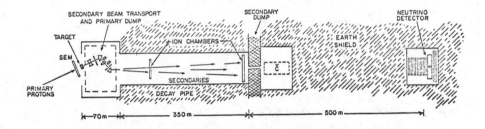

Fig. 1:  E21 Dichromatic Beam and Apparatus.

## BEAM AND APPARATUS

The dichromatic beam[5] and apparatus[6] which are
shown in Fig. 1 have been discussed elsewhere. The
400 GeV proton beam is targeted on an aluminum block
1/2" W x 3/4" H x 12" L. The total intensity of secondary charged particles was measured with ionization
chambers at two different points in the decay pipe
(430 ft. and 1130 ft.). Both of these ion chambers
were calibrated relative to each other and absolutely

308

to a foil irradiation and secondary emission monitor
(SEM) by passing a 200 GeV primary proton beam through
them simultaneously.  The SEM was calibrated absolutely
with single turn extracted beam from the Fermilab Main
Ring and a toroid.  The downstream ion chamber was
separately calibrated in a low intensity, $10^6$ ppp, 200
GeV proton beam where individual particles were counted.  The agreement between the calibrations is 2%.
During neutrino event data runs the two ion chambers
tracked each other to better than 5% and disagreed
to this extent because of beam containment within the
ion chambers.  Corrections up to 10% have been made
for the specific ionization of the gas in the ion
chambers for different species of particles ($\pi$, K,
p) relative to the 200 GeV protons used for calibration.

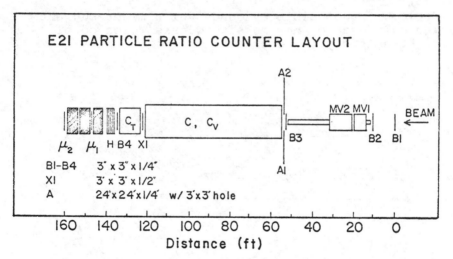

Fig. 2:  Particle Ratio Measurement Apparatus.

To know the number of $\nu_K$ and $\nu_\pi$ at the apparatus,
the $\pi^\pm$, $K^\pm$, p fractions of particles in the beam were
measured.  The counter arrangement is shown in Fig.
2.  These counters were located immediately downstream
of a steel dump at the end of the decay pipe.  The
secondary beam of reduced intensity ($10^6$ particles/
pulse) entered the detectors through  a 4" square hole
in the dump.  Both a sixty foot long differential
counter and ten foot long threshold counter (set to
count pions) were used in the measurement.  The differential counter focussed 1.84 ± .016 mr light onto
an iris of width 0.040".  A pressure curve for -230

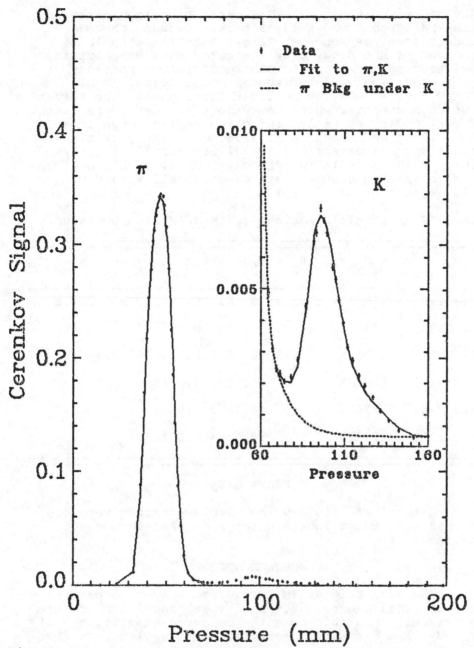

Fig. 3: Cerenkov pressure curve for the dichromatic beam tuned to -230 GeV. The inset shows the measured pion background under the K peak.

GeV is shown in Fig. 3. The inset shows the measured pion background under the K peak. Because the secondary beam was somewhat larger than the 4" hole at the end of the decay pipe, Cerenkov data were obtained by steering the beam, to sample different portions of the vertical and horizontal phase space. Table I and Fig. 4 show the particle fractions at the target for mean energies where neutrino event data were taken. The secondary particle production angles range from 0 to ±2 mr and the $\Delta p/p$ is ±16% ($\sigma$). The particle fractions are determined to ±(1-5)% for $\pi$'s, ±(3-7)% for K's and ±(1-2)% for protons. This, combined with the ionization chamber data give overall $\nu_\pi$ and $\nu_K$ flux errors of ±(5 to 9)%.

Table I   Particle Fractions vs. Mean Hadron Momentum

| $< P_\pi >$ | $< P_k >, < P_p >$ | K/$\pi$ | P/$\pi$ |
|---|---|---|---|
| +122±19, | 125±18, 131±19 | 0.103 ±0.004 | 0.989±0.018 |
| +170±23, | 173±23, 182±25 | 0.1295±0.005 | 3.076±0.065 |
| +215±29, | 218±28, 235±33 | 0.1618±0.013 | 8.029±0.406 |
| −128±23, | −128±23 | 0.0567±0.0019 | |
| −179±28, | −177±26 | 0.0472±0.0020 | |
| −202±31, | −179±26 | 0.0458±0.0011 | |

## EVENT DATA

A total of 18,000 $\nu$ and 12,000 $\bar{\nu}$ events were obtained with two triggers which are labeled muon and hadron.

## MUON TRIGGERS

The muon trigger (MT) required a muon to penetrate the toroidal magnet (~2.4 m of steel and 1.5 m of target steel). In this case the measured quantities are $E_h$, $E_\mu$, and $\theta_\mu$, with $\theta_\mu$ restricted to angles less than 110 mr. These events are used to check the energy

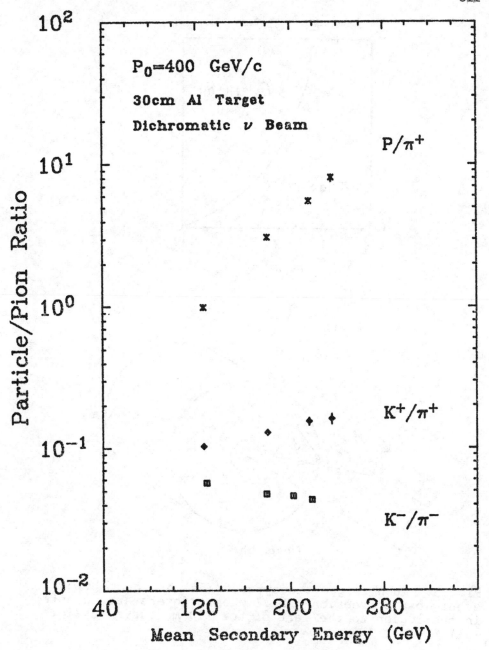

Fig. 4:  Particle ratios vs. mean hadron energy.

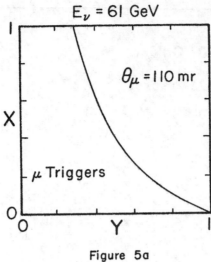

Figure 5a

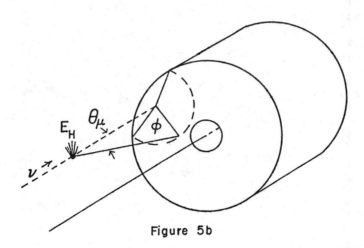

Figure 5b

Fig. 5: Muon trigger acceptance (5a) and geometrical event loss correction (5b). Muon triggers are accepted in the region to the left of the 110 mr curve in Fig. (5a).

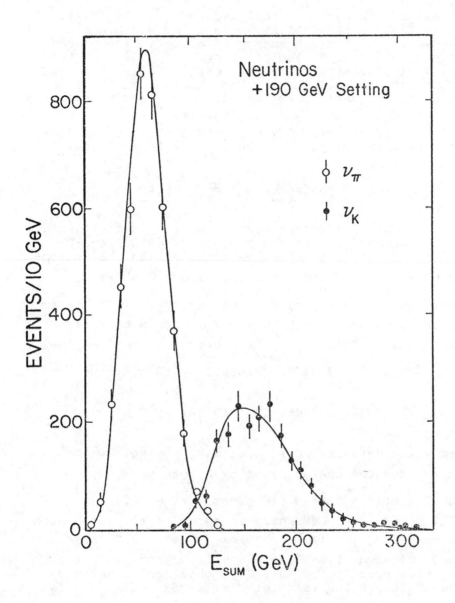

Fig. 6: Total energy distribution for muon trigger events (+190 GeV).

calibration of the calorimeter and muon spectrometer
for both $\nu_K$ and $\nu_\pi$ at the six different beam momenta.
The calibration agrees at all twelve energies to 4%.
     In Fig. 5a the scaling variable MT acceptance
is shown for a neutrino energy of 61 GeV.  The y acceptance begins to be limited above y = 0.28.  In the
low y region geometrical event loss is fully corrected
by the $\phi$ rotation defined in Fig. 5b.  Since the total
energy of the events is measured, the $\nu_K$ and $\nu_\pi$ data
are separated.  An example of this separation is shown
in Fig. 6 for the +190 GeV data.  The y = 0 cross sections are then obtained in the y $\leq$ 0.2 region from:

$$\sigma_o = \left.\frac{d\sigma}{dy}\right|_{y=0} = \frac{CN}{FT\,\Delta y} \quad \text{where}$$

N is the number of events for y $\leq$ 0.2, F is the flux,
T is the number of target nucleons, $\Delta y$ is 0.2 and C
is a correction factor for the shape of the y distribution between y = 0 and y = 0.2.  Figure 7 shows $\sigma_o^{\nu}$
and $\sigma_o^{\bar{\nu}}$ as a function of neutrino energy.  If the form
of $\sigma_o/E_\nu$ is assumed to be a + b $E_\nu$, then a best fit
to the data of Fig. 7 yield (in units of $10^{-38}$ cm$^2$/GeV):

$$\sigma_o^{\nu}/E_\nu = (0.77\pm0.06) - (0.66\pm0.65)\times10^{-3}\ E_\nu \quad \text{and}$$

$$\sigma_o^{\bar{\nu}}/E_{\bar{\nu}} = (0.75\pm0.06) - (0.24\pm0.50)\times10^{-3}\ E_{\bar{\nu}}.$$

Clearly the data are consistent with a single value
of $\sigma_o/E_\nu$ and the charge symmetry hypothesis: $\sigma_o^{\nu} = \sigma_o^{\bar{\nu}}$.  The best combined fit gives $\sigma_o/E = (0.719 \pm 0.035)\times10^{-38}$ cm$^2$/GeV which in terms of the integrated
structure function is $\int F_2^{\nu n}(x)\,dx = 0.46 \pm 0.02$ at a
mean $Q^2$ of about 5 GeV$^2$.  This value agrees with the
Gargamelle result of 0.49 $\pm$ 0.05 obtained at energies
below 12 GeV.[7]  The value of $\int F_2(x)\ dx$ obtained this
way, from the low y cross section, is independent of
the Callen-Gross assumption of spin 1/2 partons.
The integral $F_2^{\nu N}(x)\,dx$ is consistent with the expected
mean squared charge of the three quark model from

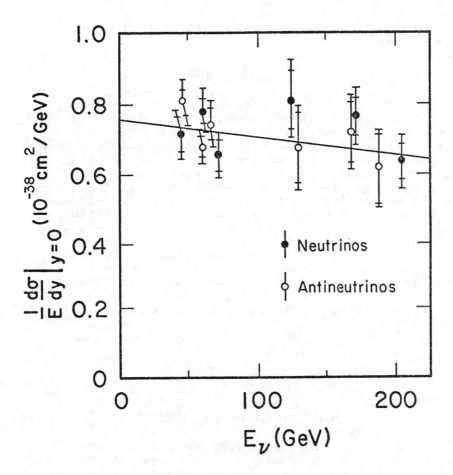

Fig. 7:  Cross Sections at y=0 for the test of charge
symmetry.  The inner error bars are statistical while
the outer limits include the systematics which are
mainly due to the flux errors.

predictions which relate electron or muon-deuteron scattering and neutrino nucleon scattering:

$$\int F_2^{\nu N}(x)\,dx \leq 18/5 \int F_2^{\mu d}(x)\,dx,$$

where $\int F_2^{\mu d}(x)\,dx$ is measured to be $0.153 \pm 0.005$ for muons[8] and $0.147 \pm .06$ to $0.178 \pm 0.04$ for electrons.[9]

## HADRON TRIGGERS

To measure total cross sections, neutrino data were also obtained with a hadron trigger. The hadron trigger required an average energy deposition in the calorimeter of at least 6 GeV. To be certain the event was a charged current neutrino interaction, a penetration requirement of 1.5 meters of steel (~9 absorption lengths) was imposed on the outgoing muon in the analysis. This condition implies muon angular acceptance out to 360 mr. The hadron trigger data were efficiency corrected by azimuthal weighting as were the muon trigger data.

Figure 8a shows the acceptance regions for both the muon and hadron triggers at a neutrino energy of 61 GeV ($< E_{\nu_\pi} >$ at 190 GeV). The substantial region of overlap for the two triggers allows a determination of efficiencies:

$$\epsilon_\mu = 97 \pm 1\% \text{ and}$$

$$\epsilon_H = 95 \pm 1\% \text{ for } E_H > 10 \text{ GeV}.$$

## SEPARATION OF $\nu_\pi$, $\nu_K$ EVENTS

In Fig. 8b a similar acceptance plot is shown for the mean $\nu_K$ energy, scaled by the mean energy ratio $< E_{\nu_K} >/< E_{\nu_\pi} >$. From a comparison of figures 8a and 8b it can be seen that all of the $\nu_K$ events, which have hadron energy in an $E_H$ region ambiguous with $\nu_\pi$ data, are muon trigger events and are consequently identified as $\nu_K$ data from total energy. Events with

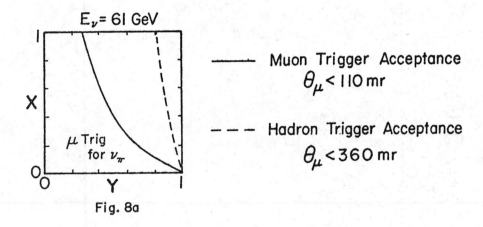

Fig. 8a

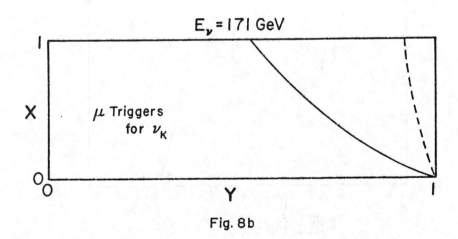

Fig. 8b

Fig. 8: Muon and hadron trigger acceptances for
neutrinos of energy 61 and 171 GeV.

318

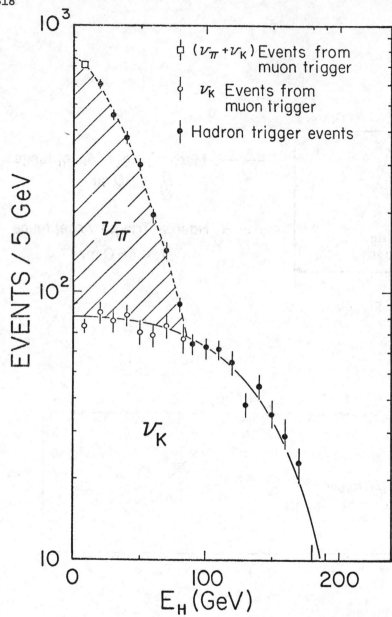

Fig. 9:  Hadron energy histogram for all data at +190 GeV.  Except for small corrections mentioned in the text this distribution is used to obtain both the number of events and the mean hadron energy for the $\nu_\pi$ and $\nu_K$ event samples.

hadron energy greater than $E_{\nu_\pi}$ can only result from $\nu_K$. Figure 9 shows the hadron energy histogram for the +190 GeV running and demonstrates the separation of $\nu_K$, $\nu_\pi$ data.

There is a small loss of data at large angles, or large y, due to the 360 mr acceptance cut. An estimate of the number of missing events is made by histogramming the scaled muon angle variable, $K_\theta \equiv 2M/E_\nu \theta^2$, which shows a small loss at low $K_\theta$. Table II gives the estimated loss as a function of energy. The correction is less than 2% for antineutrinos and less than 7% for neutrinos.

Table II  Estimated Event Loss Due to 360 mr Cut

| $E_\nu$ (GeV) | Loss | $E_{\bar\nu}$ (GeV) | Loss |
|---|---|---|---|
| 45.2 | 0.066±.021 | 45.9 | 0.016±.005 |
| 61.3 | 0.052±.015 | 60.6 | 0.000±.005 |
| 72.4 | 0.042±.013 | 65.7 | 0.013±.005 |
| 125.0 | 0.020±.007 | 129.0 | 0.00 ±.005 |
| 171.0 | 0.014±.005 | 168.0 | 0.00 ±.005 |
| 205.0 | 0.00 ±.005 | 188.0 | 0.00 ±.005 |

## TOTAL CROSS SECTIONS

The values of the cross section and mean energies are given in Table III. They are shown graphically in Fig. 10. Included in the plot are the lower energy results from Gargamelle (GGM).[7,10] The GGM $\bar\nu$ slope of 0.28 goes through the $\bar\nu_\pi$ data but undershoots the $\bar\nu_K$ points. The $\nu$ slope of 0.74, which fits the GGM data well, is an overestimate for our high energy data. The best fit slopes, in units of $\sigma/E$ are;

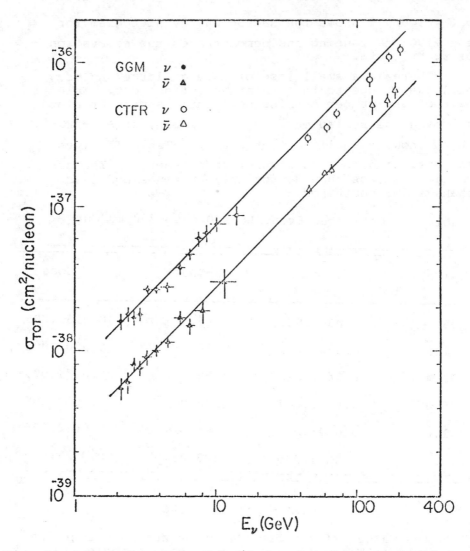

Fig. 10:  Neutrino and antineutrino total cross sections.
The lower energy results ($E_\nu$ < 12 GeV) from Gargamelle
(GGM) are also shown.  The curves are the GGM best fits
at low energy extrapolated to the higher energy region
of our measurements.  See text for details.

$$S_\nu = 0.61 \pm 0.03 \times 10^{-38} \ cm^2/GeV,$$

$$S_{\bar\nu} = 0.29 \pm 0.015 \times 10^{-38} \ cm^2/GeV$$

with $\chi^2$ of 3.8 and 6.5, respectively, for 5 degrees of freedom. It is possible the decrease in the slope for neutrinos is consistent with scaling violations observed in ep and μp deep inelastic scattering. The cross sections reported here are in good agreement with BEBC data obtained in a recent CERN narrow band neutrino run.[11] The cross section ratio $\sigma_{\bar\nu}/\sigma_\nu$, of Fig. 11, shows a (20 ± 10)% rise from 45 GeV to the high energy points, 125 to 205 GeV. Our data do not show the more substantive rise reported by other authors.[2] This rise does not allow a right handed coupling quark (B quark) of mass 5-7 GeV which was suggested to explain the original measurements of HPWF.[12] It is possible to fit the cross section ratio with a combination of scaling violations and asymptotic freedom effects.[13]

Table III   Cross Sections vs. Mean Neutrino Energy

| E (GeV) | $\sigma^\nu$ ($10^{-38} \ cm^2$) | E (GeV) | $\sigma^{\bar\nu}$ ($10^{-38} \ cm^2$) |
|---------|----------|---------|----------|
| 45.2 | 30.1±2.0 | 45.9 | 13.2±0.7 |
| 61.3 | 35.3±1.8 | 60.0 | 16.9±0.8 |
| 72.4 | 44.4±3.0 | 65.7 | 18.1±1.1 |
| 125.0 | 76.3±9.3 | 129.0 | 51.0±7.4 |
| 171.0 | 109.2±7.6 | 168.0 | 54.6±5.7 |
| 205.0 | 122.4±9.8 | 188.0 | 63.3±8.6 |

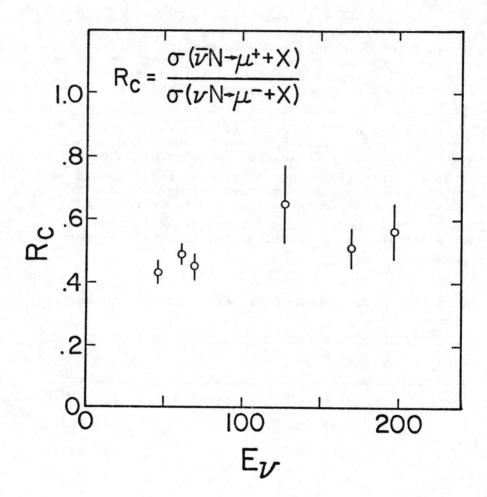

Fig. 11:  Ratio of antineutrino and neutrino cross
sections.

# MEAN ELASTICITY

From the acceptance considerations discussed earlier it is clear the y distribution has missing events at high y because the muon energy is unmeasured. On the other hand the $\nu_K$ and $\nu_\pi$ hadron energy distribution is measured. Using our knowledge of the mean beam energy with position in the detector we can establish a mean neutrino energy which allows a calculation of $< y > = < E_H >/< E_\nu >$. The equivalence of $< E_H >$ with $< y > * < E_\nu >$ follows from the central limit theorem and integration of the hadron energy resolution function over a complete sample of data in our apparatus. A small correction to $< y >$ is necessary for the data which escapes both the muon and hadron trigger conditions ($\theta_\mu > 350$ mr). Figure 12 shows the $< y >$ versus neutrino energy and the values are given in Table IV. Again it is noted that we measure no unusually significant increase as was found by previous authors (the high y anomaly). Averaging over the six energies one finds $< y_\nu > = 0.47 \pm 0.01$ and $< y_{\bar\nu} > = 0.32 \pm 0.01$. These values are consistent with the BEBC data and our previously published data.[14] For the antineutrino data the 10% increase in $< y >$ from the lower energies ( 65 GeV) to the higher energies (150 GeV) may be related to the $(20 \pm 10)\%$ increase in $\sigma_{\bar\nu}$ discussed earlier.

Table IV  $< y >$ vs. Mean Neutrino Energy

| E (GeV) | $< y >^\nu$ | E (GeV) | $< y >^{\bar\nu}$ |
|---------|-------------|---------|-------------------|
| 45.2 | 0.487±.026 | 45.9 | .306±.014 |
| 61.3 | 0.461±.022 | 60.0 | .314±.016 |
| 72.4 | 0.452±.023 | 65.7 | .314±.016 |
| 125.0 | 0.503±.026 | 129.0 | .343±.020 |
| 171.0 | 0.473±.020 | 168.0 | .350±.022 |
| 205.0 | 0.476±.022 | 188.0 | .339±.029 |

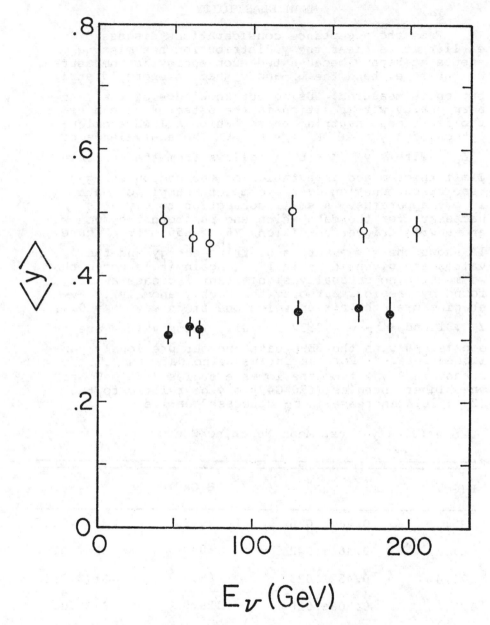

Fig. 12: Mean Inelasticity, < y >, for neutrinos (open circles) and antineutrinos (closed circles).

# INTEGRATED STRUCTURE FUNCTIONS

Integration of Eq. (3) and calculation of the first moment of the y distribution yields:

$$\sigma^{\nu,\bar{\nu}} = \frac{G^2 ME}{\pi} \left\{ \frac{1}{3} f_1 + \frac{1}{2} f_2 \pm \frac{1}{3} f_3 \right\}, \qquad (4)$$

$$\sigma^{\nu,\bar{\nu}} < y >^{\nu,\bar{\nu}} = \frac{G^2 ME}{\pi} \left\{ \frac{1}{4} f_1 + \frac{1}{6} f_2 \pm \frac{5}{24} f_3 \right\}. \qquad (5)$$

Since there are three structure functions and four measured moments, the consistency of the structure functions can be tested. One way to see this is to define the ratios:

$$S_{\pm} = \frac{\sigma^{\nu} < y >^{\nu} \pm \sigma^{\bar{\nu}} < y >^{\bar{\nu}}}{\sigma^{\nu} \pm \sigma^{\bar{\nu}}}. \qquad (6)$$

The $S_-$ ratio, displayed in Fig. 13 as a function of energy shows the consistency of the $f_3$ determination. The assumption of spin 1/2 for the partons (i.e., Callen-Gross, $2f_1 = f_2$) implies $S_+ = 7/16$ and the data of Fig. 13 are reasonably consistent with this value. If, however, one lets $\Delta \equiv (f_2 - 2f_1)/f_2$ then $S_+ = (7-3\Delta)/(16-4\Delta)$. The measured mean of $S_+ = 0.424 \pm 0.007$ gives $= 0.17 \pm 0.09$ which implies a small violation of Callen-Gross at the two standard deviation level.

With the assumption, $2f_1 = f_2$, the integrated structure functions, $f_2$ and $f_3$, have been fitted to the zeroth and first moments of the y distributions. These are displayed in Fig. 14 where it is evident there is a gradual decrease in the value of $f_3$ with energy. This effect which may be associated with scale breaking or other phenomena is equivalently seen as a decrease in the B parameter ($B \equiv f_3/f_2$) or as an increase in the fractional amount of momentum carried by the antiquark in the nucleon as $E_{\nu,\bar{\nu}}$ is increased.

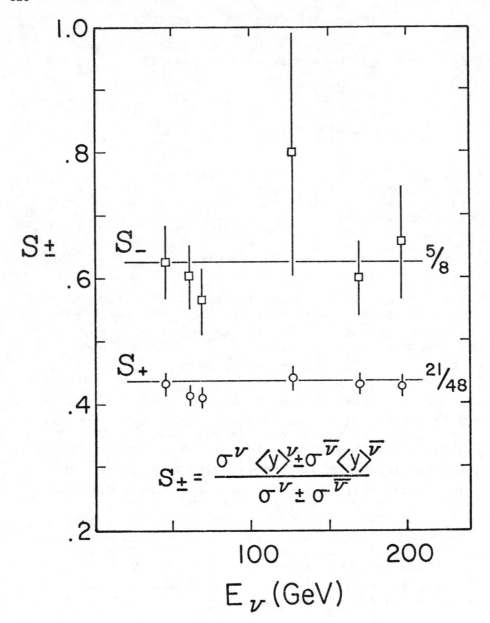

Fig. 13: Integrated structure function ratios.

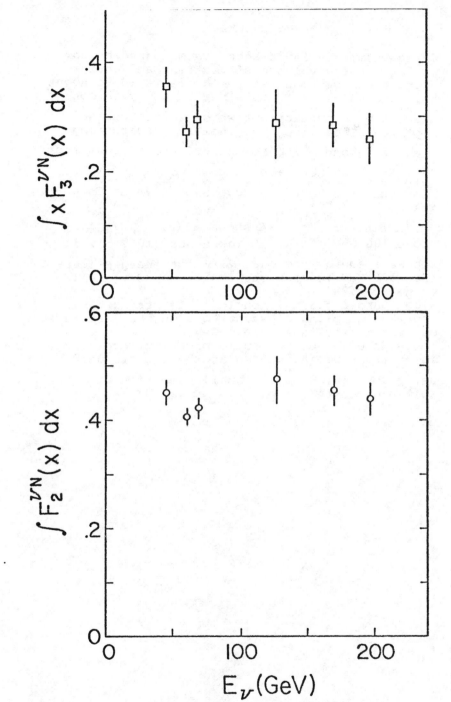

Fig. 14: Integrated structure functions.

## CONCLUSIONS

(1) Charge symmetry invariance, as measured by comparing the equivalence of neutrino and antineutrino cross sections at y = 0, is valid to about 5%. The value of $f_2$ obtained at low y is consistent with predictions which relate charged lepton and neutrino deep-inelastic scattering.

(2) The mean y and $\sigma^\nu/E$, for neutrinos, is flat from $45 \leq E_\nu \leq 205$ GeV. The value of $\sigma^\nu/E = 0.61$ is less than the lower energy value 0.74 of Gargamelle.

(3) For antineutrinos $\sigma^{\bar\nu}/E$ shows a $(20 \pm 10)\%$ rise above 100 GeV and $< y >$ increases $(10 \pm 5)\%$ in the same energy region. $\sigma^{\bar\nu}/E$ for energies less than 100 GeV is consistent with the Gargamelle value of 0.28.

(4) Right-handed quarks, coupled full strength via the weak coupling constant to u quarks, are ruled out for masses less than 8 GeV.

(5) There is evidence for scale breaking, especially in $f_3$ and some indication at the two standard deviation[3] level, for a small violation of the Callen-Gross relation.

REFERENCES

(a) Present address:  Fermi National Accelerator Laboratory, Batavia, Illinois  60510.

(b) Present address:  University of Rochester, Rochester, New York  14627.

(c) Present address:  Northwestern University, Evanston, Illinois  60201.

1. C. H. Llewellyn-Smith, Physics Reports $\underline{36}$, 263 (1972).

2. A. Benvenuti et al., Phys. Rev. Lett. $\underline{37}$, 189 (1976).

3. A. Benvenuti et al., Phys. Rev. Lett. $\underline{36}$, 1478 (1976).

4. M. Holder et al., Phys. Rev. Lett. $\underline{39}$, 433 (1977).

5. P. Limon et al., Nucl. Instrum. Methods $\underline{116}$, 317 (1974).

6. B. C. Barish et al., Phys. Rev. Lett. $\underline{35}$, 1316 (1975).
   B. C. Barish et al., Nucl. Instrum. Methods $\underline{130}$, 49 (1975).

7. D. H. Perkins, Reports on Progress in Physics $\underline{40}$, 409 (1977).

8. H. L. Anderson et al., Phys. Rev. Lett. $\underline{37}$, 4 (1976).

9. J. I. Friedman and H. W. Kendall, Annual Review of Nucl. Sci. $\underline{22}$, 203 (1972).

10. T. Eichten et al., Phys. Lett. $\underline{46B}$, 281 (1973).

11. P. C. Bosetti et al., Phys. Lett. $\underline{70B}$, 273 (1977).

12. R. M. Barnett, Phys. Rev. Lett. $\underline{36}$, 1163 (1976) and R. M. Barnett, H. Georgi, and H. D. Politzer, Phys. Rev. Lett. $\underline{37}$, 1313 (1976).

13. F. J. Sciulli, Private Communication. See e.g., I. Hinchliffe and C. H. Llewellyn Smith, Phys. Lett. $\underline{70B}$, 247 (1977).

14. B. C. Barish et al., Phys. Rev. Lett. $\underline{38}$, 314 (1977).

# RECENT RESULTS FROM NEUTRINO
## INTERACTIONS IN HEAVY NEON *

M. Kalelkar, C. Baltay, D. Caroumbalis, H. French,
M. Hibbs, R. Hylton, and W. Orance
Columbia University, New York, N.Y. 10027

A.M. Cnops, P.L. Connolly, S.A. Kahn, H.G. Kirk,
M.J. Murtagh, R.B. Palmer, N.P. Samios, and M. Tanaka
Brookhaven National Laboratory, Upton, N.Y. 11973

## ABSTRACT

We report recent results from an analysis of
100,000 pictures of the Fermilab 15 ft bubble chamber
filled with heavy neon and exposed to the double horn
focused, wideband $\nu_\mu$ beam. We have found 164 dilepton
($\mu^- e^+$) events with 33 vees, in good agreement with the
GIM model of charm production. We have also observed
the production of the charmed $D^0$ meson, followed by the
decay $D^0 \rightarrow K^0\pi^+\pi^-$, at a rate of $(0.7 \pm 0.2)\%$ of all
charged current events. From a subsample of our film,
we have found four events of the purely leptonic
neutrino-electron elastic scattering process.

## INTRODUCTION

We present recent results from a study of $\nu_\mu$
interactions in heavy neon. The experiment was carried
out at Fermilab using the two-horn focused wideband
muon neutrino beam and the 15 ft chamber filled with a
heavy neon-hydrogen mixture (64 atomic % neon). A
total of 150,000 pictures was taken with an average of
$10^{13}$ 400 GeV protons per pulse on the neutrino target.
The interaction length for hadrons is 125 cm, so that
hadrons typically interact, while muons leave the
chamber without interaction, and can thus be identified
on the scan table. Electrons are easily identifiable
through visible bremsstrahlung, since the radiation
length is 40 cm.

## DILEPTON PRODUCTION

We have previously published[1] results on dilepton
production from the first 50,000 pictures of our
exposure. We have now analyzed 100,000 pictures,
corresponding to about 60,000 charged current neutrino
interactions. In this sample, we have found 164 events
with a $\mu^-$, an $e^+$ and anything else. The $e^+$ is required
to have two signatures and a momentum over 300 MeV/c.
With these cuts, the background from asymmetric Dalitz

pairs is a few percent. The $\mu^-$ is identified as the fastest negative leaving track. No momentum cut is made. From a comparison of interacting and non-interacting tracks of both signs, the background due to fake $\mu^-$ (hadron punchthrough) is determined to be about 10%. After correcting for these backgrounds, scan efficiency ($\sim$ 90%), and $e^+$ identification efficiency ($\sim$ 85%), we obtain a dilepton rate of

$$R = \frac{\nu_\mu + Ne \rightarrow \mu^- + e^+ + \ldots}{\nu_\mu + Ne \rightarrow \mu^- + \ldots} = (0.5 \pm 0.15)\% \quad .$$

This rate is calculated using half of our events for which we have an accurate normalization. Figure 1 shows the momentum distribution of the $e^+$ and $\mu^-$, and also the total visible energy.

We have examined the 164 $\mu^- e^+$ events for associated $K_S \rightarrow \pi^+\pi^-$ and $\Lambda \rightarrow p\pi^-$ decays. We find 33 such vees (25 events with a single vee, 4 with a double vee), consisting of 20 unambiguous $K^0$'s, 3 unambiguous $\Lambda$'s, and 10 $K/\Lambda$ ambiguities. From half of our events, this corresponds to a neutral strange particle rate of $0.6 \pm 0.2$ per dilepton event, in good agreement with the GIM model of charm production by neutrinos. From our 60,000 charged current events, we find that 6% have a visible vee. At that rate, we would expect 10 vees in 164 $\mu^- e^+$ events, whereas we actually see 33.

Figure 2 shows the $K^0 e^+$ effective mass from 19 $\mu^- e^+$ events with a single $K^0$. The data are not in good agreement with the distribution expected from the $K^0 e^+ \nu$ decay of a spin zero $D^+$ meson at 1868 MeV. However, the distribution is consistent with a calculation by Barger et al.[2] assuming a $K\pi e\nu$ decay.

## OBSERVATION OF $D^0 \rightarrow K^0 \pi^+ \pi^-$

We have measured all events with vees in about 80,000 pictures, corresponding to 46,000 charged current events with a muon momentum over 2 GeV/c. We obtain good 2 or 3 constraint fits for 1815 $K_S \rightarrow \pi^+\pi^-$ and 1367 $\Lambda \rightarrow p\pi^-$. Correcting for branching ratios and detection efficiencies, this corresponds to a ($K^0 + \bar{K}^0$) rate of $(13.6 \pm 1.5)\%$ of all charged current events, and a ($\Lambda^0 + \bar{\Sigma}^0$) rate of $(5.0 \pm 0.5)\%$.

Figure 3 shows the $K_S \pi^+ \pi^-$ mass distribution, indicating a peak in the mass region of the charmed $D^0$ meson seen at SPEAR.[3] The best fit to a polynomial background plus a Gaussian, shown by the curve, gives the following parameters:

$M = 1850 \pm 15$ MeV, $\quad \sigma = 20 \pm 8$ MeV

corresponding to 64 events above a background of 180, with a statistical significance of four standard deviations. The width is consistent with our experimental mass resolution of 20 MeV. No corresponding peak is apparent near the D mass in the events without a $\mu^-$. This is consistent with the prediction of the GIM model that the charm changing neutral current interactions are absent. If the peak were due to $K^*$ production, then one might expect it to be present in events both with and without a $\mu^-$.

Correcting for branching ratios and detection efficiencies, we obtain a rate

$$\frac{\nu_\mu + Ne \to \mu^- + D^0 + \ldots, \ D^0 \to K^0 \pi^+ \pi^-}{\nu_\mu + Ne \to \mu^- + \ldots} = (0.7 \pm 0.2)\% \ .$$

Figure 4 shows the distribution in Z, the fraction of the hadronic energy carried by the $D^0$. We have used the visible hadronic energy for each event, correcting for our estimate of the energy lost due to missing neutrals or charged tracks that interact close to the vertex and therefore fail to reconstruct. The solid lines represent all of the events in the $D^0$ region of the $K^0 \pi^+ \pi^-$ mass distribution, while the dashed lines give the contribution from the background under the $D^0$, obtained by using control regions above and below the $D^0$.

## NEUTRINO-ELECTRON SCATTERING

From an analysis of half of our pictures, we have found four events of the purely leptonic neutral current process

$$\nu_\mu + e^- \to \nu_\mu + e^- \ .$$

The signature for this process is a single high energy $e^-$ making a very small angle with the neutrino direction. The energies and angles of the four events are as follows:

| E (GeV) : | 31.5 | 34.6 | 9.0 | 8.9 |
| θ (mrad) : | $8 \pm 2$ | $4 \pm 2$ | $8 \pm 5$ | $4 \pm 2$ |

There are two backgrounds to this process. One comes from $\gamma$'s originating outside the chamber and converting asymmetrically. By scanning for energetic forward $\gamma$'s and measuring the probability for asymmetry, we find this background to be on the order of 1%. The other background comes from the reaction $\nu_e + n \to e^- + p$, where the proton is too slow to be visible. From the known cross section for this process, the $\nu_e$ flux in the beam, and the estimated $q^2$ distribution, we expect a background of about 10%.

REFERENCES

1. C. Baltay et al, Phys. Rev. Lett. <u>39</u>, 62 (1977).
2. V. Barger et al, Phys. Rev. D<u>16</u>, 746 (1977).
3. G. Goldhaber et al, Phys. Rev. Lett. <u>37</u>, 255 (1976).

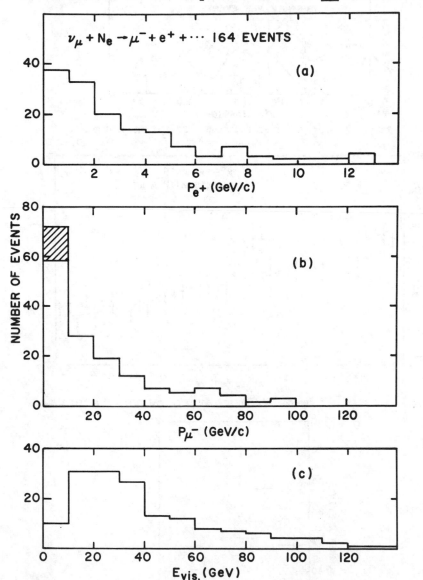

Fig. 1 Momentum of a) the $e^+$, and b) the $\mu^-$, the dilepton sample. The shaded events are the background from hadron punchthrough. c) The total visible energy.

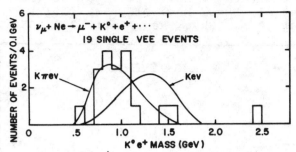

Fig. 2  K$^O$ e$^+$ mass from dilepton sample.

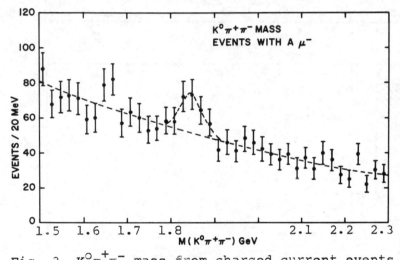

Fig. 3  K$^O\pi^+\pi^-$ mass from charged current events.

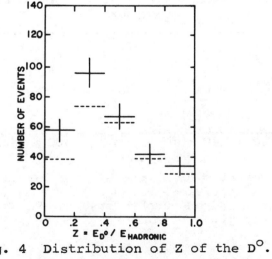

Fig. 4  Distribution of Z of the D$^O$.

# EVIDENCE FOR SCALING VIOLATION IN $\nu$N AND $\bar{\nu}$N INCLUSIVE SCATTERING AND TEST OF QCD[*]

Aachen-Bonn-CERN-London-Oxford-Saclay Collaboration[**]


presented by J.H.Mulvey

Department of Nuclear Physics, Oxford.

---

[*]  This report was based on preliminary results of an analysis now completed and published as Oxford pre-print 16/78 which should be consulted for final results and further details. It has been submitted to Nuclear Physics B.

[**] P.C.Bosetti, H.Deden, M.Deutschmann, P.Fritze, H.Grässler, F.J.Hasert, J.Morfin, H.Seyfert, R.Schulte and K.Schultze, Aachen.

K.Böckmann, H.Emans, C.Geich-Gimbel, R.Hartmann, A.Keller, T.P.Kokott, W.Meincke, B.Nellen and R.Pech, Bonn.

D.C.Cundy, J.Figiel, A.Grant, D.Haidt, P.O.Hulth, D.J.Kocher, D.R.O.Morrison, E.Pagiola, L.Pape, C.H.Peyrou, P.Porth, P.Schmid, W.G.Scott, H.Wachsmuth and K.L.Wernhard, CERN.

S.Banerjee, K.W.J.Barnham, R.Beuselinck, I.Butterworth, E.F.Clayton, D.B.Miller, K.J.Powell, Imperial College London.

C.L.Davis, P.Grossmann, R.McGow, J.H.Mulvey, G.Myatt, D.H.Perkins, R.Pons, D.Radojicic, P.Renton and B.Saitta, Oxford.

V.Baruzzi, M.Bloch, M.De Beer, W.Hart, Y.Sacquin, B.Tallini and D.Vignaud, Saclay.

ABSTRACT

An analysis of the nucleon structure functions determined from inclusive neutrino and anti neutrino scattering is presented. The dependence of the moments of the structure functions on $q^2$ is investigated and QCD predictions are tested. The QCD parameter, $\Lambda$, is determined with the result: $\Lambda = (0.74 \pm 0.05)$ GeV.

## 1. INTRODUCTION

This report describes the results of an investigation of scaling violation in neutrino and anti-neutrino inclusive scattering over a range in $q^2$ (the 4-momentum transfer squared) from 0.1 to 100 GeV$^2$. The data are derived from two bubble chamber experiments: Gargamelle filled with freon and using the wide-band neutrino beam of the CERN-PS [1,2] and new data from BEBC, filled with a 74% (mole) neon-hydrogen mixture, exposed to the CERN-SPS narrow-band beam. The numbers of events used in this analysis were 3,000 neutrino and 2,000 anti neutrino events from Gargamelle with energy, E, in the range 2 to 20 GeV and 1120 neutrino and 270 anti neutrino events in the energy range 20 to 200 GeV from BEBC.

In view of the limited time for this presentation details of the experimental method will be omitted; some results of the BEBC experiment have been published [3] and the analysis of the nucleon structure functions is dealt with fully elsewhere [4].

## 2. THE INCLUSIVE CHARGED-CURRENT CROSS SECTION

The inclusive charged-current cross section for neutrino (anti-neutrino) scattering on an isoscalar target can be written:

$$\frac{d^2\sigma(\nu,\bar{\nu})}{dxdy} = \frac{G^2ME}{\pi} \left[ (1-y-\frac{Mxy}{E}) \cdot F_2(x,q^2) \right.$$

$$+ (y^2/2) \cdot 2xF_1(x,q^2)$$

$$\left. \pm y(1-y/2) \cdot xF_3(x,q^2) \right] \qquad (1)$$

where $x = q^2/2M\nu$; $\nu = E-E_\mu$; $y = \nu/E$ and we have assumed charge symmetry, $M_W \to \infty$, and $\cos\theta_c = 1$ (i.e. neglected $\Delta S=1$ and $\Delta C=1$ contributions).

The form of this expression is determined by the (V-A) character of the charged current. Exact scaling behaviour requires that the nucleon structure functions, $F_1$, $F_2$ and $F_3$, be independent of $q^2$ and only functions of the Bjorken variable x; this leads to the following predicted behaviour of the total cross section $\sigma$ and average $q^2$ as E is changed:

$$\sigma/E = \text{constant}$$

and

$$\langle q^2 \rangle /E = 2M \langle x.y \rangle = \text{constant}.$$

Tests of these relationships are given in figures 1 and 2 [3].

First, the high energy neutrino data suggest a fall in $\sigma(\nu N)/E$; the BEBC data alone are not sufficient but taken together with that of Gargamelle the trend is clear. On the other hand the ratio $\sigma(\bar{\nu} N)/E$ seems rather constant. In both cases the data of the CTF experiment at Fermilab are in good agreement with BEBC and the results from ANL, although with rather large errors, tend to confirm the Gargamelle low energy neutrino cross section.

Figure 2 shows a clear fall with energy of the quantity $\langle q^2 \rangle /E$, again taking Gargamelle and BEBC data together. The lines shown on figure 2 are not fits to the data, they have slopes derived from an empirical fit [5] to inclusive eN and $\mu N$ scattering: $\langle q^2 \rangle /E \propto E^{-0.14}$.

Thus there is evidence of a departure from exact scaling and, in the case of figure 2, evidence that this is similar in character to the scaling violation seen in electron and muon scattering at SLAC and Fermilab.

## 3. THE NUCLEON STRUCTURE FUNCTIONS

To investigate the nature of possible scale violating effects the structure functions must be evaluated as functions of x and $q^2$.

This has been done via the following quantities, computed using the known $\nu$ and $\bar{\nu}$ fluxes* $\phi$ (E) and $\bar{\phi}$ (E):

--------------------------------------------------------------

* These have been determined [3] from measurements of the muon flux in the iron shielding at the end of the decay tunnel; there is a $\pm 7\%$ uncertainty in calibration and a possible 15% uncertainty in the K/$\pi$ ratio however these are not large enough to have a significant effect on the main results of this analysis [4].

$$N_1(x,q^2) = A \iint \phi(E) \cdot \frac{G^2ME}{\pi} (y^2/2) \, dy \, dE$$

$$N_2(x,q^2) = A \iint \phi(E) \cdot \frac{G^2ME}{\pi} (1-y-\frac{Mxy}{2E}) \, dy \, dE$$

$$N_3(x,q^2) = A \iint \phi(E) \cdot \frac{G^2ME}{\pi} y(1-y/2) \, dy \, dE$$

and similarly $\overline{N}_1$, $\overline{N}_2$, $\overline{N}_3$ for $\overline{y}$. $N_1(x,q^2)$ is the expected number of events in a given $x,q^2$ bin if $2xF_1=1$ and $F_2=xF_3=0$; and correspondingly for $N_2$ and $N_3$.

The observed numbers of events are therefore predicted to be, for given values of $F_1$, $F_2$ and $F_3$:

$$N_{obs}(x,q^2) = (2xF_1)N_1 + (F_2)N_2 + (xF_3)N_3 \qquad (2a)$$

$$\overline{N}_{obs}(x,q^2) = (2xF_1)\overline{N}_1 + (F_2)\overline{N}_2 - (xF_3)\overline{N}_3 \qquad (2b)$$

where the $N_{obs}(\overline{N}_{obs})$ correspond to the actual experimental numbers corrected for the effects of energy resolution, experimental selection procedures etc.

Two equations are not sufficient to solve for three structure functions so as a first step the Callan-Gross relation, $2xF_1 = F_2$, was checked. $F_3$ can be eliminated and by integrating over intervals of $y$ the contributions of the $2xF_1$ and $F_2$ terms were compared with the $y$ and $\overline{y}$ y-distributions; this was done for four regions of x and $q^2$ with the result (for BEBC data) shown in figure 3(a). The dashed curves correspond to equal contributions from $2xF_1$ and $F_2$ to the weighted sum of dN/dy and $d\overline{N}/dy$ (obtained through the elimination of $F_3$ in equations 2(a) and 2(b)) and this case is in good agreement with the data. By fitting the data best values of $A=2xF_1/F_2$ can be obtained and these are shown (BEBC data only) in figure 4 as a function of $q^2$. The average value obtained is $A=0.89\pm0.12$ (systematic error $\pm0.08$).* This result is consistent with unity and

---

\* A is related to the ratio $R=\sigma_S/\sigma_T$: $R=(1+\frac{4M^2x^2}{q^2} -A)/A$ and the result for A corresponds to an average value $R=0.12 ^{+0.18}_{-0.14}$.

this has been used for all subsequent analysis. Figure 3(b) shows good agreement of the data, in the form of the weighted difference of dN/dy and $\overline{dN}$/dy, with the predicted contribution due to the $xF_3$ term obtained from equations 2(a) and 2(b) by taking A=1.

Putting for $N_{obs}(\overline{N}_{obs})$ the corrected experimental numbers, equations 2(a) and 2(b) were solved for $F_2(x,q^2)$ and $xF_3(x,q^2)$; the results for different intervals of x are shown as functions of $q^2$ in figures 5(a) and 5(b).

The Gargamelle and BEBC data together show clear evidence of scaling violation; where the two sets overlap (for small x) the same trend is apparent in both. The continuity of the two sets is a demonstration of the reliability of the data and a striking confirmation of the validity of the current-current form of the weak interaction: the same values of the structure functions at a given x and $q^2$ are obtained from data from interactions made by neutrinos differing in energy by a factor of about 30.

The form of scaling violation - the structure functions rise with $q^2$ at low x and fall at high x - is the same as that observed in electron and muon scattering. The dashed lines shown in figure 5(a) are derived from the fit to eN and $\mu$N data by Perkins Schreiner and Scott [5]; the magnitude of $F_2$ is also in good agreement with the quark model prediction $F_2^{\nu N}(x,q^2)=(9/5)F_2^{ed}(x,q^2)$ and the values so derived from electron and muon scattering on deuterium are also shown in figure 5(a).

The structure function $xF_3$ (figure 5(b)) shows the same general behaviour as $F_2$ but is significantly smaller (the dashed lines are the same as those in figure 5(a)).

The effect of radiative corrections has been examined, using the calculations of Kiskis [6] and also of R.Barlow and S.Wolfram [7], but are expected to be small and have little $q^2$ dependence.

Integrals of the structure functions are shown as functions of $q^2$ in figures 6(a), 6(b) and 6(c). $\int F_2 dx$ is the momentum fraction carried by quarks and anti quarks ($Q+\overline{Q}$, or valence plus sea); its complement is thus the contribution due to gluons. As figure 6(a) shows ($Q+\overline{Q}$) falls to about 40% as $q^2$ rises to 25 GeV$^2$. Figure 6(b) shows $\int xF_3 dx$, which measures the valence quark momentum fraction, $Q_v$; this falls from ~1 at low $q^2$ (where the elastic contribution is dominant) to ~30% at high $q^2$. Figure 6(c) shows the ratio $B(q^2)=\int F_3 dx/\int F_2 dx$ which measures the ratio ($Q_v/Q+\overline{Q}$); this also falls slowly with rising $q^2$, to about 0.65 at 25 GeV$^2$ implying a rise in the momentum fraction carried by the quark-antiquark sea, as well as by the gluons.

The momentum fractions carried by quarks and anti-quarks can be determined as a function of x from the structure functions:

$$Q = \frac{1}{2} [F_2 + xF_3]$$

and

$$\overline{Q} = \frac{1}{2} [F_2 - xF_3]$$

The results are shown in figure 7(a), averaged over the $q^2$ range 0.3 to 3 GeV$^2$ and in 7(b) for the $q^2$ range 3 to 30 GeV$^2$. The shrinkage of the quark x-distribution at higher $q^2$ is evident.

At this stage we may summerize by saying that there is clear evidence of scaling violation and that this follows a pattern very similar to that found in eN and $\mu$N scattering. Production thresholds (e.g. strange and charmed particles) must make some contribution to this behaviour, but at small x; target mass and quark-gluon interaction effects may also play an important part indeed there are clear qualitative indications of effects which are expected in quantum-chromodynamics (QCD), for example the increasing importance of gluons and the quark-antiquark sea at high $q^2$.

## 4. HIGHER MOMENTS OF THE STRUCTURE FUNCTIONS AND TESTS OF QCD

QCD makes quantitative predictions about the $q^2$ dependence of the moments of the structure functions.

The analysis is performed in terms of the Nachtmann [8] moments using the variable $\xi$, rather than x, since $\xi$ includes the effect of target (nucleon) mass and moments in this variable are believed to be more reliably predicted by QCD, at finite $q^2$ [9].

$$\xi = \frac{2x}{[1 + \sqrt{1+4M^2x^2/q^2}]}$$

and the Nachtmann moments for $F_2$ and $xF_3$ are [10]:

$$M_2(N,q^2) = \int_0^1 \frac{\xi^{N+1}}{x^3} F_2(x,q^2) \frac{[(N^2+2N+3)+3(N+1)\sqrt{1+4M^2x^2/q^2} +N(N+2)4M^2x^2/q^2]dx}{(N+2)(N+3)}$$

and

$$M_3(N_1q^2) = \int_0^1 \frac{\xi^{N+1}}{x^3} \cdot xF_3(x,q^2) \frac{[1+(N+1)\sqrt{1+4M^2x^2/q^2}]dx}{(N+2)}$$

In QCD [10] the effective quark-gluon coupling is:

$$\alpha_{s/\pi} = [12/(33-2m)] /\ln(q^2/\Lambda^2)$$

where m = number of quark flavours and $\Lambda$ is a parameter to be determined by experiment.

The expressions for the moments in general contain three terms each varying as a known power of $\alpha_s$ in the region of $q^2$ where $\alpha_s$ is small ($q^2 \gg \Lambda^2$). These predictions are derived in 1st order perturbation theory and correspond to QCD "radiative corrections":

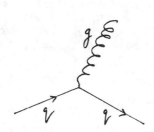

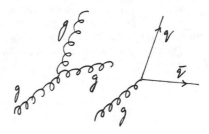

Flavour:   non-singlet                                   Singlet

## 4.1 MOMENTS OF $xF_3$

The function $xF_3$ measures the valence quark momentum distribution and so is a non-singlet in flavour which leads to a relatively simple expression for its $q^2$-dependence in 1st order:

$$M_3(N,q^2) = \text{Constant} / [\ln(q^2/\Lambda^2)]^{d^{NS}}$$

where

$$d^{NS} = \frac{4}{(33-2m)} \left[ 1 - \frac{2}{N(N+1)} + 4 \sum_{2}^{N} 1/j \right]$$

for $q^2 \gg \Lambda^2$ and $q^2 \gg$ quark masses and quark transverse momentum.

This prediction can be tested by noting that, for example:

$$\frac{M_3(5,q^2)}{[M_3(3,q^2)]^{1.456}} = \text{Constant}$$

since $d^{NS}(N=5)/d^{NS}_{(N=3)} = 1.456$, independent of m and $\Lambda$.

Figure 8 shows two such moment ratios as a function of $q^2$. They are independent of $q^2$, at least for $q^2 > 1$ GeV$^2$ and this is a direct, quantitative confirmation of a QCD prediction. Below $q^2 \sim 1$ GeV$^2$ second order effects do become important [11].

Alternatively, since

$$\ln[M_3(N_1)] = \frac{d^{NS}(N_1)}{d^{NS}(N_2)} \ln[M_3(N_2)] + Const.$$

ratios of the $d^{NS}$ can be determined; figure 9 shows the logarithm of one moment plotted against another. In the case of exact scaling all the data for a given pair would lie at the same point; instead the data are distributed along lines with slopes in very good agreement with the QCD predictions. This result is summarized in table I.

TABLE I

Test of QCD using moments of $xF_3$

| Pair of N-values | Observed Slope | QCD prediction |
|---|---|---|
| 5,3 | $1.45 \pm 0.12$ | 1.456 |
| 8,3 | $1.92 \pm 0.24$ | 1.882 |
| 6,2 | $3.0 \pm 0.4$ | 2.528 |

The slope predictions depend only on the square bracket in the definition of $d^{NS}$ and the agreement with experiment is a confirmation of the theory and proves the vector nature of the gluons. Fixed point theories (in which $\alpha_s \to$ const. for $q^2 \to \infty$) also give a power law dependence for the moments but the values of the indices are not predicted. Thus the results of this test must be regarded as a significant success for QCD.

The data can next be used to determine $\Lambda$. Thus, for $N=3$ as example:

$$[M_3(3,q^2)]^{-1.62} = Const.(\ln q^2 - \ln \Lambda^2) \quad \text{for } m=3;$$

$\Lambda$ can be determined from the intercept of this linear relation. The data are shown in figure 10. From this, and other moments of $xF_3$, the results in Table II were obtained, for $q^2 > 1$ GeV$^2$.

TABLE II

Values of the QCD parameter $\Lambda$ (Gev).

| N | 3 | 3 | 5 | 6 |
|---|---|---|---|---|
| m | 3 | 4 | 3 | 3 |
| $\Lambda$ | 0.73 ±0.07 | 0.69 ±0.07 | 0.77 ±0.07 | 0.73 ±0.06 |

The data follows the predicted linear relationship and the values obtained for $\Lambda$ from the different moments are stable. The contribution from c-quarks is expected to be small and so the m=3 values are preferred. The effects of 2nd order corrections and flux uncertainties are of the same order as the statistical errors shown.

The best value, from a simultaneous fit to the different N-moments, is $\Lambda$=0.74±0.05 GeV (neglecting second order corrections).

This determination of $\Lambda$ has only a weak dependence on A, $xF_3$ being insensitive to values of A in the range 0.5 to 1.0. The values of $\Lambda$ obtained from analysis of $F_2$ using eN and $\mu$N data tend to be smaller, in the range 0.3 to 0.6 GeV, but depend on the parameter $R = \sigma_S / \sigma_T$ which is not well known.

4.2 MOMENTS OF $F_2$ AND THE GLUON MOMENTUM DISTRIBUTION

$M_2(N,q^2)$ contains singlet as well as non-singlet terms. The rather complex expression can be cast in the form:

$$Y. \, M_2(N,q^2) = M_2(N,q_0^2) + X. \, G(N,q_0^2)$$

where $\lambda$ and X are known functions [4] of $N, q^2, q_0^2, \Lambda$ and m and $q_0^2$ is a reference value of $q^2$ chosen in this analysis to be $q_0^2$=5 GeV$^2$.

Thus QCD predicts a linear relationship between $Y.M_2(N,q^2)$ and X with a slope, $G(N,q_0^2)$, which is the $N^{th}$ moment of the gluon momentum distribution at $q^2=q_0^2$. Figure 11 shows that the data follow the QCD prediction, lying on straight lines, and the values of $G(N,q_0^2)$ are given in Table III, with the valence quark moments, $M_3(N,q_0^2)$, for comparison.

The momentum sum rule: $G(2,q_0^2)=1-M_2(2,q_0^2)$ provides another estimate for the N=2 gluon moment and this is also given in Table III.

344

The results show values of gluon moments, at $q^2=5$ GeV$^2$, which are comparable to those of the valence quarks.

TABLE III

Moments of the Gluon Distribution at $q_0^2=5$ GeV$^2$
($\Lambda = 0.74$ GeV)

| N | G(N,5) | Momentum Sum-rule | M$_3$(N,5) |
|---|--------|-------------------|-----------|
| 2 | 0.62 ± 0.15 | 0.45 ± 0.03 | 0.45 ± 0.07 |
| 3 | 0.12 ± 0.05 | - | 0.12 ± 0.02 |
| 4 | 0.03 ± 0.02 | - | 0.045 ± 0.01 |
| 5 | 0.02 ± 0.01 | - | 0.027 ± 0.007 |

## 5. SUMMARY OF CONCLUSIONS

(a) Neutrino and anti-neutrino inclusive charged current scattering in the $q^2$ range 0.1 to 100 GeV$^2$ can be described in terms of the (V-A) theory with three nucleon structure functions depending on x and $q^2$; there is no evidence of anomalous behaviour.

(b) The violation of Bjorken scaling observed in the behaviour of the structure functions is similar in character to that found in eN and $\mu$N inclusive scattering.

(c) The average value of $A=2xF_1/F_2$ is $0.89\pm0.2$, consistant with the Callan-Gross relationn:  A=1.  This result corresponds to a value for R($=\sigma_S/\sigma_T$) of $0.12^{+0.18}_{-0.18}$.

(d) The $\int F_2dx$ falls from ~1 at low $q^2$ to ~0.4 at $q^2 \approx 25$ GeV$^2$.

(e) $\int xF_3dx$ also falls and $B(q^2)=\int xF_3dx/\int F_2dx$ decreases from ~1 at low $q^2$ to ~0.65 at $q^2 \approx 25$ GeV$^2$.

(f) The behaviour of the Nachtmann moments of xF$_3$ is in good quantitative agreement with the predictions of first order QCD for $q^2>1$ GeV$^2$ and provides direct support for the vector nature of the gluon.

(g)  From the $xF_3$ moments the QCD parameter $\Lambda$ is determined to be
$(0.74 \pm 0.05)$GeV, somewhat higher than has been found from the $F_2$
moments in eN and $\mu$N scattering.

(h)  With this value of $\Lambda$ the $q^2$ dependence of the $F_2$ moments leads
to values of the gluon moments, at $q^2 = 5$ GeV$^2$, which are
comparable with the valence quark moments.

## 6. ACKNOWLEDGEMENTS

I wish to acknowledge the help of my colleagues in the ABCLOS
collaboration and especially D.H.Perkins and W.Scott for their work
on which this report is based. The collaboration of the Gargamelle
P.S. experiment is thanked for making their data available and we
are also indebted to the operating crews of BEBC and the SPS.

## REFERENCES

1.   T.Eichten et al, Phys. Lett. <u>46B</u> (1973) 281.

2.   H.Deden et al, Nucl. Phys. <u>B85</u> (1975) 269.

3.   P.Bosetti et al, Phys. Lett. <u>70B</u> (1977) 273.

4.   ABCLOS Collaboration "Analysis of Nucleon Structure Functions in
     CERN Bubble Chamber Neutrino Experiments" Oxford Preprint 16/78.
     Submitted to Nuclear Physics.

5.   D.H.Perkins, P.Schreiner and W.G.Scott, Phys. Lett. <u>67B</u> (1977)
     347.

6.   J.Kiskis Phys. Rev. <u>D8</u> (1973) 2129.

7.   R.Barlow and S.Wolfram, Oxford pre-print in preparation.

8.   O.Nachtmann, Nucl. Phys. <u>B63</u> (1973) 237
     and <u>B78</u> (1974) 455.

9.   S.Wandzura, Nucl. Phys. <u>B112</u> (1977) 412.

10.  D.G.Gross and F.Wilczek, Phys. Rev. <u>D8</u> (1973) 3633;
     and <u>D9</u> (1974) 980
     also H.Georgi and G.D.Politzer, Phys. Rev. <u>D9</u> (1974) 416.

11.  I.G.Floratos et al, Nucl. Phys. <u>B129</u> (1977) 66
     A.J.Buras et al, Nucl. Phys. <u>B131</u> (1977) 308.

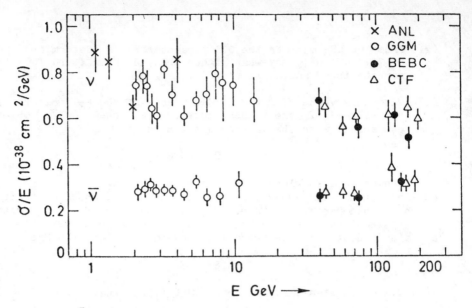

**Figure 1**   $\sigma/E$ for inclusive neutrino and anti-neutrino charged current interaction [3].

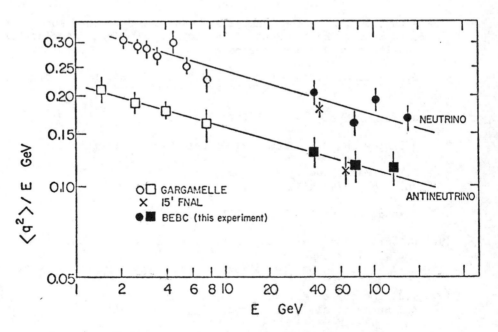

**Figure 2**   $\langle q^2 \rangle /E$ [3]. The slopes of the lines are from an empirical fit [5] to eN and μN scattering data.

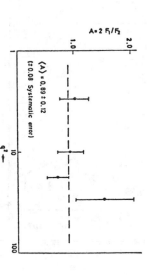

$$\omega_1\left(\frac{dN}{dy}\right) + \omega_2\left(\frac{d\bar{N}}{dy}\right) \rightarrow$$

a) X=0-0.1  $Q^2$=1-10

b) X=0.1-0.2  $Q^2$=3-30

c) X=0.2-0.4  $Q^2$=3-100

d) X=0.4-1.0  $Q^2$=10-100

Figure 3(a)

Sum of weighted y-distributions for neutrino and antineutrino. The solid curve is the sum of equal contributions (A = 1) from the terms $F_2(x, q^2)$, shown and $2xF_1(x,q^2)$ shown

Figure 4

Values of A obtained from fit to y distribution (Figure 3(a)).

$A = 2 F_1/F_2$

$\langle A \rangle = 0.89 \pm 0.12$
($\pm 0.08$ Systematic error)

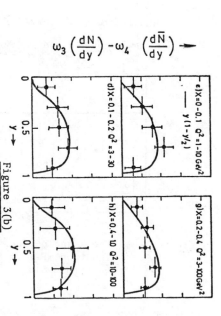

$$\omega_3\left(\frac{dN}{dy}\right) - \omega_4\left(\frac{d\bar{N}}{dy}\right) \rightarrow$$

e) X=0-0.1  $Q^2$=1-10 GeV$^2$
— y(1-y/2)

d) X=0.1-0.2  $Q^2$=3-30

g) X=0.2-0.4  $Q^2$=3-100 GeV$^2$

h) X=0.4-1.0  $Q^2$=10-100

Figure 3(b)

Difference of weighted y-distributions for neutrino and antineutrino. The curve is that predicted for the contribution of the $xF_3(x,q^2)$ term, assuming A = 1.

348

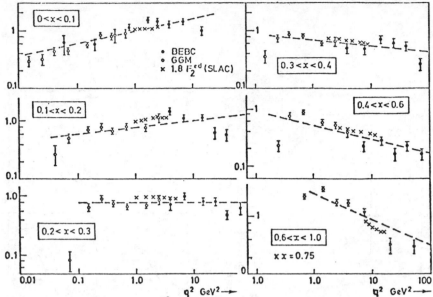

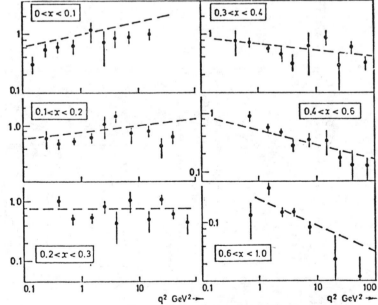

**Figure 5(a)** The x and $q^2$ dependence of the nucleon structure function $F_2(x, q^2)$. The slopes of the dashed lines are from an empirical fit [5] to the eN and μN scattering data.

**Figure 5(b)** The X and $q^2$ dependence of the nucleon structure function $xF_3(x, q^2)$. The lines are those drawn on Figure 5(a).

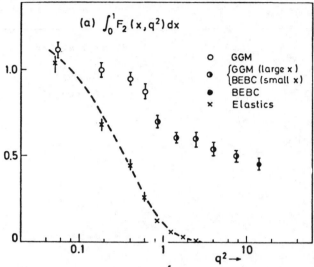

Figure 6(a)  The $q^2$ dependence of $\int F_2(x,q^2)\,dx$ which measures the momentum fraction carried by quarks and anti-quarks $(Q + \bar{Q})$

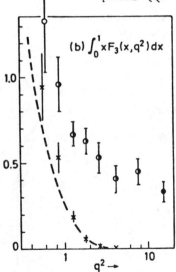

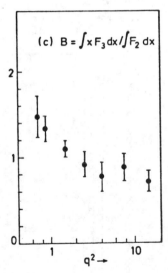

Figure 6(b)

The $q^2$ dependence of $\int xF_3(x,q^2)\,dx$ which measures the momentum fraction carried by valence quarks, $Q_v$.

Figure 6(c)

The $q^2$ dependence of the ratio $B = \int xF_3(x,q^2)\,dx / \int F_2(x,q^2)\,dx$. In terms of quark momentum fractions $(1-B)/2 \simeq \bar{Q}/(Q + \bar{Q})$.

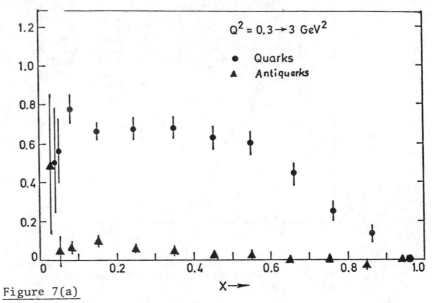

Figure 7(a)

The momentum fraction carried by quarks and antiquarks as a function of x. Data from the Gargamelle experiment.

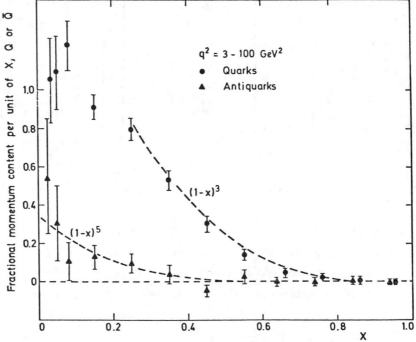

Figure 7(b)    The momentum fraction carried by quarks (and antiquarks) as a function of x. Data from the BEBC experiment.

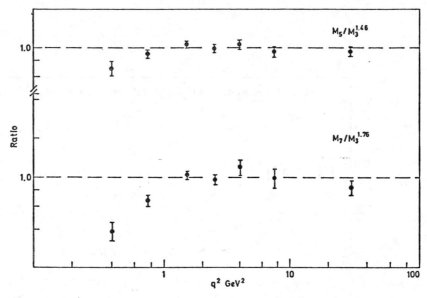

Figure 8     The ratios $M_3(N = 5)/[M_3(N = 3)]^{1.46}$ and $M_3(N = 7)/[M_3(N = 3)]^{1.76}$ as a function of $q^2$. The constancy of these ratios is a demonstration that QCD correctly predicts the power law for the $q^2$ dependence of the moments of $xF_3(x,q^2)$, at least for $q^2 > 1$ GeV$^2$ and so provides evidence for the vector nature of the gluon.

Figure 9

An alternative test of the QCD power law prediction for the moments of $xF_3(x,q^2)$. The slopes of the lines (and the numbers quoted) are the QCD predictions; these are compared with the experimental slopes in Table II.

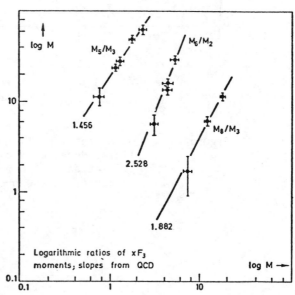

352

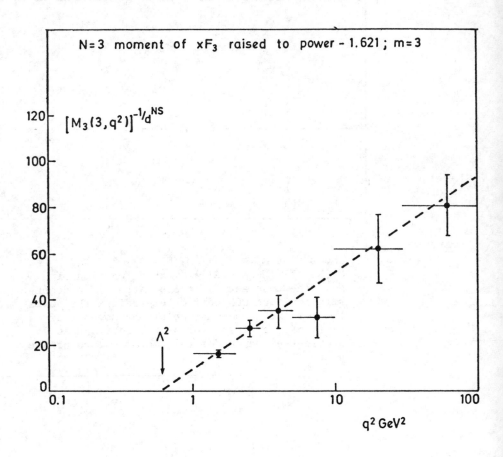

N=3 moment of xF$_3$ raised to power – 1.621; m=3

$\left[M_3(3,q^2)\right]^{-1/d\ ^{NS}}$

$\Lambda^2$

q$^2$ GeV$^2$

Figure 10    Determination of the QCD parameter $\Lambda$. The
linear dependence of $[M_3(3,q^2)]^{-1.62}$ is a
prediction of QCD and $\Lambda$ is obtained from the
intercept on the q$^2$ axis.

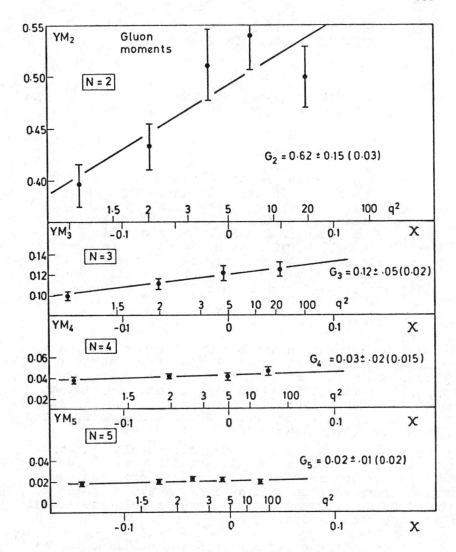

Figure 11

Values of the function $M_2(N,q_0^2).Y(q^2)$ plotted against $X(q^2)$ (see text). QCD predicts a linear relation with a slope equal to the gluon moment $G(N,q_0^2)$; the value of $q_0^2$ is 5 GeV$^2$ and $\Lambda$ is taken to be 0.74 GeV.

# MEASUREMENT OF THE PRODUCTION OF μ e EVENTS IN ANTINEUTRINO-NUCLEON INTERACTIONS

D. Sinclair

University of Michigan
Ann Arbor, Michigan  48109

## ABSTRACT

We have measured the ratio of $\mu^+e^-$ events to all $\mu^+$ events in an antineutrino hydrogen-neon experiment using the Fermilab 15 ft. bubble chamber. Based on 12 events with $p_\mu > 4.0$ GeV/c and $p_e > 0.8$ GeV/c we find this ratio to be .22±.07%. Using a model of charmed particle production to correct for the cuts on $p_e$ and $p_\mu$ we obtain a ratio of .35±.11%. The data indicate a higher ratio for $E_{\bar{\nu}} > 60$ GeV.

## I  INTRODUCTION

The experiment on which I shall report was performed by a collaboration[1] of four laboratories: Fermilab, Institute of High Energy Physics at Serpukhov, Institute of Theoretical and Experimental Physics at Moscow, and The University of Michigan.  In this experiment we continue[2] the search for μ e events produced in an antineutrino beam.

The production of μ e pairs in neutrino interactions at high energies is well established[3,4,5] . Their production rate is ~ .5% of the total charged current cross section, and a higher yield of strange particles is observed for these events relative to all charged current events.

Corresponding experiments in antineutrino beams[2,6] have had considerably fewer events to study and the question of μ e production could not be answered on the same level of sensitivity.  Furthermore, it is important to compare the data on μ e production in neutrino and antineutrino beams with the production of dimuons observed in counter experiments,[7,8,9] where all data strongly support the quark model[10] of charmed particle production.

ISSN:   0094-243X/78/354/$1.50   Copyright 1978 American Institute of Physics

## II THE EXPERIMENT

The new data come from an exposure of 85,000 pictures using the Fermilab 15 ft. bubble chamber filled with a hydrogen-neon mixture containing 64 at. % of neon. The density of this mixture is 0.77g cm$^{-3}$ and the radiation length is 39 cm.

The chamber was exposed to a broad-band double-horn-focused antineutrino beam. An absorptive plug downstream of the target was used to suppress the neutrino contamination to about 5% of the flux (that is ~ 12% of the event rate). The proton energy was 400 GeV and the average intensity was 1.0 x 10$^{13}$ protons/pulse. The energy spectrum of antineutrino events is shown in figure 1.

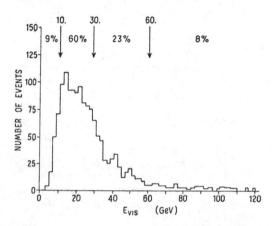

Fig. 1. Distribution of visible energy for events with a $\mu^+$.

The film was divided equally among the four laboratories and scanned for neutral-induced events with visible momenta along the beam direction greater than about 1 GeV/c. Events consisting of only a single charged track were not included. A total of about 20,000 events were found.

Scanners examined all tracks coming from the primary interaction vertex for evidence of electrons or positrons; that is spiralization, sudden change of curvature, bremstrahlung, trident formation, annihilation (for positrons), or large $\delta$ rays. All primary vertices were examined by two scanners independently and possible electron or positron tracks were noted. Physicists studied all events with such tracks provided they passed a momentum cut of > 300 MeV/c, measured with

a template. Obvious pairs were rejected. Single electron or positron tracks were retained if they exhibited
at least two of the criteria mentioned above.

To detect the muons we relied mainly on the external muon identifier (EMI) which consists of approximately 600g cm$^{-2}$ of absorber inside the vacuum vessel
of the bubble chamber followed by 23 m$^2$ of multiwire
proportional chambers.[11] The EMI detected approximately 80% of the muons with momentum ($p_\mu$) greater than
4.0 GeV/c. About 70% of the remainder were detected by
a kinematic method which selects as the muon that track
with large (> 1.6 GeV/c) transverse momentum ($p_T$) relative to the total momentum of the other tracks. Figure
2 shows the distribution of $p_T$ for muons and for hadrons and demonstrates that this selection does not mis-
identify any hadrons.

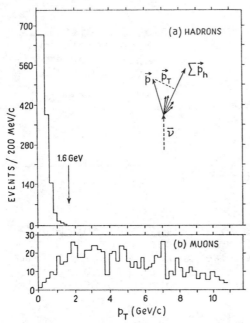

Fig. 2. Distribution
of events vs. $p_T$
which is the trans-
verse momentum of a
selected track rela-
tive to the direction
defined by the sum of
the momenta of the
hadrons ($\Sigma \vec{p}_h$). In (a)
the selected track
is a hadron. In (b)
it is the muon.

We estimated the number of charged current events
in our fiducial volume by measuring 25% of the events
found by the scanners. However, all events with single
electrons or positrons were measured. The data are summarized in Table 1. The results shown are for $p_\mu$> 4.0
GeV/c. For muons below this momentum both the EMI geometric acceptance and background problems are important.
The single electron events have $p_e$> .8 GeV/c. Below
this momentum, background from electromagnetic processes

(mainly Compton scattering) becomes relatively large. The picture quality is such that a close-in Compton electron vertex (within about 2.5 cm of the interaction vertex) may not be resolved.

The number of charged current events in Table I has been corrected for missing single-track events.[12] The estimates of antineutrino energy ($E_\nu$) were made using an average correction for neutral energy loss characteristic of the total event sample.[13] No correction was made for random scanning losses which affect both $\mu$ e events and other charged current events equally. Also no correction was applied for EMI acceptance which is assumed to be the same for $\mu$ e events as for all other charged current events.

TABLE I.  Uncorrected data for charged current and $\mu$ e events.

| $E_{\bar{\nu}}$ (GeV) | $\bar{\nu}{\to}\mu^+X$ | $\bar{\nu}{\to}\mu^+e^-X$ | $\nu{\to}\mu^-X$ | $\nu{\to}\mu^-e^+X$ |
|---|---|---|---|---|
| 10 - 30 | 4243 | 4 | 276 | 2 |
| 30 - 60 | 1536 | 3 | 296 | 1 |
| 60 -150 | 540 | 5 | 196 | 3 |
| All | 6319 | 12 | 768 | 6 |

### III   CORRECTIONS TO THE DATA

In order to obtain the relative rate of $\mu$ e events to charged current events it is necessary to correct the data in Table I for electron detection efficiency; for background processes which simulate electrons from the primary vertex; and for the signal loss due to the cuts imposed on $p_\mu$ and $p_e$. We now discuss these corrections.

(1)  Electron detection efficiency.

This is the product of two factors. The first, referred to as the pick efficiency, is the probability that the scanner detects an electron track from the primary vertex and thereby brings the event to the attention of a physicist. The second is the identification efficiency which is the probability that the track picked by the scanner passes the identification test applied by the physicist, namely, that it have at least two signatures characteristic of an electron. All primary vertices were examined by two scanners independently and approximately 2000 were flagged by scanners for examination by physicists. Most of these contained positron electron pairs, not single electrons. From an analysis of these data we estimate that the pick efficiency for single electrons is .85±.05. The electron identification efficiency was measured by examining a sample of gamma pairs close to the primary vertex. It was found to be .83±.05 for electrons with $p_e$> 0.8 GeV/c. We therefore estimate the electron detection efficiency to be .72±.07.

(2) Background processes

The most important of these are close-in Compton electrons. If the Compton electron vertex should occur within ~ 2.5 cm of the interaction vertex it probably would not be resolved. A less important source of electromagnetic background are asymmetric gamma conversions within 2.5 cm of the primary vertex or assymetric Dalitz pairs having an undetected electron or positron. To estimate the size of these backgrounds we computed the energy spectrum of gammas under the assumption that the gammas all came from neutral pion decay. Other sources of background considered were: electron neutrino events in which a hadron is misidentified as a muon by the EMI; δ-rays close to the primary vertex; and small angle $K_{e3}$ decays in flight. After correction for electron detection efficiency, the background from all sources mentioned above was estimated to be 2.0 events for electrons and less than 0.3 events for positrons with $p_e$ > 0.8 GeV/c.

(3) Signal loss due to momentum cuts

This correction depends on the model assumed to produce the μ e events. We believe that the kinematic properties of these events are consistent with a model in which the muon is the leading lepton and the electron comes from the decay of a charmed particle produced at the hadron vertex. Hence we have used such a model to correct for our momentum cuts. Specifically, we used a variation of the standard quark model suggested by Barnett.[14] The electrons were assumed to come entirely from the decays of D̄ mesons and we chose the momentum spectrum expected for a four body decay of the D. Figure 3 shows the spectrum of $p_e$ estimated from the model along with the experimental distribution for the 12 μ+e- events. Figure 4(a) shows the effect of the momentum cuts, estimated from the model, as a function of antineutrino energy. The effect of the cut on muon momentum is quite severe in the lowest energy

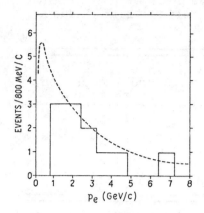

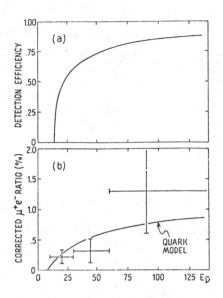

Fig. 3 Distribution of electron momentum for the 12 $\mu^+e^-$ events. The broken curve is calculated assuming the electrons come from D decay.

Fig. 4 (a) Estimated effect of momentum cuts on detection efficiency for $\mu^+e^-$ events. (b) Corrected $\mu^+e^-$ ratio.

interval ($10 < E_\nu < 30$ GeV) because near threshold the production of charmed particles is excluded, for kinematic reasons, from the region of low y. The effect of the cut on $p_\mu$, however, is to exclude events with high y, thereby restricting charmed particle production to the central region of the y distribution. The effect of both momentum cuts, averaged over the entire range of our antineutrino energy spectrum, was to eliminate approximately 40% of the $\mu$ e events.

## IV   RESULTS AND CONCLUSIONS

The corrected results are shown in Table II. Other features of the $\mu^+e^-$ events are shown in Table III.

TABLE II.  Corrected values for the rate of $\mu^+ e^-$ events
as a percentage of all antineutrino induced
charged current events.

| $E_{\bar{\nu}}$ (GeV) | $\sigma(\mu^+ e^-)/\sigma(\mu^+)$ with cuts on $p_e, p_\mu$ | $\sigma(\mu^+ e^-)/\sigma(\mu^+)$ with model dependent corrections |
|---|---|---|
| 10-30 | 0.11±.06% | 0.22±.15% |
| 30-60 | 0.23±.14% | 0.32±.19% |
| 60-150 | 1.1 ±.6% | 1.3 ±.7% |
| All | 0.22±.07% | 0.35±.11% |

TABLE III. Some features of the 12 $\mu^+ e^-$ events.  x and
y are the scaling variables.  W is the in-
variant mass of the hadrons.  These quantities have not been corrected for missing
neutral energy.

| $E_{VIS}$ (GeV) | $p_\mu$ (GeV/c) | $p_e$ (GeV/c) | $x_{VIS}$ | $y_{VIS}$ | V's | $W_{VIS}$ |
|---|---|---|---|---|---|---|
| 69. | 54. | 2.3 | ~.01 | .22 | K° | 5.3 |
| 20. | 8.6 | 1.2 | ~.03 | .57 | | 4.5 |
| 103. | 81. | 3.2 | .79 | .21 | K° | 2.9 |
| 143. | 87. | 6.6 | .11 | .39 | $\Lambda/K°, \bar{\Lambda}/K°$ | 9.6 |
| 29. | 10. | 1.8 | .08 | .64 | $\Lambda$ | 5.6 |
| 24. | 7.6 | 3.8 | .08 | .67 | $\Lambda$ | 5.3 |
| 16. | 6.2 | 3.4 | .31 | .61 | $K°_{e3}$ | 3.5 |
| 44. | 36. | 1.2 | .64 | .17 | | 2.2 |
| 22. | 5.1 | 1.0 | .17 | .77 | | 5.1 |
| 93. | 77. | 4.6 | .14 | .17 | $\Lambda, K°$ | 5.0 |
| 116. | 104. | 2.7 | .08 | .11 | K° † | 4.8 |
| 26. | 5.9 | 1.7 | .28 | .76 | | 5.2 |

†1 C fit. located ~ 3 mean lives from $\bar{\nu}$ interaction,
hence could be a regenerated $K°_s$ .

The visible energy ($E_{VIS}$) of these events is significantly greater than that of typical $\mu^+$ events.
$\langle E_{VIS} \rangle$ = 60 GeV for $\mu^+e^-$ events.
$\langle E_{VIS} \rangle$ = 28 GeV for all $\mu^+$ events.

The features of the events in Table III strongly
support the interpretation that the electrons come from
the semi-leptonic decays of charmed particles. According to the model of Glashow, Iliopoulos and Maiani,[10]
production of charmed particles by $\bar{\nu}$ is mainly off the
strange sea.

$$\bar{\nu} + \begin{pmatrix} \bar{s} \\ s \end{pmatrix} \rightarrow \bar{c} + \mu^+ + s$$
$$\hookrightarrow e^- + \bar{\nu} + \bar{s}$$

Hence each $\mu^+e^-$ event should be accompanied by two
strange particles, at least. In addition, since the
process involves the interaction of anti fermions, the
y distribution should be flat, except for threshold
effects.

After correction for neutral decays, the data show
1.5±.5 neutral strange particles per $\mu^+e^-$ event. This
is quite consistent with 2 strange particles (charged
as well as neutral) per event, and is significantly
higher than the average strange particle content of
ordinary charged current events. The distribution of
$y_{VIS}$ for the $\mu^+e^-$ events is shown in figure 5, where
it is contrasted with the corresponding distribution
for ordinary charged current events of comparable
energy.

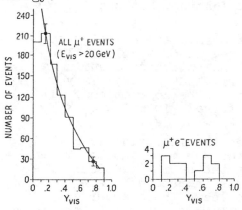

Fig. 5. Comparison of y-distributions between $\mu^+e^-$ events
and all $\mu^+$ events of comparable energy.

The conclusions we
draw from this experi-
ment are the following:
1. There is definite
evidence for direct
production of electrons
by antineutrinos at a
rate of ~ .4% of normal
charged current events.
However, the rate may
be limited by threshold
effects.
2. The features of
direct electroproduc-
tion by antineutrinos
strongly support the
GIM(10) model.

References and Footnotes

1.  J.P. Berge, D. Bogert, R. Endorf, R. Hanft, J.Malko,
    G. Harigal, G. Moffatt, F. Nezrick, W. Smart, and
    J. Wolfson, FERMILAB;   V. Ammosov, A. Denisov, P.
    Ermolov, V. Gapienko, V. Klukhin, V. Koreshev, A.
    Mukhin, P. Pitukhin, Y. Rjabov, E. Slobodjuk, and
    V. Sirotenko, IHEP, USSR;   V. Efremenko, A. Fedotov,
    P. Gorichev, V. Kaftanov, G. Kliger, V. Kolganov,
    S. Krutchinin, M. Kubantsev, I. Makhlueva, V.
    Shekeljan, and V. Shevchenko, ITEP, USSR;   J. Bell,
    C.T. Coffin, W. Louis, B.P. Roe, R.T. Ross, A.A.
    Seidl, D. Sinclair, and E. Wang, U. of MICHIGAN.

2.  J.P. Berge et al., Phys. Rev. Lett. 38, 266 (1977)

3.  H. Deden et al., Phys. Lett. 58B, 361 (1975)

4.  (a) J. von Krogh et al., Phys. Rev. Lett. 36, 710
        (1976)
    (b) P. Bosetti et al., Phys. Rev. Lett. 38, 1248
        (1977)

5.  C. Baltay et al., Phys. Rev. Lett. 38, 62 (1977)

6.  H.C. Ballagh et al., Phys. Rev. Lett. 39, 1650 (1977)

7.  A. Benvenuti et al., Phys. Rev. Lett. 38, 1183 (1977)

8.  B.C. Barish et al., Phys. Rev. Lett. 39, 981 (1977)

9.  M. Holder et al., Phys. Lett. 69B, 377 (1977)

10. S.L. Glashow, J. Iliopoulos, and L. Maiani, Phys.
    Rev. D2, 1285 (1970)

11. R.J. Cence et al., Nucl. Instrum. Methods 138, 245
    (1976)

12. The missing single track events are expected to be
    confined largely to the region of small y.  The
    number of missing events is estimated assuming a
    y distribution of the form $dN/dy = (1-y+1/2\ y^2)$ -
    $yB(1-1/2\ y)$ with $B = 0.8$ .  The correction is 19%
    for the lowest energy interval (10 GeV < $E_{\bar{\nu}}$ < 30
    GeV) and negligible for $E_{\bar{\nu}}$ > 30 GeV.

13. The energy of the incident antineutrino is estimated
    by summing the momentum of the muon and the momen-
    tum of the hadrons along the beam direction.  The
    momentum of the hadrons is increased by 25% to ac-
    count for neutrals which leave the chamber unde-
    tected.

14. R.M. Barnett, PR 14D, 70 (1976)

# INCLUSIVE PRODUCTION OF HADRONS IN $\nu_\mu$Ne AND $\bar{\nu}_\mu$Ne INTERACTIONS

T.H. Burnett[*], S.E. Csorna, D. Holmgren, G. Kollman,
H.J. Lubatti, K. Moriyasu, H. Rudnicka[†], G.M. Swider, B.S. Yuldashev[††]
Visual Techniques Laboratory, Department of Physics
University of Washington, Seattle, Washington 98195

H.C. Ballagh, H.H. Bingham, P. Bosetti, W.B. Fretter, D. Gee,
G. Lynch, J.P. Marriner, J. Orthel, M.L. Stevenson, G.P. Yost
University of California and Lawrence Berkeley Laboratory
Berkeley, California 94720

(Presented by Bekhzad S. Yuldashev)

## ABSTRACT

Data on the inclusive production of hadrons in $\nu_\mu$Ne and $\bar{\nu}_\mu$Ne interactions are presented and compared with the corresponding quantities obtained in $\pi^\pm$Ne interactions. With 224 $\nu$Ne and 219 $\bar{\nu}$Ne events, no differences are seen which would distinguish the hadronic states created by pions and neutrinos.

## INTRODUCTION

In this paper we present preliminary data on the inclusive production of hadrons in $\nu_\mu$Ne and $\bar{\nu}_\mu$Ne interactions. We compare our results with those obtained in $\pi^\pm$Ne interactions and find that within statistics we can discern no differences. A similar conclusion was previously reported for $\nu_e$Ne and $\bar{\nu}_e$Ne interactions.[1]

The data reported here come from a study of neutrino interactions in the Fermilab 15-foot bubble chamber filled with a 64% neon - 36% hydrogen (atomic) mixture. Thus $\gtrsim$ 96% of the interactions occur on Ne. We neglect a small hydrogen background. The radiation and interaction lengths are 39 cm and $\sim$ 1.4 m, respectively. For each interaction we require a noninteracting leaving track with momentum greater than 4 GeV/c that is identified as a muon by the EMI with likelihood,[2] $\mathcal{L} > 5$, and that the total visible energy be greater than 10 GeV. From measurements of the charged secondaries, the neutrino energy was reconstructed using the method of $p_T$ balance in the $\nu$-$\mu$ plane to correct the total

---

[*] A.P. Sloan Foundation Research Fellow.
[†] Visitor from the Institute of Nuclear Physics and Technology of the Academy of Mining and Metallurgy, Cracow, Poland.
[††] Visitor from the Physical Technical Institute, Tashkent, USSR.

hadronic momentum.[3]   We can reliably identify protons in the
momentum range $0.2 \lesssim p \lesssim 1.0$ GeV/c.  Protons with $p \gtrsim 1.0$ GeV/c
will be included in the positive ("minimum-ionizing, non-muon")
secondary tracks, $N_+$.  Thus, assuming that $K^{\pm}$ contamination is
negligible,[4,5] the $N_+$ distribution contains pions and fast
protons, while the negative track distribution, $N_-$, contains pions
only.  We also require that there be $\geq 1$ pion (i.e., one or more
tracks which cannot be identified as protons).  Since there are
few events with hadronic invariant mass $W > 10$ GeV, we restrict ourselves to $1 \leq W \leq 10$ GeV.  This leaves 224 $\nu_\mu$ and 219 $\bar{\nu}_\mu$ events.
In comparing with $\pi^{\pm}$Ne interactions at 10.5 GeV/c[5] ($\sqrt{s} = 4.4$ GeV)
we make the further restriction $3 < W < 6$ GeV, leaving 112 $\nu_\mu$ and
100 $\bar{\nu}_\mu$ events.

RESULTS

In Fig. 1 and Table I we give the average multiplicity of
positive and negative particles combined, $< N_\pm > = < N_+ > + < N_- >$

as a function of W and $Q^2$. Within the statistical significance of these data we observe no dependence of the average multiplicity on $Q^2$ for fixed W intervals. Similar conclusions have been obtained in ep, μp, νp and $\bar{\nu}$p interactions.[6] In Fig. 2a we show the dependence of $< N_\pm >$ on W for $\nu(\bar{\nu})$Ne interactions and compare it with $\pi^{\pm}$Ne interactions [5,7] and note that within statistics the W dependence is similar. In Fig. 2b we give the dependence of the average multiplicity of produced negative particles, $< N_-^{Pr} >$, where

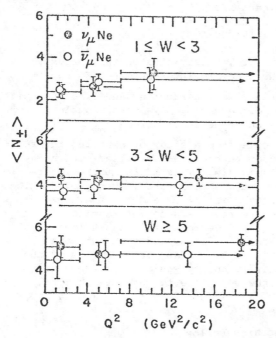

Fig. 1.  The average multiplicity as
a function of $Q^2$ for different
values of W.

hadronic momentum.[3]   We can reliably identify protons in the momentum range $0.2 \lesssim p \lesssim 1.0$ GeV/c.  Protons with $p \gtrsim 1.0$ GeV/c will be included in the positive ("minimum-ionizing, non-muon") secondary tracks, $N_+$.  Thus, assuming that $K^{\pm}$ contamination is negligible,[4,5] the $N_+$ distribution contains pions and fast protons, while the negative track distribution, $N_-$, contains pions only (i.e., one or more tracks which cannot be identified as protons).  We also require that there be $\geq 1$ pion.  Since there are few events with hadronic invariant mass $W > 10$ GeV, we restrict ourselves to $1 \leq W \leq 10$ GeV.  This leaves 224 $\nu_\mu$ and 219 $\bar{\nu}_\mu$ events. In comparing with $\pi^{\pm}$Ne interactions at 10.5 GeV/c[5] ($\sqrt{s}$ = 4.4 GeV) we make the further restriction $3 < W < 6$ GeV, leaving 112 $\nu_\mu$ and 100 $\bar{\nu}_\mu$ events.

RESULTS

In Fig. 1 and Table I we give the average multiplicity of positive and negative particles combined, $< N_\pm > = < N_+ > + < N_- >$

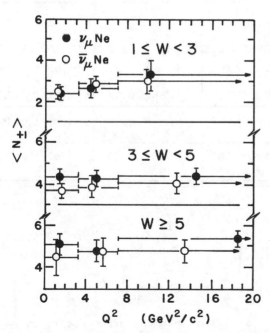

Fig. 1.  The average multiplicity as a function of $Q^2$ for different values of W.

as a function of W and $Q^2$.  Within the statistical significance of these data we observe no dependence of the average multiplicity on $Q^2$ for fixed W intervals.  Similar conclusions have been obtained in ep, μp, νp and $\bar{\nu}$p interactions.[6]  In Fig. 2a we show the dependence of $< N_\pm >$ on W for $\nu(\bar{\nu})$Ne interactions and compare it with $\pi^{\pm}$Ne interactions[5,7] and note that within statistics the W dependence is similar.  In Fig. 2b we give the dependence of the average multiplicity of produced negative particles, $< N_-^{Pr} >$, where

Table I.  The average multiplicity, $< N_\pm >$, as a function of $Q^2$ and W for    a) $\nu_\mu$ Ne    and    b) $\bar{\nu}_\mu$ Ne interactions.

a)

| $Q^2$ (GeV$^2$/c$^2$)  /  W (GeV) | 1.0 - 3.0 | 3.0 - 5.0 | $\geq 5$ |
|---|---|---|---|
| 0 - 3 | 2.39 ± 0.15 | 4.25 ± 0.37 | 5.05 ± 0.57 |
| 3 - 7 | 2.71 ± 0.40 | 4.15 ± 0.38 | 4.63 ± 0.50 |
| $\geq 7$ | 3.33 ± 0.76 | 4.27 ± 0.24 | 5.66 ± 0.25 |
| $\geq 0$ | 2.56 ± 0.18 | 4.23 ± 0.18 | 5.34 ± 0.21 |

b)

| $Q^2$ (GeV$^2$/c$^2$)  /  W (GeV) | 1.0 - 3.0 | 3.0 - 5.0 | $\geq 5$ |
|---|---|---|---|
| 0 - 3 | 2.42 ± 0.16 | 3.61 ± 0.32 | 4.45 ± 0.86 |
| 3 - 7 | 2.80 ± 0.33 | 3.90 ± 0.37 | 4.61 ± 0.63 |
| $\geq 7$ | 3.00 ± 0.71 | 4.05 ± 0.37 | 4.78 ± 0.36 |
| $\geq 0$ | 2.49 ± 0.14 | 3.81 ± 0.20 | 4.67 ± 0.30 |

$$< N_-^{Pr} > = \left[ \sum_{N_- \geq 1} N_- \cdot \sigma_{N_-} \right] \bigg/ \left[ \sum_{N_- \geq 1} \sigma_{N_-} \right] \quad \text{for } \nu \text{ and positive}$$

hadron beams and $$< N_-^{Pr} > = \left[ \sum_{N_- \geq 2} (N_- - 1) \cdot \sigma_{N_-} \right] \bigg/ \left[ \sum_{N_- \geq 2} \sigma_{N_-} \right]$$

for $\bar{\nu}$  and negative hadron beams.  This has the effect of removing the negative charge given to the hadron system by the beam particle, which we assume to be a $W^-$ for $\bar{\nu}_\mu$ interaction.  In most hadron-nucleus interaction experiments, the inelastic interactions in which there is only one charged particle have considerable elastic contamination.  Thus, the above definition uses only data which do not have such background. From Fig. 2b, it is seen that $< N_-^{Pr} >$ in neutrino-neon interactions has the same W dependence as $< N_-^{Pr} >$ in hadron-nucleus interactions.[5,7,8] We note that $< N_-^{Pr} >$ is free of possible proton contamination which might be present in $N_{\pm}$.

In an earlier publication[7] we showed that there is a universal relationship between the dispersion $D_-^{Pr} = \left[ < N_-^{Pr^2} > - < N_-^{Pr} >^2 \right]^{1/2}$ versus $< N_-^{Pr} >$ for hadron-nucleus interactions, and we see from

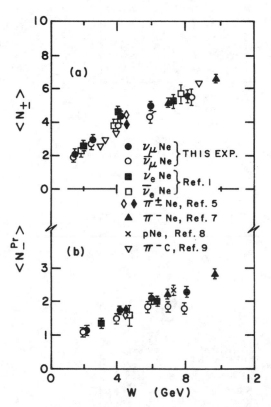

Fig. 3 that our neutrino-neon data are consistent with this conclusion. While it is not generally understood, it appears that the universal relationship between $D_-^{Pr}$ and $< N_-^{Pr}>$ holds for various projectiles and targets.

Several authors[5,7,9] have pointed out that hadron-nucleus interactions approximately satisfy KNO scaling.[10] In Fig. 4 we add our $\nu(\bar{\nu})$Ne data to a compilation of $\pi$Ne and $\pi$C interactions[7] and find that $\nu$Ne interactions are also consistent with approximate KNO scaling.

We now turn to the production of protons in neutrino-neon interactions. In Fig. 5 we give the invariant inclusive cross section of protons observed in $\nu_\mu$Ne and $\bar{\nu}_\mu$Ne interactions (combined) and note that there is good agreement with $\pi^\pm$Ne and $\pi^-$C inter-

Fig. 2. Comparison of average multiplicities of charged hadrons as a function of W (or $\sqrt{s}$) for neutrino (antineutrino) and hadron-induced interactions.

actions.[5,11] The average multiplicity of slow protons $(0.2 \lesssim p \lesssim 0.8$ GeV/c) for $\nu$Ne and $\bar{\nu}$Ne combined in the range $3 < W < 6$ GeV is $< N_p > = 0.68 \pm 0.08$ which agrees well with $< N_p > = 0.76 \pm 0.04$ for $\pi^\pm$Ne interactions at 10.5 GeV/c. Thus the

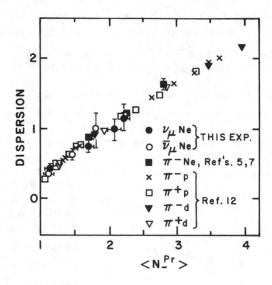

Fig. 3.   Comparison of dispersion versus average multiplicity of
produced negative particles for $\nu(\bar{\nu})$Ne, $\pi^{\pm}$Ne, $\pi^{\pm}$d, and
$\pi^{\pm}$p interactions.

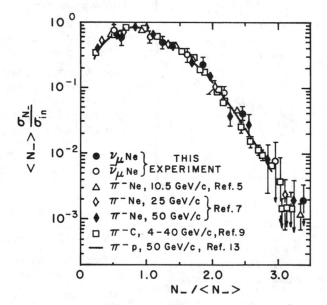

Fig. 4.   Comparison of KNO scaling variable for $\nu(\bar{\nu})$Ne interactions
and hadron interactions.

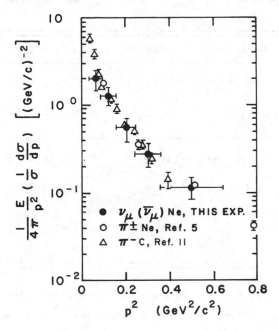

Fig. 5.  Inclusive cross section of identified protons in $\nu_\mu(\bar\nu)$Ne, $\pi^\pm$Ne and $\pi^-$C interactions.

slow protons produced in $\nu$Ne and $\pi$Ne interactions are similar.  In $\pi^\pm$Ne interactions it has also been observed[5] that there is an excess of positive charge for fast minimum-ionizing secondaries and this has been interpreted as resulting from unidentified fast protons. The positive excess has been determined by applying charge symmetry and taking the difference $<N_\pm>_{\pi^+ Ne} - <N_\mp>_{\pi^- Ne} \equiv <N_p^f>_\pi$.  We can see from Table II that a similar result is obtained for $\nu_\mu(\bar\nu_\mu)$Ne interactions for the region $3 < W < 6$ GeV, by using

$<N_\pm>_{\nu Ne} - <N_\mp>_{\bar\nu Ne} = <N_p^f>_\nu$.  This suggests that there is also an excess of fast protons in neutrino-neon interactions which is quantitively similar to the excess observed in $\pi$Ne interactions.[5]

In Table III we give the average transverse momentum of negative pions produced in $\nu$Ne and $\pi^\pm$Ne interactions.[5]  For neutrino interactions, the average transverse momentum is defined with respect to the direction of the total visible hadronic momentum.  Within errors, we obtain good agreement between $\nu$Ne and $\pi$Ne interactions.

TABLE II.  Difference $< N_\pm >_{\nu Ne} - < N_\mp >_{\bar\nu Ne}$ as compared to $\pi^\pm Ne$ data.

| $\nu_\mu (\bar\nu_\mu) Ne$ Interactions, $3 \leq W \leq 6$ GeV | |
|---|---|
| $< N_+ >_{\nu Ne} - < N_- >_{\bar\nu Ne}$ | $= 0.69 \pm 0.13$ |
| $< N_+ >_{\bar\nu Ne} - < N_- >_{\nu Ne}$ | $= 0.13 \pm 0.14$ |
| $\pi^\pm Ne$ Interactions, $\sqrt{s} = 4.4$ GeV [5] | |
| $< N_+ >_{\pi^+ Ne} - < N_- >_{\pi^- Ne}$ | $= 0.78 \pm 0.04$ |
| $< N_+ >_{\pi^- Ne} - < N_- >_{\pi^+ Ne}$ | $= 0.46 \pm 0.03$ |

TABLE III.  Average transverse momenta of negative pions in $\nu_\mu (\bar\nu_\mu) Ne$ interactions at $3 < W < 6$ GeV and $\pi^\pm Ne$ interactions at 10.5 GeV/c. [5]

| Reaction | $< p_\perp >$ (MeV/c) |
|---|---|
| $\nu_\mu Ne \to \pi^- + \ldots$ | $315 \pm 16$ |
| $\pi^+ Ne \to \pi^- + \ldots$ | $309 \pm 4$ |
| $\bar\nu_\mu Ne \to \pi^- + \ldots$ | $307 \pm 16$ |
| $\pi^- Ne \to \pi^- + \ldots$ | $348 \pm 4$ |

## CONCLUSION

We find that within statistics the inclusive characteristics of hadrons produced in $\nu_\mu Ne$ and $\bar\nu_\mu Ne$ interactions are similar to those observed in $\pi^\pm Ne$ interactions.

## ACKNOWLEDGMENTS

We are grateful for the assistance of the staff of the Neutrino Laboratory and Accelerator Division of Fermilab.  This work is supported in part by the DOE and NSF.

# REFERENCES

1. T.H. Burnett et al., "Neutrino-77", ed. M. A. Markov et al. (Nauka-Moscow), Vol. II, p. 132 (1978).

2. R.J. Cence et al., Nucl. Inst. and Meth. 138, 245 (1976); G. Lynch, LBL Phys. Note 808 (1975).

3. The total visible hadronic momentum is increased in magnitude to balance the momentum transverse to the $\nu$ direction in the $\mu\nu$ plane. We have verified that no distortion which could affect any of our conclusions results from this procedure. See H. Rudnicka, University of Washington Internal Report VTL-HEP-58 (1978).

4. M. Derrick et al., Phys. Rev. D17, 1 (1978).

5. W.M. Yeager et al., Phys. Rev. D16, 1294 (1977).

6. See, for example, J.W. Chapman et al., Phys. Rev. Lett. 36, 124 (1976); C. del Papa et al., Phys. Rev. D13, 2934 (1976); C.K. Chen et al., Preprint DESY 77/69 (1977); K. Bunnell et al., Preprint SLAC-PUB-2061 (1977), and Ref. 4.

7. B.S. Yuldashev et al., VTL-PUB-48 (1978), to be published in Acta Phys. Pol. B9 (1978).

8. D.J. Miller and R. Nowak, Lett. Nuovo Cimento 13, 39 (1975).

9. S.A. Azimov et al., Nucl. Phys. B107, 45 (1976).

10. Z. Koba, H.B. Nielsen and P. Olesen, Nucl. Phys. B40, 317 (1970).

11. S.A. Azimov et al., Sov. J. Nucl. Phys. 23, 519 (1976).

12. Data for $\pi^{\pm}p$ and $\pi^{\pm}d$ interactions are taken from A. Wroblewski, Proc. of VIII Inst. Symposium on Multiparticle Reactions, Kayserberg (1977).

13. G.A. Akopjanov et al., Nucl. Phys. B75, 401 (1974).

CHARGED AND NEUTRAL CURRENT COUPLINGS OF QUARKS

(As Seen By Neutrinos)[*]

R. Michael Barnett
Stanford Linear Accelerator Center
Stanford University, Stanford, California 94305

ABSTRACT

Charged and neutral current coupling of u and d quarks are examined and compared with recent data.

[*]Research supported by the Department of Energy.

There is now considerable information about the couplings of quarks to gauge bosons in theories of the weak and electromagnetic interactions. Much has been learned about charged-current couplings from the energy dependence of neutrino cross-sections and of y distributions, and also from the lack of t and b quark production in neutrino scattering. Knowledge of neutral-current couplings has also come mainly from neutrino experiments. In this talk I will first discuss aspects of charged-current scattering including the question of whether the cross-sections are "anomalous" in any sense and the subject of limits on the production of new heavy quarks. In the second part, I will discuss a new, unique determination of neutral-current couplings, using data from deep-inelastic and elastic neutrino scatterings and from neutrino-induced exclusive and inclusive pion production.

The Weinberg-Salam (WS) theory[1] of weak and electromagnetic interactions with the Glashow-Iliopoulos-Maiani (GIM) quark structure[2] has been a remarkable phenomenological success when compared with all other weak interaction models. It is worthwhile, therefore, to discuss charged-current couplings in that context before in more general contexts. The WS-GIM model has the couplings:

$$\begin{pmatrix} u \\ d' \end{pmatrix}_L \quad \begin{pmatrix} c \\ s' \end{pmatrix}_L \quad \begin{pmatrix} t \\ b' \end{pmatrix}_L$$

where the primes indicate that the weak interaction eigenstates $(d', s', b')$ do not coincide with the mass eigenstates $(d, s, b)$. This mixing among the quark states is indicated with the weak coupling matrix[3]:

$$
\begin{array}{c}
u \\
c \\
t
\end{array}
\begin{pmatrix}
C_c & -S_c C_2 & -S_c S_2 \\
S_c C_1 & C_c C_1 C_2 - S_1 S_2 e^{i\delta} & C_c C_1 S_2 + S_1 C_2 e^{i\delta} \\
S_c S_1 & C_c S_1 C_2 + C_1 S_2 e^{i\delta} & C_c S_1 S_2 - C_1 C_2 e^{i\delta}
\end{pmatrix}
$$

where $C_1 \equiv \cos \Theta_1$, $S \equiv \sin \Theta_1$, etc. and the rows and columns correspond to the quarks indicated. In the discussion which follows, I use the notation $\overline{ud}_L \equiv \bar{u} \gamma_\mu (1+\gamma_5) d$.

J. Ellis et al.[4] have noted that the universality of quark and lepton couplings requires (in the WS-GIM model) that the ratio of coupling constants squared for $\overline{ub}_L$ to $\overline{ud}_L$ (which is $\tan^2 \Theta_c \sin^2 \Theta_c$) be less than 0.003. Since the ratio of $\overline{us}_L$ to $\overline{ud}_L$ is 0.05, one finds that $\sin^2 \Theta_2 \approx \tan^2 \Theta_2 < 0.1$ and that in this six-quark model $\Theta_c$ is equal to the usual Cabibbo angle $(\Theta_c \approx 13^\circ)$.

The determination of limits on $\overline{td}_L^c$ is more complicated. J. Ellis et al.[4] have also noted that an estimate can be obtained following a procedure analogous to that of Gaillard and Lee[5] for a four-quark model. These latter authors used the $K_L$-$K_S$ mass difference to estimate the charmed-quark mass (given the Cabibbo angle). When this procedure is extended to the six-quark model, one finds that the results depend on the accuracy of the Gaillard-Lee estimate of the $K^0 \bar{K}^0$ transition amplitude (they suggested that their

estimate of the amplitude was good within an order of magnitude).
If f is defined as the multiplicative deviation from their estimate,
then the ratio of coupling constants squared for $\bar{t}d_L$ to $\bar{u}d_L$ ($\tan^2\theta_c$
$\sin^2\theta_1$) can be found as a function of f and the mass of the t quark
(with $m_c = 1.5$ GeV):

|  | f=1 | f=2 | f=5 |
|---|---|---|---|
| $m_t = 5$ GeV | 0 | 0.012 | 0.031 |
| $m_t = 15$ GeV | 0 | 0.0035 | 0.013 |

While these limits are not as severe as for $\bar{u}b_L$, the coupling $\bar{t}d_L$ is
nonetheless quite small.

One can say that $\sin^2\theta_1 \lesssim 0.4$ even for a 5 GeV quark. Therefore, the coupling $\bar{c}d_L$ is always much smaller than $\bar{c}s_L$, and the
coupling $\bar{t}b_L$ is large. Since the signs of $\sin\theta_1$ and $\sin\theta_2$ are not
known, one cannot make definitive statements about the relative
magnitudes of $\bar{t}s_L$ to $\bar{t}d_L$ or of $\bar{c}b_L$ to $\bar{u}b_L$. However, for most (but
not all) angles, the coupling constants squared for $\bar{t}s_L$ and $\bar{c}b_L$ are
much larger than those for $\bar{t}d_L$ and $\bar{u}b_L$, respectively.

In summary, for the WS–GIM model, the t quark should couple
dominantly to the b quark, with (in most cases) a secondary coupling
to s quarks and with a relatively small coupling to d quarks. The
b quark, if it is lighter than the t quark, is likely to decay into
c quarks. The b quark should have a very small coupling to u quarks.

In order to consider more general limits on charged-current
couplings (outside the context of the WS–GIM model), one can study
the energy dependence of $\sigma_{tot}$ and $<y>$ in neutrino scattering,
$\nu_\mu N \rightarrow \mu^- + X$ (where $y \equiv (E_\nu - E_\mu)/E_\nu$). The applicability of limits
from neutrino experiments is restricted to couplings via those gauge
bosons which also couple to $\nu_\mu$; in most models, this means the usual
W boson. The requirement used in determining limits is that consistency with all available data be obtained. Some care must be
given, since each experiment has different cuts, efficiencies and
corrections to the data. All curves shown below were calculated in
the context of QCD (i.e.– they contain scaling violations).

In Fig. 1 the data from neutrino scattering for the average
value of y versus energy are shown. The line is the prediction (with
QCD corrections included) for the WS–GIM model. The CERN-Dortmund-

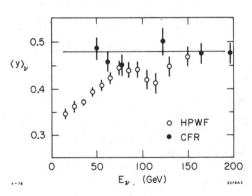

Fig. 1. The average value of y
in deep-inelastic neutrino scattering versus energy. The line
is the QCD prediction for the
standard four-quark model. The
data are from Refs. 7 and 8.

Heidelberg-Saclay (CDHS) data[6] which are not shown have no energy dependence consistent with the Cal Tech-Fermilab-Rockefeller (CFR) data.[7] The Harvard-Pennsylvania-Wisconsin-Fermilab (HPWF) data[8] contain efficiencies and cuts which reduce $\langle y \rangle$ at low energies. It is clear that the WS-GIM model is consistent with the data. For neutrinos, $\langle y \rangle$ does not set significant limits on any couplings.

The cross-section for charged-current neutrino scattering[9] is shown in Fig. 2. The lowest solid curve is the prediction of the

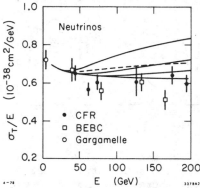

Fig. 2. The total charged-current cross-section for neutrino scattering vs. energy. The solid curves are the QCD predictions for the standard four-quark model with (from bottom to top): 1) no $\bar{t}d_L$, 2) $\bar{t}d_L$ added and $m_t$=9 GeV, 3) $m_t$ =5 GeV. The dotted curve has $\bar{t}d_R$ added with $m_t$=5 GeV. The data are from Ref. 9.

WS-GIM model. The three solid curves show the effect of a $\bar{t}d_L$ coupling equal in magnitude to $\bar{u}d_L$ for $m_t$=5, 7, and 9 GeV (from top). One can say roughly that $m_t \gtrsim 8$ GeV. The dotted curve is the result of a $\bar{t}d_R$ coupling for a 5 GeV t quark. In this case there is not a strict limit, and one can conclude only that $m_t \gtrsim 6$ GeV. Of course, one learns more about $\bar{t}d_R$ couplings from neutral-current interactions as is discussed later.

Since the upsilon meson, $T(9.4)$, seems to imply the existence of a quark with mass around 5 GeV, it is interesting to examine the limits for quarks of such mass. If $m_t$ = 5 GeV and if the ratio of coupling constants squared for $\bar{t}d_L$ to $\bar{u}d_L$ is 0.2, then the results are similar to those for $m_t$ = 9 GeV shown in Fig. 2 (or for a ratio of 0.4, similar to the dotted curve). Therefore, a $\bar{t}d_L$ coupling squared must be about 0.3 or less of that for $\bar{u}d_L$ (while for $\bar{t}d_R$ the limit is only 0.8).

The limits obtained for $\bar{u}b_L$ couplings are not quite as strict as for $\bar{t}d_L$. As long as $m_b \geq 7$ GeV, the coupling could be as strong as for $\bar{u}d_L$. If $m_b \approx 5$ GeV, then the coupling squared for $\bar{u}b_L$ can be 0.7 or less of that for $\bar{u}d_L$. Therefore, a substantial admixture of $\bar{u}b_L$ is allowed (although much stronger limits were found for the WS-GIM model earlier).

To study the possibility of a $\bar{u}b_R$ coupling, one can examine $\sigma_{tot}$ and $\langle y \rangle$ in antineutrino scattering which are shown[6-9] in Figs. 3 and 4. Clearly there is absolutely no <u>need</u> for any $\bar{u}b_R$ coupling. Any energy dependence present in the data is probably just that expected from scaling violations resulting from QCD corrections.[10] Since the data from different collaborations have different cuts, efficiencies and corrections, some experimentalists prefer to use the variable B which is determined by fitting to

$$\frac{d^2\sigma^{\nu,\bar{\nu}}}{dxdy} = \frac{G^2 M E}{\pi} F_2(x) \left[ \frac{(1\pm B)}{2} + \frac{(1\mp B)}{2} (1-y)^2 \right]$$

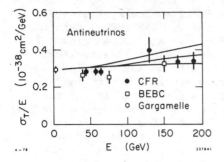

Fig. 3. The total charged-
current cross-section for antineutrino scattering vs. energy.
The solid curves are the QCD
predictions for the standard
four-quark model with (from
bottom to top): 1) no $\bar{u}b_R$, 2)
$\bar{u}b_R$ added and $m_b$=9 GeV, 3) $m_b$
=7 GeV. The data are from Ref. 9.

where

$$B = -xF_3(x)/F_2(x).$$

Fig. 5 shows the data[6-9] and
the QCD predictions without
and with a coupling $\bar{u}b_R$ (for
$m_b$ = 7, 9, 11 GeV). It is
evident from this figure (and

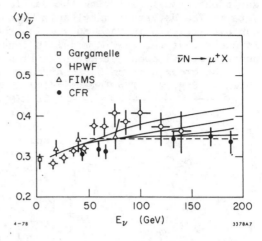

Fig. 4. The average value of y in
deep-inelastic antineutrino scattering vs. energy. The solid curves
are the QCD predictions for the
standard four-quark model with (from
bottom to top): 1) no $\bar{u}b_R$, 2) $\bar{u}b_R$
added and $m_b$=11 GeV, 3) $m_b$=9 GeV,
4) $m_b$=7 GeV. The dashed line represents the values for CDHS data
reported by J. Steinberger.[9] The
data are from Ref. 9.

from Figs. 3 and 4) that $m_b \gtrsim 11$ GeV if $\bar{u}b_R$ has the same strength as
$\bar{u}d_L$. Also shown in Fig. 5 are the curves for $\bar{u}b_L$ which lead to the
limits discussed earlier. If one examines the effects of $m_b \approx 5$ GeV
(motivated by the existence of $T(9.4)$), then the ratio of couplings
squared for $\bar{u}b_R$ to $\bar{u}d_L$ must be 0.1 or less, as seen in Fig. 6.
These are very strict limits. While the results given here (and
shown in the figures) include the scaling violations of QCD, little
change in the limits results even if all scaling violation is ignored.
    The data shown in Figs. 3-5 indicate (for each experiment) that
there is very little variation with energy above E = 50 GeV; this
result is consistent with the expectations from QCD. However, below
50 GeV the situation is not at all clear. Leaving aside the dispute
over the existence of a "high-y anomaly", the question of energy
dependence at "low" energies is very interesting. An important test
of QCD would be obtained with a careful measurement by a single
experiment of the energy dependence of <y> and $\sigma_{tot}$ between 10 and
50 GeV. Of course, another excellent test is the $q^2$ dependence of
the structure function $F_2(x)$ or, similarly, the E dependence of <x>.
At present, QCD is the only theory of the strong interactions, and
it is vital, therefore, to test its predictions.
    The question of the possible existence of $\bar{u}b_R$ or $\bar{t}d_R$ couplings
can be addressed from a completely different perspective. In gauge
theories of the weak and electromagnetic interactions, the charged-

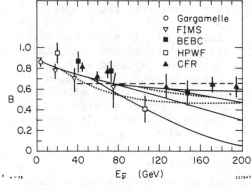

Fig. 5. The B parameter for anti-neutrinos vs. energy. The solid curves are the QCD predictions for the standard four-quark model with (from top to bottom): 1) no $\bar{u}b_R$, 2) $\bar{u}b_R$ added and $m_b$=11 GeV, 3) $m_b$=9 GeV, 4) $m_b$=7 GeV. The two dotted curves have $\bar{u}b_L$ added with (from top to bottom) $m_b$=9 GeV and $m_b$=5 GeV. The dashed line represents the value for CDHS data reported by J. Steinberger.[9] The data are from Refs. 6–9.

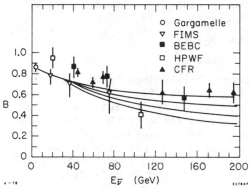

Fig. 6. The B parameter for anti-neutrinos vs. energy. The solid curves are the QCD predictions for the standard four-quark model with (from top to bottom): 1) no $\bar{u}b_R$, 2) $\bar{u}b_R$ added, $m_b$=5 GeV, and couplings squared 0.1 of $\bar{u}d_L$, 3) 0.2 of $\bar{u}d_L$, 0.3 of $\bar{u}d_L$. The data are from Refs. 6–9.

current and neutral-current interactions are intimately related. For example, in $SU(2) \times U(1)$ models,[1] the weak neutral currents can be obtained by adding a term found by an isospin rotation of the charged currents to a term proportional to the electromagnetic current.

   Larry Abbott and I have recently completed an analysis[11] of neutral-current couplings. Our analysis is independent of models, but the conclusions are, of course, applicable to any model. As will be shown, we have obtained a unique determination of the neutral-current couplings of u and d quarks which shows that in $SU(2) \times U(1)$ models, there can be no $\bar{t}d_R$ or $\bar{u}b_R$ couplings.

   Assuming a V, A structure and starting with the effective neutral-current Lagrangian:

$$\mathcal{L} = \frac{G}{\sqrt{2}}\, \nu\gamma_\mu(1+\gamma_5)\nu \left[ u_L\, \bar{u}\gamma_\mu(1+\gamma_5)u + u_R\, \bar{u}\gamma_\mu(1-\gamma_5)u \right.$$

$$\left. + d_L\, \bar{d}\gamma_\mu(1+\gamma_5)d + d_R\, d\gamma_\mu(1-\gamma_5)d \right] \;,$$

the allowed values of the coefficients $u_L$, $u_R$, $d_L$, and $d_R$ were determined from four types of neutrino experiments. There is always an ambiguity in the overall sign of the four coefficients (couplings); we chose a sign convention by requiring $u_L$ to always be positive.

   The first input is data for deep-inelastic scattering ($\nu N \rightarrow \nu X$). We chose to use CDHS data[12] since it is at high energy (<E> $\approx$ 100 GeV) and has small error bars. These data give limits on the values of $u_L^2 + d_L^2$ and $u_R^2 + d_R^2$. When plotted on the $u_L$-$d_L$ and $u_R$-$d_R$ planes, the regions allowed at the 90% confidence level are, therefore, annuli as shown in Fig. 7.

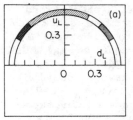 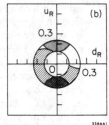

Fig. 7. The left (a) and right (b) coupling planes. Annular regions are allowed by deep-inelastic data. The region shaded with lines is allowed by deep-inelastic, elastic and exclusive-pion data. The regions shaded with dots are allowed by deep-inelastic and inclusive-pion data.

Since the radii are well-determined by the deep-inelastic data, it is useful in analyzing other data to plot the allowed values of the angles $\Theta_L$ and $\Theta_R$ which are defined as

$$\Theta_L \equiv \arctan (u_L/d_L)$$

$$\Theta_R \equiv \arctan (u_R/d_R) \ .$$

This plot has the advantage of showing correlations between the left and right planes which are not evident in Fig. 7.

The next input is data for elastic neutrino-proton scattering ($\nu p \rightarrow \nu p$). New data[13] from the HPW collaboration for both neutrinos and antineutrinos has been reported at this conference and is used for this analysis. Significant portions of the possible angular regions are eliminated by this data. The allowed region (at the 90% confidence level) is contained within the dotted line in Fig. 8.

A crucial new input into this analysis is data for exclusive pion production[14] for which six channels are available now:

$$\nu p \rightarrow \nu p \pi^0 \qquad \bar{\nu}N \rightarrow \bar{\nu} N \pi^0$$
$$\nu n \rightarrow \nu n \pi^0 \qquad \bar{\nu}n \rightarrow \bar{\nu} p \pi^-$$
$$\nu p \rightarrow \nu n \pi^+$$
$$\nu n \rightarrow \nu p \pi^-$$

To analyze these data, the detailed model of Adler[15] was used. This model includes non-resonant production, incorporates excitation of the $\Delta(1232)$ resonance, and satisfies current algebra constraints. Because we ignore resonances with mass above 1.4 GeV and use soft-pion theorems, we must require that the invariant mass of the pion-nucleon system (W) be less than 1.4 GeV. The data are not available with this cut (and cannot be obtained when a neutron is in the final state). Fortunately most of the cross-section is below W = 1.4 GeV, and indications are that application of the cut would strengthen our conclusions. Since we consider only neutral to charged current ratios, the effect of the cut is minimized. Since there is some uncertainty in the model used for analysis, the allowed region is defined as that region within a factor of two of the data (about 3 standard deviations). Application of the limits from these data reduced the allowed region to that shown by shading with lines in both Figs. 7 and 8. In Fig. 8 for $\Theta_L \approx 135°$, any value of $\Theta_R$ was allowed by the elastic data, but now with the exclusive pion data the limits are greatly improved.

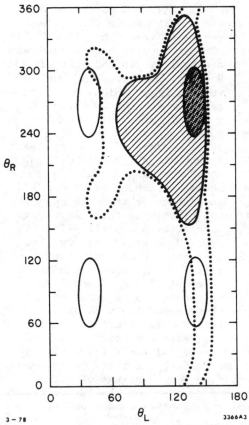

360

300

240

$\theta_R$

180

120

60

0

0          60          120          180

$\theta_L$

3 – 78                                         3366A3

Fig. 8. Allowed angles for specific radii. The dotted curve indicates the area allowed by elastic
data. The region shaded with lines
is allowed by elastic and exclusive-pion data. The elliptical
regions are allowed by inclusive-
pion data. The area shaded with
dots is the only region allowed by
all data.

A final input is data for
inclusive pion production[16]
($\nu N \to \nu \pi \dot{X}$). The analysis[17] of
these data requires significant
parton-model assumptions; in
particular, it is assumed that
pions produced in the current
(W-boson) fragmentation region
are decay products of the struck
quarks. The ratios of $\pi^+$ to $\pi^-$
production for $0.3 < z < 0.7$ where
$z \equiv E_\pi / E_{hadron}$ were measured.
The data were taken at low ener-
gies where one could question
parton-model assumptions. However, our determination of
neutral-current couplings is
unique even without these data;
its inclusion serves only to
further reduce the one allowed
region. In Figs. 7 and 8 the
final allowed region (at the
90% confidence level) is shown
by shading with both lines and
dots.

This allowed region cor-
responds to the couplings

$u_L = 0.33 \pm 0.07$     $u_R = -0.18 \pm 0.06$

$d_L = -0.40 \pm 0.07$     $d_R = 0.0 \pm 0.11$

where the errors are 90% confidence limits and an overall sign
convention has been assumed.

Our results are compared
with the predictions of various
models in Fig. 9. $SU(2) \times U(1)$
models[1] all have the same prediction for the left plane,
which is shown with a line indicating the results for different
values of $\sin^2 \theta_W$. The WS model[1,2]
is shown on the right plane with a similar line, whereas for other
$SU(2) \times U(1)$ models[18-20] (A, B and C) only the point corresponding to
$\sin^2 \theta_W = 0.3$ (determined from the left plane) is shown on the right
plane. Note that on both planes the WS model agrees with all data
for $\sin^2 \theta_W$ between 0.22 and 0.30. In fact it is only for the $m_Z/m_W$
ratio found from the minimal Higgs boson structure[1] that agreement
is obtained. This is a remarkable result.

For all other $SU(2) \times U(1)$ models, the predictions are far from
the allowed region for any value of $m_Z/m_W$. Those models are unequiv-
ocably ruled out. Also shown in Fig. 9 are the predictions of two

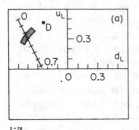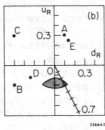

Fig. 9. Various gauge models compared with the allowed coupling constant region. The line marks the WS model for values of $\sin^2\theta_W$ from 0.0 to 0.7. A,B and C indicate $SU(2)\times U(1)$ with A) a $\bar{u}b_R$ coupling[18], B) a $\bar{t}d_R$ coupling[19] and C) both couplings[20] (vector model). D and E indicate $SU(3)\times U(1)$ models with: D) u in a right-handed singlet[21] and E) u in a right-handed triplet.[22] For A,B,C and E, $u_L$ and $d_L$ lie within the shaded region.

$SU(3) \times U(1)$ models[21,22]; they too fail to enter the allowed region in the right plane (their parameters were chosen to obtain the best fit in the left plane). However, there are $SU(2)^L \times SU(2)^R \times U(1)$ models[23] which can very closely mimic the results of the WS model, and therefore cannot be ruled out by our analysis. In fact, Georgi and Weinberg[24] have shown that any $SU(2) \times U(1) \times G$ models meeting certain requirements can mimic the WS model. In such cases other tests must be sought such as those involving $\nu e$ scattering, parity-violation, etc. However, it is now clear that only models with neutral-current couplings very similar to those of the WS model should be considered in constructing theories of the weak and electromagnetic interactions.

Although neutrino experiments are very difficult, they provide powerful tools for analyzing the structure of the weak interactions. In the last few years, there has been enormous progress in determining the couplings of both charged and neutral currents of u and d quarks. In the future we can expect to learn more about the couplings of other quarks and of the leptons.

I would like to acknowledge contributions to this work from S. Adler, J. Bjorken, F. Gilman, A. Mann, C. Matteuzzi, Y. J. Ng, J. Strait, L. Sulak and especially Larry Abbott and François Martin.

## REFERENCES

1. S. Weinberg, Phys. Rev. Lett. <u>19</u>, 1264 (1967); A. Salam, in <u>Elementary Particle Physics: Relativistic Groups and Analyticity</u>, edited by N. Svartholm (Almquist and Wiskell, Stockholm, 1968), p. 367.
2. S.L. Glashow, J. Iliopoulos and L. Maiani, Phys. Rev. <u>D2</u>, 1285 (1970).
3. M. Kobayashi and K. Maskawa, Prog. Theor. Phys. <u>49</u>, 652 (1973).
4. J. Ellis et al., Nucl. Phys. <u>B131</u>, 285 (1977); J. Ellis, M.K. Gaillard, and D.V. Nanopoulos, Nucl. Phys. <u>B109</u>, 213 (1976).
5. M.K. Gaillard and B.W. Lee, Phys. Rev. <u>D10</u>, 897 (1974).
6. M. Holder et al., Phys. Rev. Lett. <u>39</u>, 433 (1977).
7. F. Sciulli et al., in <u>Proc. 1977 Int. Symposium on Lepton and Photon Interactions at High Energies</u>, edited by F. Gutbrod (DESY, Hamburg, 1977), p. 239.

8. A. Benvenuti et al., Phys. Rev. Lett. 36, 1478.

9. B.C. Barish et al., (CFR), Phys. Rev. Lett. 39, 1595 (1977); A. Benvenuti et al., (HPWF), Phys. Rev. Lett. 37, 189 (1976); J.P. Berge et al., (FIMS), Phys. Rev. Lett. 39, 382 (1977); P.C. Bosetti et al., (BEBC), Phys. Lett. 70B, 273 (1977), K. Kleinknecht (CDHS), in Proc. 1977 Int. Symposium on Lepton and Photon Interactions at High Energies, edited by F. Gutbrod (DESY, Hamburg, 1977) p. 271; J. Steinberger, invited talk at the 1977 Irvine Conference.

10. G. Altarelli, G. Parisi, and R. Petronzio, Phys. Rev. Lett. 63B, 183 (1976); G. Altarelli and G. Parisi, Nucl. Phys. B126, 298 (1977); M. Barnett, H. Georgi and H.D. Politzer, Phys. Rev. Lett. 37, 1313 (1976); J. Kaplan and F. Martin, Nucl. Phys. B115, 333 (1977); M. Barnett and F. Martin, Phys. Rev. D16, 2765 (1977); V.I. Zakharov, in Proc. of the XVIII Int. Conf. on High Energy Physics, Tbilisi, July 1976, edited by N.N. Bogoliubov et al. (Joint Institute for Nuclear Research, Dubna, 1977) Vol. II, p. B69; A. Zee, F. Wilczek and S.B. Treiman, Phys. Rev. D10, 2881 (1974); A.J. Buras and K.J.F. Gaemers, Phys. Lett. 71B, 106 (1977); A.J. Buras, Nucl. Phys. B125, 125 (1977); A.J. Buras et al., Nucl. Phys. B131, 308 (1977); G.C. Fox, Caltech Report CALT-68-619 (1977).

11. L.F. Abbott and R.M. Barnett, report SLAC-PUB-2097 (1978).

12. M. Holder et al., Phys. Lett. 72B, 254 (1977); see also J. Blietschau et al., Nucl. Phys. B118, 218 (1977); A. Benvenuti et al., Phys. Rev. Lett. 37, 1039 (1976); B.C. Barish et al., Caltech report no. CALT-68-601 (1977).

13. J. Strait, Harvard University Ph.D. thesis (1978), W. Kozanecki, Harvard University Ph.D. thesis (1978), and D. Cline et al., Phys. Rev. Lett. 37, 252 and 648 (1976); see also W. Lee et al., Phys. Rev. Lett. 37, 186 (1976); M. Pohl et al., report no. CERN/EP/PHYS 77-53 (1977).

14. W. Krenz et al., report no. CERN/EP/PHYS 77-50 (1977); O. Erriques et al., CERN report (1978); see also S.J. Barish et al., Phys. Rev. Lett. 33, 448 (1974); W. Lee et al., Phys. Rev. Lett. 38, 202 (1977).

15. S.L. Adler, Ann. Phys. 50, 189 (1968) and Phys. Rev. D12, 2644 (1975); see also S.L. Adler et al., Phys. Rev. D12, 3501 (1975).

16. H. Kluttig et al., Phys. Lett. 71B, 446 (1977).

17. L.M. Sehgal, Phys. Lett. 71B, 99 (1977); P.Q. Hung, Phys. Lett. 69B, 216 (1977); P. Scharbach, Nucl. Phys. B82, 155 (1974).

18. M. Barnett, Phys. Rev. Lett. 34, 41 (1975), Phys. Rev. D11, 3246 (1975); P. Fayet, Nucl. Phys. B78, 14 (1974); F. Gürsey and P. Sikivie, Phys. Rev. Lett. 36, 775 (1976); P. Ramond, Nucl. Phys. B110, 214 (1976).

19. M. Barnett, Phys. Rev. D13, 671 (1976).

20. A. De Rújula et al., Phys. Rev. D12, 3589 (1975); F.A. Wilczek et al., Phys. Rev. D12, 2768 (1975); H. Fritzsch et al., Phys. Lett. 59B, 256 (1975); S. Pakvasa et al., Phys. Rev. Lett. 35, 702 (1975).

21. G. Segre and J. Weyers, Phys. Lett. 65B, 243 (1976); B.W. Lee and S. Weinberg, Phys. Rev. Lett. 38, 1237 (1977); B.W. Lee and

R.  Shrock,  report no. Fermilab–PUB–77/48–THY.

22.  M. Barnett and L.N. Chang, Phys. Lett. 72B, 233 (1977); M. Barnett et al., Phys. Rev. D (to be published); P. Langacker and
G. Segre, Phys. Rev. Lett. 39, 259 (1977).

23.  J. Pati and A. Salam, Phys. Rev. D10, 275 (1974); H. Fritzsch
and P. Minkowski, Nucl. Phys. B103, 61 (1976); R.N. Mohapatra
and D.P. Sidhu, Phys. Rev. Lett. 38, 667 (1977); A. De Rújula,
H. Georgi and S.L. Glashow, Annals Phys. 109, 242 and 258 (1977).

24.  H. Georgi and S. Weinberg, Phys. Rev. D17, 275 (1978).

CONFERENCE PROGRAM

Monday Morning, March 6
Welcoming Address by E. Q. Campbell, Dean of Vanderbilt University Graduate School

Hadronic Reactions - R. Diebold (ANL), Chairman
M. S. Witherell (Princeton) - Search for a Stable Dihyperon
A. Yokosawa (ANL) - Resonant-Like Structures in the pp System in the Mass Region 2100 to 2800 MeV
D. Miller (Purdue) - Survey of Baryonium Experiments
D. Green (Carnegie-Mellon) - Search for Narrow States in the Reaction $\bar{p}p \to \pi_f^+ X^-$ at 6 GeV/c
L. Price (Columbia) - Elastic Scattering Near $90^\circ$ c.m.: Tests of Parton and Resonance Models
G. Brandenburg (MIT) - Inclusive Scattering Results from the Fermilab Single Arm Spectrometer
K. Young (U. of Washington) - High p Jets Produced in High Energy pp Collisions*

Monday Afternoon, March 6

Lepton and Lepton Pair Production - A. G. S. Smith (Princeton), Chairman
R. Kephart (Fermilab) - Status Report on Upsilon and High Mass Dimuon Continuum*
W. P. Oliver (Tufts) - Dimuon Production by Protons in Tungsten
K. J. Anderson (Chicago) - Production of High Mass Muon Pairs by Pions and Protons*
R. F. Mozley (SLAC) - Production of Leptons and Lepton Pairs in $\pi^\pm p$ Interactions
M. Goldberg (Syracuse) - Production of $e^+e^-$ and $\pi^0$ Pairs at the ISR
S. M. Morse (BNL) - Properties of Prompt Muons Produced by 28 GeV Proton Interactions

Monday Evening, March 6

Hadronic Reactons (Continued)
M. Jacob (CERN) - Summary of ISR and SPS Results, Especially High p Phenomena*
A. Erwin (Wisconsin) - A Preliminary Look at Hadron Structure With a Double Arm Calorimeter

Tuesday Morning, March 7

Hadronic Reactions and Leptons (Continued)
E. L. Berger (ANL) - Massive Lepton Pair Production in Hadronic Collisions
C. Callan (Princeton) - Developments in Strong Interactions Theory*
M. H. Shaevitz (Cal Tech) - Observation of Prompt Single Muon Production by Hadrons at Fermilab

J. H. Mulvey (Oxford) - The Observation of Prompt Neutrinos
From 400 GeV Proton Nucleus Interactions

W. C. Schlatter (CERN) - Results of A Beam Dump Experiment at
the CERN SPS Neutrino Facility

G. J. Tarnopolsky (CERN) - Electron-Muon Coincidences in Proton-
Proton Collisions at the CERN Intersecting Storage Rings

A. Zylberstejn (Saclay) - $e^+e^-$ Pair Production at the ISR with
Masses from 1.4 to 6 GeV/$c^2$ and High $P\perp$ Pizero Production[*]

H. Bøggild (Niels Bohr Institute) - High $P\perp$ Results with Split
Field Magnet at CERN ISR[*]

## Tuesday Afternoon, March 7

### $e^+e^-$ Colliding Beam Interactions - K. Berkelman (Cornell), Chairman

U. Timm (DESY) - Recent Results from the PLUTO and DASP Detectors at DORIS

B. Gobbi (Northwestern) - Results from the SPEAR Magnetic Detector with the Lead Glass Addition[*]

L. J. Nodulman (UCLA) - New Results from DELCO

R. Baldini (Frascati) - Results from ADONE[*]

E. Eichten (Harvard) - Heavy Quark Systems

## Wednesday Morning, March 8

### Neutrino Experiments - C. Baltay (Columbia), Chairman

T. Y. Ling (Ohio State) - New Results on Dimuon Production by
High Energy Neutrinos and Antineutrinos

A. Mann (U. of Pennsylvania) - Inelastic Interactions of Anti-
Neutrinos[*]

W. D. Schlatter (CERN) - Characteristics of Trimuon Events
Observed in High Energy Neutrino Interactions

E. Fisk (Fermilab) - Charged Current Neutrino and Antineutrino
Cross Sections Results from the CITFR Experiment

M. Kalelkar (Columbia) - Recent Results from Neutrino Interactions in Heavy Neon

## Wednesday Afternoon, March 8

### Neutrino Experiments (Continued)

J. H. Mulvey (Oxford) - Evidence for Scaling Violation in $\nu$N
and $\bar{\nu}$N Inclusive Scattering and Test of QCD

D. Sinclair (Michigan) - Measurement of the Production of $\mu$e
Events in Antineutrino-Nucleon Interactions

B. S. Yuldashev (U. of Washington) - Inclusive Production of
Hadrons in $\nu_\mu$Ne and $\bar{\nu}_\mu$Ne Interactions

J. Strait (Harvard) - $\nu$p and $\bar{\nu}$p Elastic Scattering at Brookhaven[*]

M. Barnett (SLAC) - Charged and Neutral Current Couplings of
Quarks (As Seen by Neutrinos)

[*]Paper not prepared for publication in these proceedings.

ADVISORS

C. Baltay
B. Barish
E. L. Berger
K. Berkelman
J. Cronin
R. Diebold
T. Ferbel
K. Gottfried
B. Richter
A. J. S. Smith

LIST OF PARTICIPANTS

N. Baggett, DOE
C. Baltay, Columbia University
K. J. Anderson, University of Chicago
R. Baldini, Frascati
V. E. Barnes, Purdue University
C. L. Basham, University of Washington
M. Barnett, SLAC
E. L. Berger, ANL
K. Berkelman, Cornell
H. Bøggild, Niels Bohr Institute
G. Brandenburg, MIT
C. Callan, Princeton University
D. Carmony, Purdue University
R. Cassell, Vanderbilt University
M. Concoran, University of Wisconsin
S. E. Csorna, Vanderbilt University
R. Diebold, ANL
D. Duke, Fermilab
E. Eichten, Harvard University
A. Erwin, University of Wisconsin
H. Fenker, Vanderbilt University
T. Ferbel, University of Rochester
W. Fickinger, Case Western Reserve University
E. Fisk, Fermilab
B. Gobbi, Northwestern University
M. Goldberg, Syracuse University
S. Hagopian, Florida State University
D. Green, Carnegie-Mellon University
E. Harvey, University of Wisconsin
F. Henyey, University of California at San Diego
K. Hidaka, ANL
M. Jacob, CERN
S. Jones, Vanderbilt University
M. Kalelkar, Columbia University
R. Kephart, Fermilab
Y. S. Kim, University of Maryland
D. Lichtenberg, Indiana University
T. Y. Ling, Ohio State University
H. Lubatti, University of Washington
A. Mann, University of Pennsylvania
J. Marraffino, Vanderbilt University
R. Mickens, Fisk University
D. Miller, Purdue University
W. Morse, BNL
R. Mozley, SLAC
J. Mulvey, Oxford University
L. Nodulman, UCLA
W. Oliver, Tufts University

D. Perrin, Neuchatel/CERN  
J. A. Poirier, NSF  
J. S. Poucher, Vanderbilt University  
E. Predazzi, Indiana University  
L. Price, Columbia University  
C. Roos, Vanderbilt University  
R. Ruchti, University of Notre Dame  
J. Russ, Carnegie-Mellon University  
F. Sannes, Rutgers University  
P. Schmidt, Brandeis University  
D. Scott, University of Wisconsin  
W. Selove, University of Pennsylvania  
M. Shaevitz, Caltech  
E. Shibata, Purdue University  
D. Sinclair, University of Michigan  
A. J. S. Smith, Princeton University  
S. Stone, Vanderbilt University  
J. Strait, Harvard University  
W. D. Schlatter, CERN  
L. R. Sulak, Harvard University  
R. Talman, Cornell University  
G. Tarnopolsky, CERN  
U. Timm, DESY  
F. Turkot, Fermilab  
J. Vande Velde, University of Michigan  
M. Webster, Vanderbilt University  
R. Whitman, Syracuse University  
M. Witherell, Princeton University  
R. Yamamoto, MIT  
T. C. Yang, University of Massachusetts  
A. Yokosawa, ANL  
K. Young, University of Washington  
B. S. Yuldashev, University of Washington  
A. Zylberstejn, Saclay

AIP Conference Proceedings

| | | L.C. Number | ISBN |
|---|---|---|---|
| No.1 | Feedback and Dynamic Control of Plasmas (Princeton) 1970 | 70-141596 | 0-88318-100-2 |
| No.2 | Particles and Fields - 1971 (Rochester) | 71-184662 | 0-88318-101-0 |
| No.3 | Thermal Expansion - 1971 (Corning) | 72-76970 | 0-88318-102-9 |
| No.4 | Superconductivity in d- and f-Band Metals (Rochester, 1971) | 74-18879 | 0-88318-103-7 |
| No.5 | Magnetism and Magnetic Materials - 1971 (2 parts) (Chicago) | 59-2468 | 0-88318-104-5 |
| No.6 | Particle Physics (Irvine, 1971) | 72-81239 | 0-88318-105-3 |
| No.7 | Exploring the History of Nuclear Physics (Brookline, 1967, 1969) | 72-81883 | 0-88318-106-1 |
| No.8 | Experimental Meson Spectroscopy - 1972 (Philadelphia) | 72-88226 | 0-88318-107-X |
| No.9 | Cyclotrons - 1972 (Vancouver) | 72-92798 | 0-88318-108-8 |
| No.10 | Magnetism and Magnetic Materials - 1972 (2 parts) (Denver) | 72-623469 | 0-88318-109-6 |
| No.11 | Transport Phenomena - 1973 (Brown University Conference) | 73-80682 | 0-88318-110-X |
| No.12 | Experiments on High Energy Particle Collisions - 1973 (Vanderbilt Conference) | 73-81705 | 0-88318-111-8 |
| No.13 | $\pi$-$\pi$ Scattering - 1973 (Tallahassee Conference) | 73-81704 | 0-88318-112-6 |
| No.14 | Particles and Fields - 1973 (APS/DPF Berkeley) | 73-91923 | 0-88318-113-4 |
| No.15 | High Energy Collisions - 1973 (Stony Brook) | 73-92324 | 0-88318-114-2 |
| No.16 | Causality and Physical Theories (Wayne State University, 1973) | 73-93420 | 0-88318-115-0 |
| No.17 | Thermal Expansion - 1973 (Lake of the Ozarks) | 73-94415 | 0-88318-116-9 |
| No.18 | Magnetism and Magnetic Materials - 1973 (2 parts) (Boston) | 59-2468 | 0-88318-117-7 |
| No.19 | Physics and the Energy Problem - 1974 (APS Chicago) | 73-94416 | 0-88318-118-5 |
| No.20 | Tetrahedrally Bonded Amorphous Semiconductors (Yorktown Heights, 1974) | 74-80145 | 0-88318-119-3 |
| No.21 | Experimental Meson Spectroscopy - 1974 (Boston) | 74-82628 | 0-88318-120-7 |
| No.22 | Neutrinos - 1974 (Philadelphia) | 74-82413 | 0-88318-121-5 |
| No.23 | Particles and Fields - 1974 (APS/DPF Williamsburg) | 74-27575 | 0-88318-122-3 |
| No.24 | Magnetism and Magnetic Materials - 1974 (20th Annual Conference, San Francisco) | 75-2647 | 0-88318-123-1 |